BASIC IMMUNOLOGY

BASIC IMMUNOLOGY

Dr Arun Ingale PhD, FSSc, FBIBI, FMS
Associate Professor
Department of Biotechnology, School of Life Sciences
North Maharastra University, Maharastra

New Central Book Agency (P) Ltd
LONDON
HYDERABAD ERNAKULAM BHUBANESWAR
NEW DELHI KOLKATA PUNE GUWAHATI

NCBA

REGD OFFICE
8/1 Chintamoni Das Lane, Kolkata 700 009, India
email: ncbapvtltd@eth.net

OVERSEAS
NCBA (UK) Ltd, 149 Park Avenue North, Northampton, NN3 2HY, UK
email: ncbauk@yahoo.com

EXPORT
NCBA Exports Pvt. Ltd, 212 Shahpur Jat, New Delhi 110 049
email: ncbaexports@ncbapvtltd.com

BRANCHES

2/27 Ansari Road, 1st Floor
Daryaganj, **New Delhi** 110 002
email: ncbadel@ncbapvtltd.com

Shop Nos. 3 & 4, Vinayak Towers
681/B Budhwar Peth, Appa Balwant Chowk
Pune 411 002
email: ncbapune@ncbapvtltd.com

Shop No. 15, Bharati Towers, Ground Floor
Forest Park
Bhubaneswar 751 001
email: ncbabub@ncbapvtltd.com

House No. 3-1-315, 1st Floor
Nimboliadda, Kachiguda, **Hyderabad** 500 027
email: ncbahydb@ncbapvtltd.com

GSS Shopping Complex, 1st Floor
Opposite College Post Office, Convent Road
Ernakulam 682 035
0484-4021054 / 53

Radhamayee Complex (Second Floor)
Opp. Axis Bank, Panbazar, Jaswanta Road
Guwahati 781 001
email: ncbaguwahati@ncbapvtltd.com

BASIC IMMUNOLOGY • Dr Arun Ingale

First Published: 2010
Reprinted: 2012, 2018

PUBLISHER AND TYPESETTER
New Central Book Agency (P) Ltd
8/1 Chintamoni Das Lane, Kolkata 700 009

PRINTER
New Central Book Agency (P) Ltd
Web-Offset Division, Dhulagarh, Sankrail, Howrah

TECHNICAL EDITOR
Dr Mita Sen

PROJECT TEAM
Suman Dutta and Kakoli Banerjee

ISBN: 978 81 7381 630 7

Price ₹ 495.00

Dedicated to my beloved

Father

Late Shri Govindrao D. Ingale

Mother

Late Bhagirathi G. Ingale

&

Shri Yeshwant G. Ingale

CONTENTS

PREFACE

I am very glad to present this book of *Basic Immunology*, written according to revised syllabus of BSc, MSc (Biotechnology, Microbiology) BPharm, MPharm, MSc Agriculture and Veterinary in all Indian Universities. This book is also useful for the medical students.

Basic Immunology has been written to fulfil a need for an up-to-date text on basic and molecular immunology. It is aimed primarily at a typical undergraduate and postgraduate student's audience in Microbiology and Biotechnology, audiences in India and elsewhere.

Basic Immunology covers a very wide area which has been widened further since the mid-1970s by the development of hybridoma technology and such techniques used as production of monoclonal antibodies and closely related Immunotechniques.

A new generation of immunologists will be required to grapple with these issues. Basic Immunology plays a key role in training those entering our field and in helping current immunologists to be as productive as possible.

While explaining the basic principle of immunology, due stress is given on various biotechnological application of the subject.

I extend my good wishes to the students and teachers of Biotechnology and Microbiology, sincerely hope that *Basic Immunology* will receive a warm welcome from them.

I welcome comments by readers of Basic Immunology for ways to improve the book and to increase its value. Such suggestions will be seriously considered in the preparation of subsequent editions.

I am very grateful to Dr N J Chikhale, Reader, Department of Biotechnology, Sant Gadge Baba Amravati University, Amravati for useful suggestions made by him time to time.

Mr Amol Hivrale, UGC Research Fellow helped greatly with the page proof reading and I would like to thank him very much for his careful work.

Finally, I would like to express my sincere thanks to Jija Yeshwant Ingale, Rashmi Ingale, Jayant, Ashmit, and Sneha Ingale for continuous encouragement for completion of this book.

Dr Arun Ingale

CHAPTER 1

IMMUNITY

INTRODUCTION

Immunity is defined as the state of resistance or insusceptibility exhibited by host to toxic substance, microorganism and foreign cell. Immunity–the state of protection from infectious diseases, it has both specific and nonspecific components. Immunity (derived from *immunitas:* Latin for exemption from civic duties and prosecution) means protection from disease and especially infectious disease. Cells and molecules involved in such protection constitute the **immune system** and the response to introduction of a foreign agent is known as the **immune response**. The immune response that results is specific and complex series of defensive action widely distributed throughout the animals body (Figure 1.1). Not all immune responses protect from disease; some foreign agents, such as the **allergens** found in house dust mite, cat dander or rye grass pollen, cause disease as a consequence of inducing an immune response. Likewise some individuals mount immune responses to their own tissues as if they were foreign agents. Thus, the immune response can cause the **autoimmune** diseases common to man, such as multiple sclerosis, diabetes, rheumatoid arthritis or myasthenia gravis. Most individuals do not suffer from autoimmune disease because they have developed **tolerance** towards their own **(self)** tissues.

Immunity may be innate (or nonspecific) and adaptive (acquired or specific). Immunity is a biological phenomenon, which consists in a long-term autosupport of the genetic "self" and "not-self" balance in the body under foreign surroundings. Immune system develops and releases certain mechanisms to exert the phenomenon. Understanding the basis of IMMUNOLOGY is achieved by the study both of the protective mechanisms challenged by "not-self" and of regulatory processes, which form internal homeostasis of "self" in the body. Lymphocytes are major cells of the immune system. In fact, the immune system is a hierarchic totality of lymphoid cells. There are **T-cells** and **B-cells**. Other cells (e.g., macrophages, dendritic cells, neutrophils and mast cells.) also attend many immune reactions. Antigen is a macromolecule, which contains a foreign or autologous information to launch immune responses. On the other hand, any antigen can be often used by explorers as an *immunobiological marker*. An appreciable account of various antigens is about ten. Antibody or immunoglobulin (Ig) is a molecule of the immune system. Antibodies as well as receptors can be bound to appropriate antigens.

There are two basic types of immunity, innate and adaptive.

I. INNATE (NON-SPECIFIC) OR NATURAL IMMUNITY

Innate immunity is a set of diseases resistance mechanism that is not specific to particular pathogen. The phagocytic cells such as macrophages play an important role in innate immunity. The normal individual has two levels of defense against foreign agents. The first type is present in neonatal animals and in invertebrates, namely **natural** or **innate** immunity. This type of immunity is sometimes referred to as non-specific but broadly specific would be a better description. The second type of immunity is **adaptive** or **acquired** immunity and is confined to vertebrates.

The immune system is composed of two major subdivisions, the innate or nonspecific immune system and the adaptive or specific immune system. The innate immune system is our first line of defense against invading organisms while the adaptive immune system acts as a second line of defense and also affords protection against re-exposure to the same pathogen. Each of the major subdivisions of the immune system has both cellular and humoral components by which they carry out their protective function. In addition, the innate immune system also has anatomical features that function as barriers to infection. Although these two arms of the immune system have distinct functions, there is interplay between these systems (i.e., components of the innate immune system influence the adaptive immune system and vice versa).

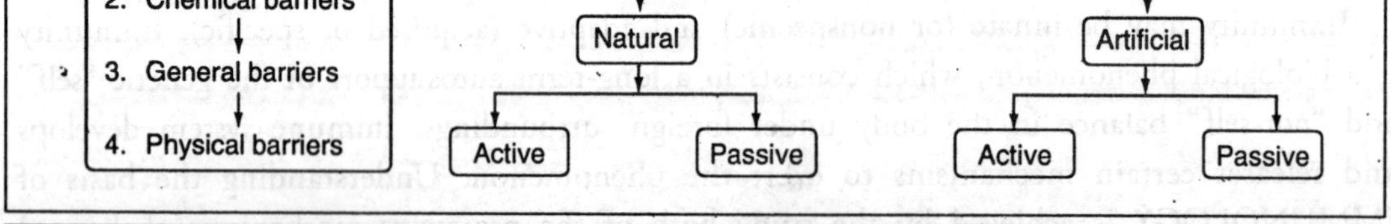

Fig. 1.1: Description of Immune response

Although the innate and adaptive immune systems both function to protect against invading organisms, they differ in a number of ways. The adaptive immune system requires some time to react to an invading organism, whereas the innate immune system includes defenses that, for the most part, are constitutively present and ready to be mobilized upon infection. Second, the adaptive immune system is antigen specific and reacts only with the organism that induced the response. In contrast, the innate system is not antigen specific and reacts equally well to a variety of organisms. Finally, the adaptive immune system demonstrates immunological memory. It "remembers" that it has encountered an invading organism and reacts more rapidly on subsequent exposure to the same organism. In contrast, the innate immune system does not demonstrate immunological memory.

The main function of the immune system is self/non-self discrimination. This ability to distinguish between self and non-self is necessary to protect the organism from invading pathogens and to eliminate modified or altered cells (e.g., malignant cells). Since pathogens may replicate intracellularly (viruses and some bacteria and parasites) or extracellularly (most bacteria, fungi and parasites), different components of the immune system have evolved to protect against these different types of pathogens. It is important to remember that infection with an organism does not necessarily mean diseases, since the immune system in most cases will be able to eliminate the infection before disease occurs. Disease occurs only when the dose of infection is high, when the virulence of the invading organism is great or when immunity is compromised. Although the immune system, for the most part, has beneficial effects, there can be detrimental effects as well. During inflammation, which is the response to an invading organism, there may be local discomfort and collateral damage to healthy tissue as a result of the toxic products produced by the immune response. In addition, in some cases the immune response can be directed towards self tissues resulting in autoimmune disease.

BARRIERS TO INFECTIONS

INNATE (OR NATURAL) IMMUNITY

The nonspecific resistance refers to those general mechanism inherited as apart of innate structure and function of each animal. These act in concert as first line defense against infectious microorganism their product or foreign macromolecule before they cause diseases. The elements of the non-specific (innate) immune system include anatomical barriers, secretory molecules and cellular components. Among the mechanical anatomical barriers are the skin and internal epithelial layers, the movement of the intestines and the oscillation of broncho-pulmonary cilia. Associated with these protective surfaces are chemical and biological agents. Innate (or natural) immunity is made up of several components are as follows:

Anatomical barriers to infections

Physical and anatomical barriers that tends to prevent the entry of pathogen are an organisms first line defense against infection. The skin and surface mucous membranes act as physical barriers are the first line of defense against infection because these are effective barrier to the entry of most organisms. The skin and mucous membranes provide a continuous surface which must be breached and back this up with mechanical protection through cilia and mucous.

The skin consists of two distinct layers, thinner outer layer epidermis and thicker inner layer dermis. The epidermis contains several layers of tightly packed epithelial cells. The dermis composed of connective tissue, contains blood vessels, hair follicles, sebaceous glands and sweat glands. The sebaceous glands produce sebum, which consist of lactic acids and fatty acids, which maintains the pH of skin between 3-5, this pH inhibits the growth of microorganisms.

Mechanical factors: The epithelial surfaces form a physical barrier that is very impermeable to most infectious agents. Thus, the skin acts as our first line of defense against invading organisms. The desquamation of skin epithelium also helps remove bacteria and other infectious agents that have adhered to the epithelial surfaces. Movement due to cilia or peristalsis helps to keep air passages and the gastrointestinal tract free from microorganisms. The flushing action of tears and saliva helps prevent infection of the eyes and mouth. The trapping effect of mucus that lines the respiratory and gastrointestinal tract helps protect the lungs and digestive systems from infection.

Physiological factors: Physiological factors that contribute to innate immunity, such as pH, temperature and various soluble factors, oxygen tension limit microbial growth. The acid environment of the stomach combined with microbial completion from the commensal flora inhibits gut infection. **Protein secretions** into external body fluids such as lysozyme also help resist invasion. Lysozyme, a hydrolytic enzyme found in mucus secretion of tears, is able to cleave peptidoglycan layer of bacterial cell wall. Soluble factors within the body such as **complement, interferons** and collectins and other "broadly specific" molecules such as C-reactive protein are of considerable importance in protection against infection. Complement protein is a group of serum protein that circulate in an inactive state. Complement proteins also form a membrane attack complex to destroy the cell of microorganism.

Chemical factors: Fatty acids in sweat inhibit the growth of bacteria. Lysozyme and phospholipase found in tears, saliva and nasal secretions can break down the cell wall of bacteria and destabilize bacterial membranes. The low pH of sweat and gastric secretions prevents growth of bacteria. Defensins (low molecular weight proteins) found in the lung and gastrointestinal tract have antimicrobial activity. Surfactants in the lung act as opsonins (substances that promote phagocytosis of particles by phagocytic cells).

Biological factors: The normal flora of the skin and in the gastrointestinal tract can prevent the colonization of pathogenic bacteria by secreting toxic substances or by competing with pathogenic bacteria for nutrients or attachment to cell surfaces.

Humoral barriers to infection: The anatomical barriers are very effective in preventing colonization of tissues by microorganisms. However, when there is damage to tissues the anatomical barriers are breached and infection may occur. Once infectious agents have penetrated tissues, another innate defense mechanism comes into play, namely acute inflammation. Humoral factors play an important role in inflammation, which is characterized by edema and the recruitment of phagocytic cells. These humoral factors are found in serum or they are formed at the site of infection.

Complement system: The complement system is the major humoral non-specific defense mechanism. Once activated complement can lead to increased vascular permeability, recruitment of phagocytic cells, and lysis and opsonization of bacteria.

Coagulation system: Depending on the severity of the tissue injury, the coagulation system may or may not be activated. Some products of the coagulation system can contribute to the non-specific defenses because of their ability to increase vascular permeability and act as chemotactic agents for phagocytic cells. In addition, some of the products of the coagulation system are directly antimicrobial. For example, beta-lysin, a protein produced by platelets during coagulation can lyse many Gram positive bacteria by acting as a cationic detergent.

Lactoferrin and transferring: By binding iron, an essential nutrient for bacteria, these proteins limit bacterial growth. Interferons are proteins that can limit virus replication in cells. Lysozyme breaks down the cell wall of bacteria. Interleukin-1–Il-1 induces fever and the production of acute phase proteins, some of which are antimicrobial because they can opsonize bacteria.

Cellular barriers to infection: Part of the inflammatory response is the recruitment of polymorphonuclear eosinophils and macrophages to sites of infection. These cells are the main line of defense in the non-specific immune system.

Neutrophils

Polymorphonuclear cells are recruited to the site of infection where they phagocytose invading organisms and kill them intracellularly. In addition, PMNs contribute to collateral tissue damage that occurs during inflammation.

Macrophages

Tissue macrophages and newly recruited monocytes, which differentiate into macrophages, also function in phagocytosis and intracellular killing of microorganisms. In addition, macrophages are capable of extracellular killing of infected or altered self target cells. Furthermore, macrophages contribute to tissue repair and act as antigen-presenting cells, which are required for the induction of specific immune responses. **Natural killer** (NK) and lymphokine activated killer (LAK) cells—NK and LAK cells can nonspecifically kill virus infected and tumor cells. These cells are not part of the inflammatory response but they are important in nonspecific immunity to viral infections and tumor surveillance.

Eosinophils

Eosinophils have proteins in granules that are effective in killing certain parasites. **Phagocytic cells** are critical in the defense against bacterial and simple eukaryotic pathogens. **Macrophages** and **Polymorphonuclear leucocytes** (PMN) can recognise bacterial and yeast cell walls through broadly specific receptors (usually for carbohydrate structures) and this recognition is greatly enhanced by activated complement (opsonin).

Artificially acquired immunity

Artificially acquired active immunity results when an animal is immunized with a vaccine. A vaccine consist of preparation of killed microorganism, living weakened microorganism or attenuated microorganism or inactivated bacterial toxin that administered to an animal to induce immunity artificially.

Artificially acquired passive immunity results when antibodies that have been produced either in an animal or by specific methods in vitro are introduced into host. Although this type of immunity is immediate, it is short lived. An example would be botulinum antitoxin produced by horse and given to human suffering from botulism food poisoning.

Types of immunity

The study of the molecular and cellular components that comprise the immune system, including their function and interaction, is the central science of immunology. The immune system is divided into a more primitive innate immune system, and acquired or adaptive immune system of vertebrates, the latter of which is further divided into humoral and cellular components.

The immune response further divide into two types are humoral mediated immunity and cell mediated immunity.

1. HUMORAL MEDIATED IMMUNITY

The humoral immune response is the aspect of immunity that is mediated by secreted antibodies, produced in the cells of the B lymphocyte lineage (B-cell). Secreted antibodies bind to antigens on the surfaces of invading microbes (such as viruses or bacteria), which flags them for destruction. Humoral immunity is called as such, because it involves substances found in the humours, or body fluids.

Humoral immunity refers to antibody production, and the accessory processes that accompany it, including: Th2 activation and cytokine production, germinal center formation and isotype switching, affinity maturation and memory cell generation. It also refers to the effector functions of antibody, which include pathogen and toxin neutralization, classical complement activation, and opsonin promotion of phagocytosis and pathogen elimination.

The concept of humoral immunity developed based on analysis of antibacterial activity of the components of serum. Hans Buchner is credited with the development of the humoral theory. In 1890 he described alexins, or "protective substances", which exist in the serum and other bodily fluids and are capable of killing microorganisms. Alexins, later redefined "complement" by Paul Ehrlich, were shown to be the soluble components of the innate response that lead to a combination of cellular and humoral immunity, and bridged the features of innate and acquired immunity.

B-cells are produced in the bone marrow and are distributed through the body in the lymph nodes. B-cells respond to the 'foreign' antigens of a pathogen by producing specific antibodies. Antibodies are complex proteins that are released into the blood and carried to the site of infection. B-cells do not fight pathogens directly. B-cells are responsible for humoral immunity. They arise from a separate population of stem cells of the bone marrow than that which gives rise to T-cells. These cells undergo multiplication and processing in lymphoid tissue elsewhere than in the thymus gland. In birds, the lymphoid tissue concerned has been located in the gut and called the bursa of Fabricius. In humans, the site is unknown although there is some evidence to suggest that such processing occurs in the bone marrow itself or in the foetal liver. Lymphocytes processed in this way are called B-lymphocytes after the bursa. B-cells, like T-cells, have surface receptors which enable them to recognise the appropriate antigen, but do not themselves interact to neutralise or destroy the antigen. On recognition of the antigen they take up residence in secondary lymphoid tissue and proliferate to form daughter lymphocytes, processed in the same way as themselves. These B-cells then develop into short-lived plasma cells. The principal function of B-cells is to make antibodies against soluble antigens. B-cell recognition of antigen is not the only element necessary for B-cell activation (a combination of clonal proliferation and terminal differentiation into plasma cells). Naive B-cells can be activated in a T-cell dependent or independent manner, but two signals are always required to initiate activation.

B-cell activation depends on one of three mechanisms

Type 1—*T-cell-independent* (polyclonal) activation, Type 2—*T-cell-dependent* activation (in which macrophages present several of the same antigen in a way that causes cross-linking of antibodies on the surface of B cells), and, T-cell-dependent activation. During T-cell-dependent activation, an antigen presenting cell (APC) presents a processed antigen to a helper T (T_h) cell, priming it. When a B-cell processes and presents the *same* antigen to the *primed* T_h *cell*, the T-cell releases cytokines that activate the B-cell. The plasma cells produce antibodies and release them into the circulation at the lymph nodes. Some of the activated B-cells do not become plasma cells instead they turn into memory cells which continue to produce small amounts of the antibody long after the infection has been overcome. As with T-cells millions of B-cells are produced each with a different antigen binding specificity. If the B-cell comes into contact with the specific type of antigen to which it is targeted it divides rapidly to form a clone of identical cells.

This antibody circulates as part of the gamma globulin fraction of the blood plasma. If the same antigens enter the body again this circulating antibody acts quickly to destroy it. At the same time memory cells quickly divide to produce new clones of the appropriate type of plasma cell. Immunoglobulins are glycoproteins in the immunoglobulin superfamily that function as antibodies. The terms *antibody* and *immunoglobulin* are often used interchangeably. They are found in the blood and tissue fluids, as well as many secretions. In structure, they are large Y-shaped globular proteins. In mammals there are five types of antibody: IgA, IgD, IgE, IgG, and IgM. Each immunoglobulin class differs in its biological properties and has evolved to deal with different antigens. Antibodies are synthesized and secreted by plasma cells that are derived from the B-cells of the immune system.

An antibody is used by the immune system to identify and neutralize foreign objects like bacteria and viruses. Each antibody recognizes a specific antigen unique to its target.

By binding their specific antigens, antibodies can cause agglutination and precipitation of antibody-antigen products, prime for phagocytosis by macrophages and other cells, block viral receptors, and stimulate other immune responses, such as the complement pathway.

An incompatible blood transfusion, causes a transfusion reaction, which is mediated by the humoral immune response. This type of reaction, called an acute hemolytic reaction, results in the rapid destruction (hemolysis) of the donor red blood cells by host antibodies. The cause is usually a clerical error (i.e. the wrong unit of blood being given to the wrong patient). The symptoms are fever and chills, sometimes with back pain and pink or red urine (hemoglobinuria). The major complication is that hemoglobin released by the destruction of red blood cells can cause acute renal failure.

2. CELL-MEDIATED IMMUNITY

T-lymphocyte cells (T-cells) originally derive from stem cells of the bone marrow. At around the time of birth, lymphocytes derived in this way leave the marrow and pass to the thymus gland in the chest, where they multiply. The lymphocytes are processed by the thymus gland, so that between them they carry the genetic information necessary to react with a multitude of possible antigens.

T-cell diversity is attained by a complex 5-stage genetic rearrangement that occurs at random in the developing T-cells in the thymus. T-cells are tested for their ability to recognise and bond to antigens in the thymus. It is thought that this occurs by positive and negative selection.The postulation is that positive selection eliminates T-cells that do not bond tightly enough to antigen type molecules produced in the thymus and negative selection eliminates T-cells that bond too tightly to self type molecules found in the thymus.

This produces a mature T-cell that is effective against antigens but is also self tolerant. It is known that approximately 95% of all developing T-cells die in the thymus. Each lymphocyte (or possibly a small number of similar T-cells) is thus able to react with one antigen. Antigens of the body's own cells are excluded so that the population of T-cells processed in this way is potentially capable of recognising 'self' and 'non-self' and only reacting against the 'non-self' molecules. Autoimmune diseases are those in which this recognition is defective, and certain of the body's cells may be recognised erroneously as 'non-self' and subsequently destroyed. Processing of T-cells occurs largely during foetal life and early childhood, after which the thymus gland shrinks in size. In adult life the active portion of the thymus is largely replaced with fatty tissue.

Mice experimentally deprived of the thymus gland immediately after birth, before they have begun to process T-cells, are incapable of cell-mediated immune responses. They fail to reject transplanted tissue. In addition, the humoral immune system works less efficiently, indicating some co-operation between the two systems. The T-cells, each processed to 'recognise' and interact with a specific antigen, circulate permanently between the blood and lymphatic systems.

There are three types of T-cells involved. Helper T-cells also known as T-helper cells, TH cells or T4 cells (these are the type of cell attacked by HIV). Killer T-cells also known as cytotoxic T-cells, TC cells or T8 cells memory T-cells another type of T8 cell.

On recognition of the antigen by a helper T-cell, one or a group of lymphocytes takes up residence in secondary lymphoid tissue (e.g. lymph glands, Peyer's patches, MALT, spleen, bone marrow) and divides to form two types of cells, memory cells, which are lymphocytes processed in the same way as themselves, and killer cells, which interact with the antigen. T-cell receptor sites sit on the surface of T-cells and provide the specificity in antigen binding. They have two components that give each different type a unique surface morphology and it is this that allows bonding to antigens. They are a Ti component which recognises antigens and a CD3 component that transfers a signal into the cell body.

T-cells respond to specific antigens. When a pathogen first infect the body, each individual antigen stimulates a single T-cell. This divides to form a clone, in the same way that B-cells do. Some of the activated T-cells become memory cells and persist in the body, ready to mount a secondary response if the pathogen attacks again. The others, however, do not produce antibodies. They develop further to become either helper T-cells, killer T-cells or suppressor T-cells. The surface of each T-cell also displays thousands of identical T-cell receptors (TCRs). The TCR binds a bimolecular complex displayed at the surface of some other cell called an antigen-presenting cell (APC). This complex consists of a fragment of an antigen lying within the groove of a histocompatibility molecule. Most of the T-cells in the body belong to one of two subsets. These are distinguished by the presence on their surface of one or the other of two glycoproteins designated: CD4 - $CD4^+$ T-cells bind an epitope consisting of an antigen fragment lying in the groove of a class II histocompatibility molecule. $CD4^+$ T-cells are essential for both the cell-mediated and antibody-mediated branches of the immune system.

These $CD4^+$ cells bind to antigen presented by antigen-presenting cells (APCs) like phagocytic macrophages and dendritic cells. The T-cells then release lymphokines that attract other cells to the area. The result is inflammation: the accumulation of cells and molecules that attempt to wall off and destroy the antigenic material (an abscess is one example, the rash following exposure to poison ivy is another).

CHAPTER

2

CELLS OF THE IMMUNE SYSTEM

HEMATOPOISIS

All blood cells are produced from specialized type of cells are called as hematopoitic stem cells. The stem cells are capable of differentiating into many other type of cell. These cells are having self renewal capacity. These types of cells are maintaining their population by cell division. The hematopoisis means the formation and development of RBC and WBC. The hematopoitic stem cells are of two types one is *unipotent* and other is *pluripotent.*

STEM CELLS

The two broad types of mammalian stem cells are: **embryonic stem cells** that are found in blastocysts, and **adult stem cells** that are found in adult tissues. In a developing embryo, stem cells can differentiate into all of the specialized embryonic tissues. .In adult organisms, stem cells and progenitor cells act as a repair system for the body, replenishing specialized cells, but also maintain the normal turnover of regenerative organs, such as blood, skin or intestinal tissues.

As stem cells can be grown and transformed into specialized cells with characteristics consistent with cells of various tissues such as muscles or nerves through cell culture, their use in medical therapies has been proposed. In particular, embryonic cell lines, autologous embryonic stem cells generated through therapeutic cloning, and highly plastic adult stem cells from the umbilical cord blood or bone marrow are touted as promising candidates. Unipotent stem cells are the cells which can differentiate into only single type of cells but *pluripotent stem* cells can differentiate into different type of cells such as erythrocyte, granulocyte, monocytes and mast cells.

PROPERTIES OF STEM CELLS

The classical definition of a stem cell requires that it possess two properties:

- **Self-renewal** the ability to go through numerous cycles of cell division while maintaining the undifferentiated state.
- **Potency** the capacity to differentiate into specialized cell types. In the strictest sense, this requires stem cells to be either **totipotent** or **pluripotent**–to be able to give rise to any mature cell type, although **multipotent** or **unipotent** progenitor cells are sometimes referred to as stem cells.

Pluripotent, embryonic stem cells originate as inner mass cells within a blastocyst. The stem cells can become any tissue in the body, excluding a placenta. Only the morula's cells are totipotent, able to become all tissues and a placenta. *Potency* specifies the differentiation potential (the potential to differentiate into different cell types) of the stem cell.

- **Totipotent** stem cells are produced from the fusion of an egg and sperm cell. Cells produced by the first few divisions of the fertilized egg are also totipotent. These cells can differentiate into embryonic and extraembryonic cell types.
- **Pluripotent** stem cells are the descendants of totipotent cells and can differentiate into cells derived from any of the three germ layers.
- **Multipotent** stem cells can produce only cells of a closely related family of cells (e.g. hematopoietic stem cells differentiate into red blood cells, white blood cells, platelets, etc.).
- **Unipotent** cells can produce only one cell type, but have the property of self-renewal which distinguishes them from non-stem cells (e.g. muscle stem cells).

All cells of the immune system have their origin in the bone marrow and they include myeloid (neutrophils, basophils, eosinophils, macrophages and dendritic cells) and lymphoid (B lymphocyte, T lymphocyte and natural killer) cells, which differentiate along distinct pathways. The myeloid progenitor (stem) cell in the bone marrow gives rise to erythrocytes, platelets, neutrophils, monocytes/macrophages and dendritic cells whereas the lymphoid progenitor (stem) cell gives rise to the NK, T-cells and B-cells (Fig. 2.1).

LYMPHOCYTE

Lymphocytes are much more common in the lymphatic system. Lymphocytes are distinguished by having a deeply staining nucleus which may be eccentric in location, and a relatively small amount of cytoplasm. Lymphocytes are derived from bone marrow stem cells. They are produced in bone marrow at very high rate (10^9 cells per day). Some lymphocyte migrate through blood or lymphatic circulatory system to secondary lymphoid tissue (thymus, spleen and lymph node) they are produce lymphocyte colonies. Lymphocyte constitute 20-40% of the bodies white blood cells in the lymph. There is approximately 10^{11} lymphocyte in the human body. These lymphocytes continually circulate in the blood and lymph and are capable of migrating into the tissue spaces and lymphoid organs.

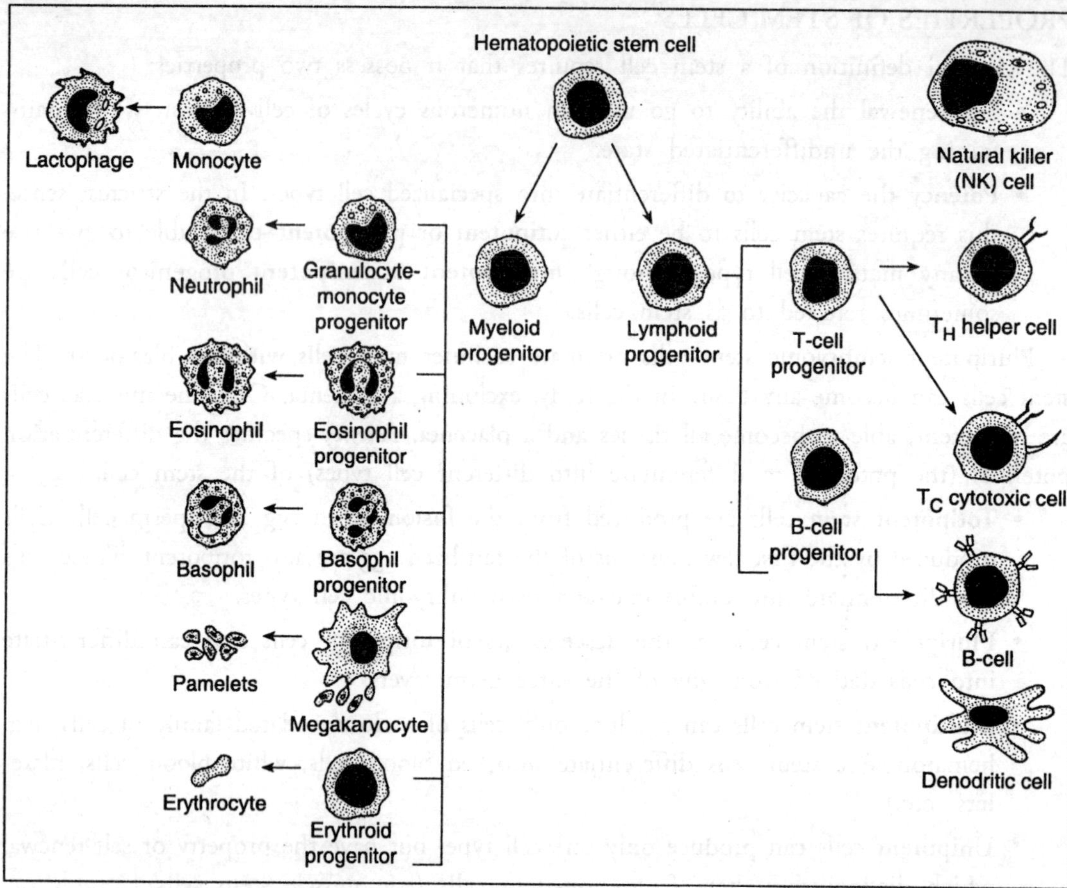

Fig. 2.1: Generation cell populations from Stem Cells

These cells have receptors for antigen and confer specificity on an immune response. Lymphocytes express receptors with varying affinity for the antigen in question. The cell with the highest affinity for the most abundant antigen will have growth advantage and will preferentially generate progeny of itself. This process is called **clonal expansion** and is antigen driven. B lymphocytes produce antibodies and some soluble mediators called cytokines. They arise in the bone marrow in adult mammals.

The lymphocyte can be broadly divided into three population B-Cells, T-Cells, and null cells.

T lymphocytes arise in bone marrow but mature in the **thymus.** The undifferentiated lymphocyte that migrate to thymus undergo special processing and become T-cells. Some T-cell are transported away from thymus and enter the blood stream where they comprise 70-80% of circulating lymphocyte. They do not produce antibody molecules but have surface receptors structurally related to Ig. T-cells recognise peptide fragments of antigen complexed with cell surface MHC glycoproteins on neighbouring cells (Fig. 2.2). The cell surface glycoproteins encoded by genes in the **Major Histocompatibility Complex** (MHC) bind fragments of antigen after it has been subjected to antigen processing.

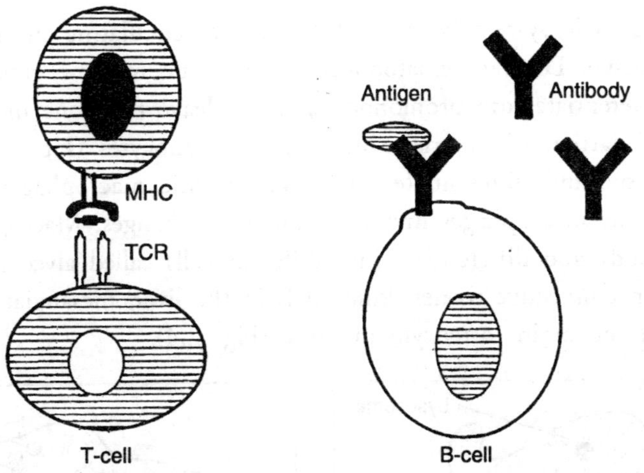

Fig. 2.2: Recognition antigens by T-cell and B-cell, T-cell recognize processed antigen and B-cell recognized native antigen

There are two sub-types of T-cell defined on the basis of function, accessory molecule expression and the type of MHC protein presenting antigen to them. This can be summarised as follows:

TABLE 2.1
Sub-types of T-cell

There are 2 basic types of T-cell

Accessory Molecule	T-cell	MHC restriction	Function
CD4	T helper	Class II	Cytokine production help for B-cells and cytotoxic T-cells
CD8	T cytotoxic	Class I	Cytokine production cytotoxic lysis of infected or tumour cells

- **B-cells:** B-cells make antibodies that bind to pathogens to enable their destruction. (B-cells not only make antibodies that bind to pathogens, but after an attack, some B-cells will retain the ability to produce an antibody to serve as a 'memory' system).

Natural killer (NK) cells are large granular lymphocytes that are cytotoxic in the absence of prior stimulation. NK cells represent a first line of defence to infections, tumour growth and other pathogenic alterations of tissue homeostasis. NK cells do not express antibodies or T-cell receptors at their cell surface. They produce cytokines and express receptors for immunoglobulin. They also possess other receptor molecules which allow them to detect some infected host cells, including tumour cells, virus, or intracellular bacteria-infected cells. Natural killer cells are able to kill cells of the body which are displaying a signal to kill them, as they have been infected by a virus or have become cancerous.

MONONUCLEAR PHAGOCYTES

Mononuclear phagocytic system consists of monocytes circulating in the blood and macrophages in the tissue. During hematopoisis in the bone marrow granulocyte, monocytes progenitor cell differentiate into promonocyte, which leave the bone marrow and enter the blood, where they farther differentiate into mature monocytes (Fig. 2.3). Monocytes circulate into the tissue and differentiate into tissue specific macrophages. Differentiation of monocytes into tissue macrophages involves number of changes. Macrophages are dispersed into throughout body and differentiate into different cells called alveolar when function in lung, histeocytes in connective tissues, kufer cells in the liver, mesangial cells in the kidney, microglial cells in the brain, osteocyte in bone (Fig. 2.4).

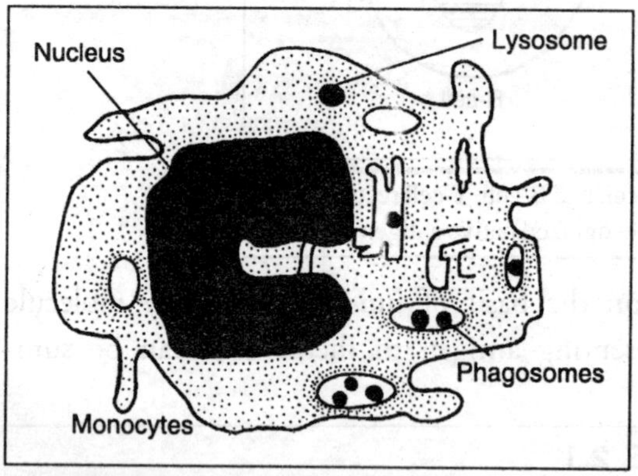

Fig. 2.3: Structure of Monocyte

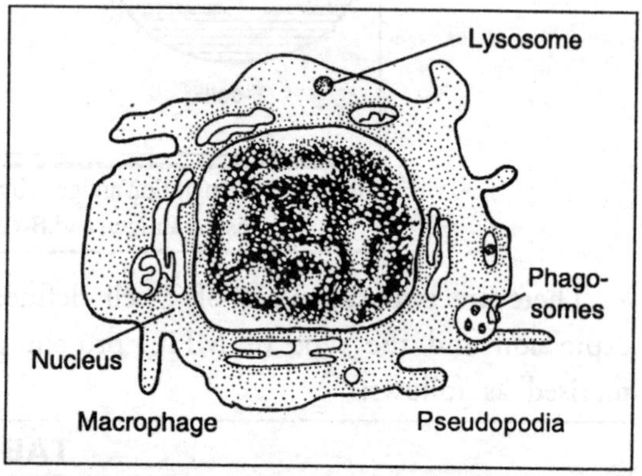

Fig. 2.4: Structure of Macrophage

If you inject "vital" dyes into experimental animals they will be taken up by various cell types including macrophages (mf), microglial cells in the CNS, endothelial cells of vascular sinusoids and reticular cells of lymphoid organs. These are the cells of the **Reticulo-Endothelial System** (RES). These cells all take up dye by pinocytosis. Only cells of the monocyte-macrophage lineage take up large particulate antigens, pieces of tissue, senescent cells, bacteria etc. by **phagocytosis.**

DENDRITIC CELLS

There are two cell types with similar names but different functions. Cells of the **dendritic cell** (DC) lineage are bone marrow derived. In the skin they are known as **Langerhans cells** (LC). These cells efficiently process antigen but cannot present it to T-cells. Langerhans cells have been shown to pick up antigen in skin and carry it via **afferent lymphatic vessels** to lymph nodes. Dendritic cells in lymph are known as "veiled" cells. In lymph nodes the cells, now known as tissue dendritic cells or **interdigitating cells,** may efficiently present antigen if they encounter the right T-cell.

Stem cells	⟶ Tissue Dendritic cells	⟶ Veiled cells	⟶ Interdigitating cells
Bone Marrow	*e.g. Langerhan's cells in the skin*	*afferent lymph*	*lymphoid tissue spleen/lymph nodes*

Follicular dendritic cells (FDC) are found in lymphoid follicles. They are called dendritic because of their morphology rather than any lineage relationship with DC. In fact, there is considerable uncertainty about their developmental origin [some evidence suggests they are long-lived bone marrow derived cells, other data that they are of epithelial origin]. FDC have receptors for immunoglobulin and complement and are able to trap antigen at their cell surface, in the form of antigen/antibody/C3d complexes, for long periods of time. They cannot present antigen to T-cells but are important in developing responses by B-cells.

GRANULOCYTES

There are three types of **granulocyte** distinguished according to their histological staining patterns.

- **Neutrophils** are multilobed nucleus and granulated cytoplasm that stains with both acid and basic dye. **Neutrophils**, also known as polymorphonuclear leukocytes, express receptors for immunoglobulin and complement and are involved in the acute inflammatory response. The cytoplasm may look transparent because of fine granules that are faintly pink in color. Neutrophils are very active in phagocytosing bacteria and are present in large amount in the pus of wounds. Neutrophils defend against bacterial or fungal infection and other very small inflammatory processes that are usually first responders to microbial infection; their activity and death in large numbers forms pus. It has life span 7-10th in circulating blood.
- **Eosinophils** carry receptors for IgE, are involved in the destruction of IgE coated parasites, such as helminths, and contribute to the response to allergens. It has bilobed nucleus and granulated cytoplasm that stains with acid dye eosin red.

Eosinophils primarily deal with parasitic infections and an increase in them may indicate such. Eosinophils are also the predominant inflammatory cells in allergic reactions. The most important causes of eosinophilia include allergies such as asthma, hay fever, and hives; and also parasitic infections. Generally their nucleus is bilobed. The cytoplasm is full of granules which assume a characteristic pink-orange color with eosin stain.

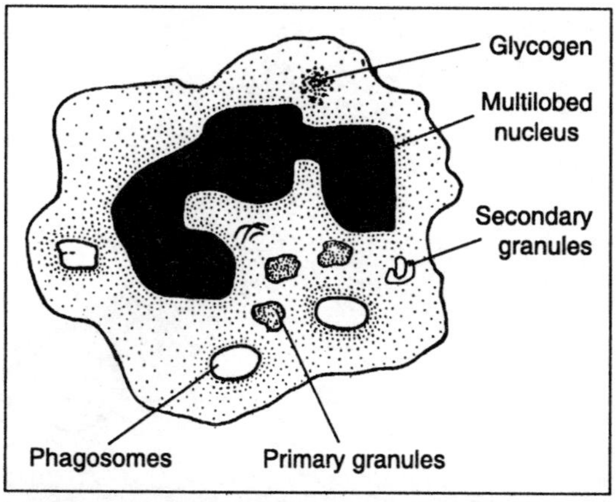

Fig. 2.5: Structure of Neutrophils

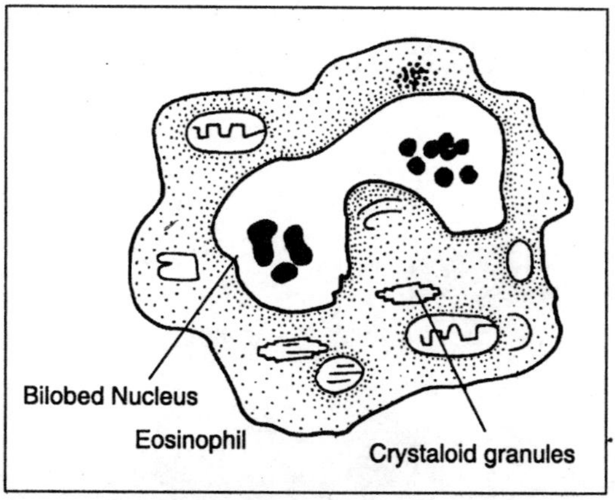

Fig. 2.6: Structure of Eosinophils

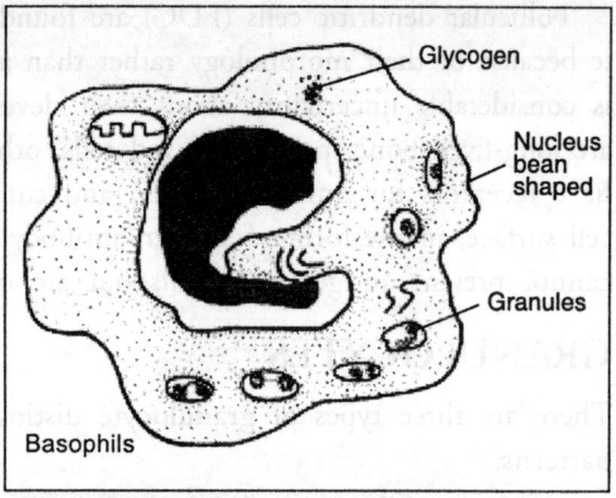

Fig. 2.7: Structure of Basophils

- **Basophils** are the circulating counterpart of tissue **mast cells.** They express high affinity receptors for IgE and are stimulated to secrete the chemicals responsible for **immediate hypersensitivity** following antigen induced aggregation of these receptors.

Basophils are chiefly responsible for allergic and antigen response by releasing the chemical histamine causing inflammation. The nucleus is bi- or tri-lobed, but it is hard to see because of the number of coarse granules which hide it. They are characterised by their large blue granules.

CHAPTER 3

ORGAN OF THE IMMUNE SYSTEM

LYMPHATIC SYSTEM

The lymphatic system is a complex network of lymphoid organs, lymph nodes, lymph ducts, lymphatic tissues, lymph capillaries and lymph vessels that produce and transport lymph fluid from tissues to the circulatory system. The lymphatic system is a major part of the immune system.

The lymphatic system has three interrelated functions: (1) removal of excess fluids from body tissues, (2) absorption of fatty acids and subsequent transport of fat, as chyle, to the circulatory system and, (3) production of immune cells such as lymphocytes (e.g. antibody producing plasma cells) and monocytes.

LYMPHATIC CIRCULATION

Unlike the blood system, the lymphatic system is not closed and has no central pump. Lymph movement occurs slowly with low pressure due to peristalsis, valves, and the milking action of skeletal muscles. Like veins, lymph travels through vessels in one way only, due to semilunar valves. This depends mainly on the movement of skeletal muscles to squeeze fluid through them, especially near the joints. Rhythmic contraction of the vessel walls through movements may also help draw fluid into the smallest lymphatic vessels, capillaries. Tight clothing can restrict this, thus reducing the removal of wastes and allowing them to accumulate. If tissue fluid builds up the tissue will swell; this is called edema. As the circular path through the body's system continues, the fluid is then transported to progressively larger lymphatic vessels culminating in the right lymphatic duct (for lymph from the right upper body) and the thoracic duct (for the rest of the body); both ducts *drain* into the circulatory system at the right and left subclavian veins. The system collaborates with white blood cells in lymph nodes to protect the body from being infected by cancer cells, fungi, viruses or bacteria. This is known as a secondary circulatory system.

Lymph vessels called lacteals are present in the lining of the gastrointestinal tract. While most other nutrients absorbed by the small intestine are passed on to the portal venous system to drain, via the portal vein, into the liver for processing, fats (lipids) are passed on to the lymphatic system, to be transported to the blood circulation via the thoracic duct. The enriched lymph originating in the lymphatics of the small intestine is called chyle. As the blood circulates, fluid leaks out into the body tissues. This fluid is important because it carries food to the cells and waste back to the bloodstream. The nutrients that are released to the circulatory system are processed by the liver, having passed through the systemic circulation. The lymph system is a one-way system, transporting interstitial fluid back to blood.

PATHOLOGY

In elephantiasis, infection of the lymphatic vessels cause a thickening of the skin and enlargement of underlying tissues, especially in the legs and genitals. It is most commonly caused by a parasitic disease known as lymphatic filariasis. Lymphedema also causes abnormal swelling, especially in the appendages (though the face, neck, and abdomen can also be affected). It occurs if the lymphatic system is damaged, or underdeveloped in some way. An estimated 170 million suffer with the disorder.

There are three stages

Stage 1: Pressing the swollen limb leaves a pit that takes a while to fill back in. Because there is little fibrosis (hardening) it is often reversible. Elevation reduces swelling.

Stage 2: Pressure does not leave a pit. Elevation does not help. If left untreated, the limb becomes fibrotic.

Stage 3: This stage of lymphedema is often called elephantiasis. It is generally only in the legs after lymphedema that has gone long untreated. While treatment can help a little, it is not reversible.

Some common causes of swollen lymph nodes include staph infections and infectious mononucleosis.

MATURATION OF LYMPHATIC TISSUES

Lymphatic tissues begin to develop by the end of the fifth week of embryonic life. Lymphatic vessels develop from lymph sacs that arise from developing veins, which are derived from mesoderm.

The first lymph sacs to appear are the paired jugular lymph sacs at the junction of the internal jugular and subclavian veins. From the jugular lymph sacs, lymphatic capillary plexuses spread to the thorax, upper limbs, neck and head. Some of the plexuses enlarge and form lymphatic vessels in their respective regions. Each jugular lymph sac retains at least one connection with its jugular vein, the left one developing into the superior portion of the thoracic duct.

The next lymph sac to appear is the unpaired retroperitoneal lymph sac at the root of the mesentery of the intestine. It develops from the primitive vena cava and mesonephric veins. Capillary plexuses and lymphatic vessels spread form the retroperitoneal lymph sac to the abdominal viscera and diaphragm. The sac establishes connections with the cisterna chyli but loses its connections with neighboring veins.

The last of the lymph sacs, the paired posterior lymph sacs, develop from the iliac veins. The posterior lymph sacs produce capillary plexuses and lymphatic vessels of the abdominal wall, pelvic region, and lower limbs. The posterior lymph sacs join the cisterna chyli and lose their connections with adjacent veins. With the exception of the anterior part of the sac from which the cisterna chyli develops, all lymph sacs become invaded by mesenchymal cells and are converted into groups of lymph nodes.

The spleen develops from mesenchymal cells between layers of the dorsal mesentery of the stomach. The thymus arises as an outgrowth of the third pharyngeal pouch.

ORGANS OF IMMUNE SYSTEM

The immune system operates throughout the body. There are, however, certain sites where the cells of the immune system are organized into specific structures. These are classified as central lymphoid tissue (bone marrow, thymus) and peripheral lymphoid tissue (lymph-nodes, spleen, mucosa-associated lymphoid tissue).

Immune system consists of lymphoid cells (both T- and B-cells) and lymphoid organs. Cells that make up the immune system are distributed throughout the body. They occur predominantly in lymphoreticular organs. These include: bone marrow, thymus, lymph nodes, spleen and mucosa associated lymphoid tissues. The cells occupy the intricacies of networks formed by reticular cells and fibers supporting frame work. The cells involved in the immune function are: lymphocytes, monocytes, macrophages of tissues, endothelial cells, mast cells, basophils, eosinophils, neutrophils etc. All cells are derived from pluripotent, self renewing stem cells of the bone marrow.

The immune system is made up of many different organs and tissues dispersed throughout the body. Based on the different roles they perform, lymphoid organs can be classified into (1) primary or central lymphoid organs and (2) secondary or peripheral lymphoid organs.

- **Primary lymphoid organs** are the sites of lymphocyte birth and/or maturation, in an **Ag independent fashion.** It also called as generative organs, where lymphocyte arise, mature and capable of recognizing self antigen.
- **Secondary lymphoid organs** are the sites of mature lymphocyte selection and expansion in an **Ag dependent fashion.** It is also called as peripheral organ, were mature lymphocyte respond to foreign antigen. Secondary lymphoid organs include lymph nodes, spleen, and mucosa associated lymphoid tissue.

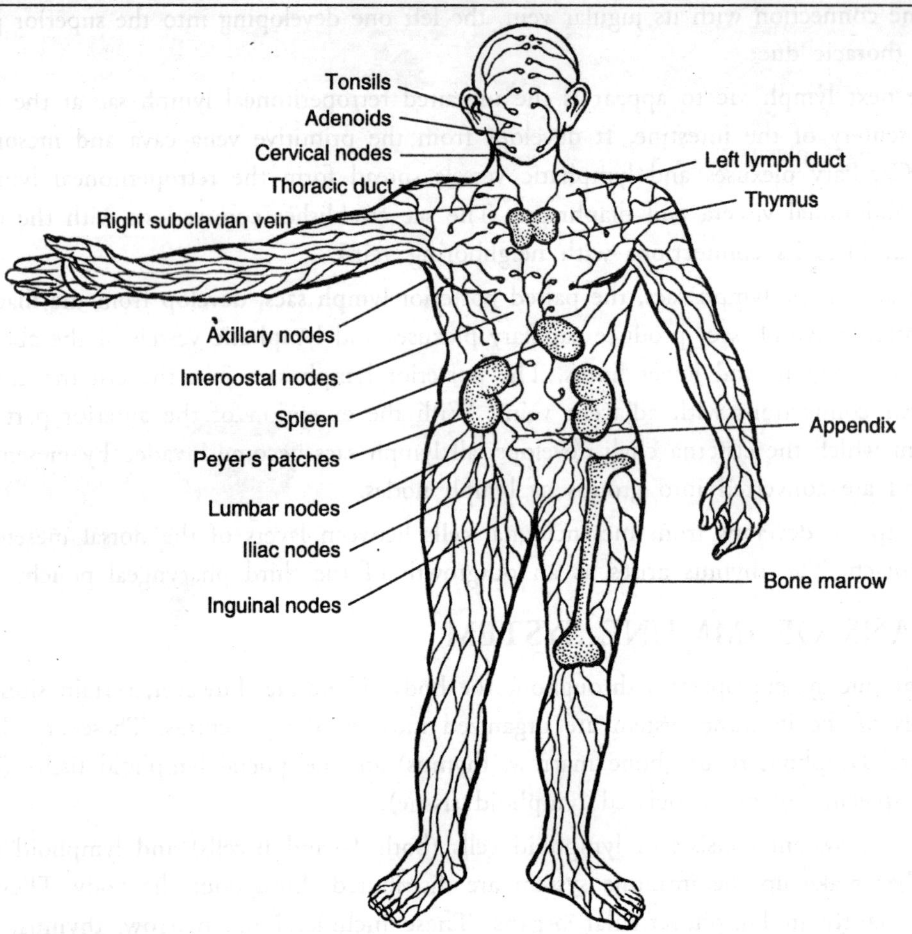

Fig. 3.1: The Human lymphoid system: Primary lymphoid organ (Bone marrow and Thymus) and secondary organ tissue are shown by labelling

PRIMARY LYMPHOID ORGANS

Primary lymphoid organs include (1) Bone marrow (2) Thymus (3) Bursa of Fabricius in birds. In the primary lymphoid organs, the T- and B-cells mature into antigen recognizing lymphocytes. Organization is simple as they are sites of growth and maturation and not sites where cells and Ag need to mix.

BONE MARROW

The bone marrow generates cells with immunological diversity for the lymphatic system. All the cells of the immune system are derived from stem cells in the bone marrow. The bone marrow is the site of origin of red blood cells, white cells (including lymphocytes and macrophages) and platelets. B-cells mature in bone marrow.

The bone marrow is the site of origin of all T- and B-cells, mononuclear phagocytes, platelets, erythrocytes, and other leukocytes in the adult. The aggregate volume and weight surpasses that of the liver. The bone marrow is divided into wedge-shaped hematopoietic compartments filled with proliferating and differentiating blood cells in connective tissue matrices bordered by venous sinuses (Figure 3.2). Monoclonal antibody-detected phenotypes and reactions for enzymes expressed at stages of differentiation show progressive hematopoietic cell maturation from immaturity in areas near bone to full differentiation at the vascular interface. Radial venous sinuses that merge with a central longitudinal vein are major components of mouse bone marrow structure. Blood cells complete maturation immediately adjacent to the dilated vascular channels into which they will emigrate.

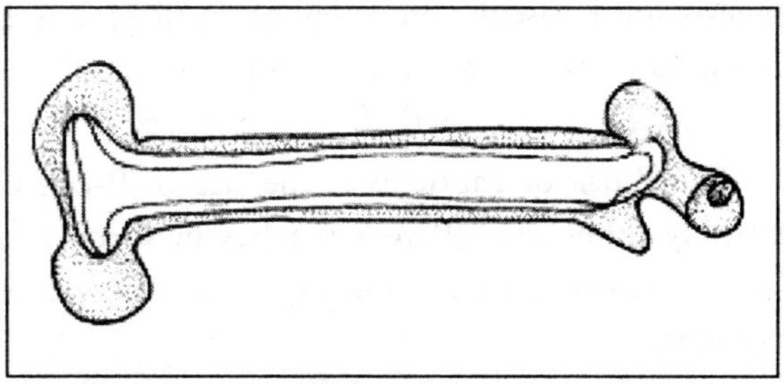

Fig. 3.2: Bone marrow

The bone marrow remains the primary site of hematopoiesis in adults until death, although some hematopoietic stem cells in connective tissue of the abdominal mesenteries can be induced to form hematopoietic colonies by cytokines released during inflammation. It is possible that the initial development of the blood-forming tissues, migration of stem cells and hematopoietic precursors may be controlled by receptor-ligand interactions and cytokines which regulate the entry, survival, proliferation and differentiation of hematopoietic progenitor cells. The microenvironment of the bone marrow is produced by a unique endothelium and connective tissue stroma combined with locally deposited cytokines that regulate compartmentalization, proliferation and differentiation of hematopoietic stem cells. Adventitial cells lining the interstitial side of venous sinuses extend cytoplasmic processes into the hematopoietic compartment, making contact with numerous cells. Stromal cells are essential for regulation of hematopoietic cell development. In situ hybridization with probes for allotypic markers proved that stromal cells are capable of transferring the hematopoietic microenvironment of the donor after allogeneic bone marrow transplantation. One way marrow stroma assists hematopoiesis is through the glycosaminoglycan-rich extracellular matrix that binds and distributes growth factors such as granulocyte-myelocyte colony stimulating factor (GM-CSF).

The hematopoietic stem cell that gives rise to all formed elements of the blood has a peripatetic life cycle. It travels to multiple sites as embryogenesis provides appropriate microenvironments. Early in development, stem cells located in the dorsal mesentery near the primitive aorta migrate through loose connective tissues to the yolk sac. The progeny rapidly divide and develop into precursors of erythroid, myeloid, monocytoid, lymphoid, megakaryocytic and endothelial cell types. Hematopoietic, lymphoreticular and endothelial cells are interdependent and intimately associated as they divide and differentiate within the organizing yolk sac. Blood cells undergo multiple obligatory interactions with endothelial cells and stromal components as they travel from the yolk sac to new sites of hematopoiesis or as they enter and leave the interstices of non-lymphoid tissues.

The location of hematophematopoiesis shifts from the yolk sac to the developing liver where clonal hematopoietic colonies expand in close proximity to endoderm-derived hepatocytes and sinus lining cells. The liver remains the major site of fetal hematopoiesis until shortly before birth when stem cells travel by blood to the spleen that supports hematopoiesis until vascular invasion of cartilaginous bones begins the process of ossification and formation of marrow cavity. Population of the developing marrow may be facilitated by chemotactic and growth factors released from bone matrix and connective tissue cells.

The Bursa of Fabricius is the site of B-cell development in birds. The bone marrow and possibly other sites (Peyer's patches in sheep and cattle) play this role in mammals, but the lack of information on the organ and how it works limits our knowledge of B-cell development.

B-CELL MATURATION

B-cell development begins in the liver and continues in the bone marrow as maturation progresses. During maturation from the primordial stem cell, B-cell progenitors contain cytoplasmic IgM. These cells do not express this protein on their surfaces although they share another surface antigen with mature B-cells. Shortly before birth, lymphocytes with detectable surface IgM appear in bone marrow and this is followed by the appearance of other cells bearing both IgM and IgD on their surface. Antigen specificity of precursor B-cells is determined by germline genes encoding the variable and hypervariable regions of the immunoglobulin light and heavy chains expressed the membranes of "precursor" B-cells. Isotype commitment has not yet occurred and each cell contains a complete set of immunoglobulin heavy chain genes for mu through alpha. The IgM+/IgDI small B-cell that leaves the bone marrow is metabolically "mature" in that it is antigen-specific, circulates in the blood and is a significant component of the recirculating pool, but it is considered a "primary" B-cell until it "switches" its membrane immunoglobulin to another isotype.

B-cells are more susceptible to tolerance induction by antigen or anti-immunoglobulins when they are in early stages of differentiation. Injection of antibodies directed against isotypic, allotypic, or idiotypic determinants of immunoglobulins in immunologically immature animals can lead to long-lasting inactivation of clonal precursors of B-cells bearing these proteins on their surfaces. This tolerant state is readily produced when the maturing B-cells express only IgM, and is more difficult to achieve after IgD is expressed. The precise mechanisms of tolerance induction following antigen or antibody exposure are unknown and theories of clonal deletion of T-cell suppression abound.

THYMUS

Thymus is the flat, bilobed organ situated above the heart. Each lobe is surrounded by capsule and is divided into lobule. In the thymus gland lymphoid cells undergo a process of maturation and education prior to release into the circulation. This process allows T-cells to develop the important attribute known as self tolerance. T-cell progenitor cells enter the thymus from bone marrow to mature and to be "educated". This lymphoepithelial organ develops from ectoderm derived from the third branchial cleft and endoderm of the third branchial pouch including mesenchymal components derived from cells of the cephalic neutral crest which all migrate from the neck to the anterior mediastinum. The rudimentary thymus enlarges in successive waves as lymphocytes and hematopoietic cells emigrate from embryonal blood vessels to infiltrate spaces between plump thymus epithelial cells. Proliferation and entry of new lymphoid precursors distend interepithelial spaces, resulting in densely packed clusters of lymphocytes in a desmosome-linked epithelial cell reticulum. Ultimately the thymus becomes an encapsulated, lobulated organ with a cortex that is densely populated with lymphocytes and a medulla which appears more epithelial because of a relative paucity of lymphoid cells. The thymus is widely thought to be an antigenically privileged site, yet antigen is essential for positive selection of specific T-cell clones for maturation and release to the periphery, acquisition of MHC haplotype restriction and tolerance induction implies that antigen must be present. A so-called blood-thymus barrier was proposed because of differences in ex vivo antibody formation in thymus fragments after injecting antigen directly into the thymus or after intravenous inoculation. The thymus gland is found in the thorax in the anterior mediastinum.

It gradually enlarges during childhood but after puberty it undergoes a process of involution resulting in a reduction in the functioning mass of the gland.

The thymus gland is arranged into an outer, more cellular, cortex and an inner, less cellular, medulla. Immature lymphoid cells enter the cortex proliferate, mature and pass on to the medulla. From the medulla mature T lymphocytes enter the circulation. The following cell types are present: lymphoid cells, epithelial cells, macrophages, other supporting cells. Thymic epithelial cells have different appearances in different locations within the gland. They form a continuous sub-capsular layer and a network in the cortex and medulla. Deep in the medulla they are also aggregated into Hassall's corpuscles.

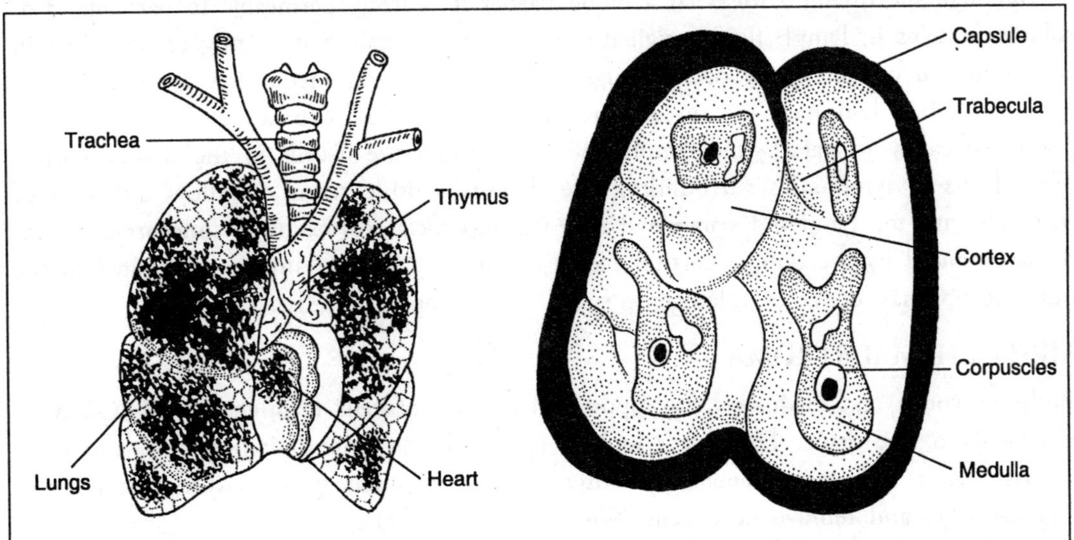

Fig. 3.3: Location and morphology of Thymus

The thymus is a flat, bilobed organ situated over the heart. Lobes are covered by a capsule and are divided into lobules by connective tissue strands called trabeculae.

The lobules are divided into 2 regions—

Cortex

Cortex comprises an outer compartment, high thymocyte concentration, site of progenitor T-cell growth, and nurse cells in outer area of cortex nourish thymocytes (Figure 3.3)

- Thymic epithelial cells in the outer cortex, act to hold cells in place and to secrete factors for differentiation and maturation including alpha 1-thymosin, beta 4-thymosin, thymopoitin, thymulin, IL-7, progenitor T-cells are double negative (cd 8-, cd 4-), start to develop TCR, as they mature and move through the cortex they become, **double positive (cd4+, cd8+)**.

Medulla

Medulla has cell density much less, cells are mature and are **single positive** either **cd 4+8- or cd 4-8+**. Studies with tracers of blood vessel permeability showed that cortical thymocytes, in contrast with those of the medulla, are protected from circulating macromolecules. Blood vessels in the thymic cortex were not permeable to the protein tracers, but the postcapillary venules of the thymic medulla permitted traces to leak along the clefts between migrating lymphocytes and endothelial cells. The tracers, which crossed medullary venule walls, had limited distribution in the thymic parenchyma because macrophages in the perivascular space ingest and retain much of the leaked tracer. The phagocytic cells and basal lamina at the border between the cortex and medulla captured any tracer that may have escaped endocytosis in the vessel wall. Thus hematogenous antigen has access, albeit restricted, to the thymic medulla.

Although the thymic cortex has a blood barrier it is freely permeable to particulates or soluble proteins in lymph fluids originating in the peritoneal cavity. Antigens deposited in the peritoneal cavity rapidly pass through diaphragmatic and mediastinal lymphatics and pass through a lymphatic plexus around the thymus. This exposes antigens from the peritoneal cavity to parathymic lymph nodes and immature T-cells in the thymic cortex. The plexus of lymphatics surrounding the thymus capsule seems to allow diffusion of materials into the lymphoid parenchyma of the superficial cortex on a concentration gradient produced by the unique cortical blood vasculature. Soluble antigens deposited in the outer cortex may deliver a tolerogenic signal to developing T-cells.

MESENCHYMAL STROMA

Both the cortex and medulla are composed of a mesh of cells forming the **STROMA** or framework of the tissue. These cells include epithelial cells, interdigitating cells, and macrophage. These cells make contact with thymocytes to stimulate growth, carry out +ve and –ve selection, and remove dead cells (90-95% of cells die).

Macrophages, Langerhans-like dendritic cells and perivascular reticular connective tissues comprise the mesenchymal stroma of the thymus. Macrophages are distributed throughout the cortex and medulla with a slight increase in density at the corticomedullary border. Immunohistology and enzymatic reactions reveal heterogeneity among thymic mononuclear cells. Nonspecific esterase-positive cells are more prevalent in the medulla but rare scattered cells are present in the corticomedullary junction and cortex. Few of the esterase-positive mononuclear phagocytes and some of the medullary macrophages contain phagocytosed nuclear debris and periodic acid Schiff (PAS) positive granules. Acid phosphatase-positive cells are concentrated at the corticomedullary junction, the location of dendritic cells. These cells express class I and II MHC antigens and stain positively with the OKt-6 monoclonal which labels Langerhans cells in skin.

Although Ia-antigen bearing epithelial cells may be important for T-cell development, it is the nonepithelial dendritic cells in the thymus that control development of helper T-cells. Monoclonal antibodies to Ia-antigen produced profound immunosuppression when injected in vivo. Repeated inoculation of newborn mice with anti-Ia results in loss of the Ia-antigen bearing Langerhans-like cells in the medulla without effecting Ia-antigen expression of thymic epithelial cells. Furthermore, such treatment selectively reduces development and release of helper T-cells into the periphery without effecting development of suppressor cells or immature thymic cortical cells. Thus helper T-cell development is entirely dependent on functional Ia-antigen bearing Langerhans cells and development of immature T-cells and suppressor cells occurs through interactions with different cells.

Noradrenergic and peptidergic fibers enter the thymus with nerve bundles and plexuses around blood vessels, penetrate into the cortex from subcapsular plexuses and branch among lymphocytes in the thymic cortex. The vasculature and lymphatic tissue of the outer and deep cortex are inervated by these fibers. The medulla is remarkably free of innervation except for fibers associated with blood vessels. Neuropeptide antigens are present in varicosities of thymic neurons. Vasoactive intestinal peptide reactive neuronal varicosities and acetylcholinesterase positive profiles are present in the deep cortex and corticomedullary regions, respectively. In vitro studies indicate that T-cell responses can be effected by neurosecretory products and anatomical studies clearly show association of nerve fibers with aggregates of lymphoid cells in vivo. However, the precise relationships of the anatomy and physiology of neuroimmune functions in T-cell development and differentiation remain to be resolved.

THYMIC MICROENVIRONMENTS

The epithelium of the thymus undergoes progressive keratinization and maturation like the skin. The epithelial reticulum of outer cortex, inner cortex, corticomedullary junction and medulla is distinguishable by monoclonal antibodies that bind cytoplasmic keratins of different molecular weight. Subcapsular thymic epithelium contains less mature keratin while the most mature keratin is found in Hassal's corpuscles in the medulla. Lectin binding patterns in the thymus also indicate corticomedullary maturation. Taken together, the epithelium of the thymus may not be static but turns over continuously like the epithelium of the skin. The Hassal's corpuscles are filled with whorles of keratinized epithelium, leukocytes and debris. They may be internal sites where exfoliated epithelium is endocytosed, degraded and carried away by leukocytes.

Thymic epithelial cells have immunomodulator and neuroendocrine functions which effect T-cell function inside and outside the thymus as indicated by the ability of thymic implants in diffusion chambers to reconstitute T-cell immune functions in thymusless mice. Granular deposits of IL-l are present in thymic cortical epithelial and/or mononuclear cells. Thymosin Al and serum thymic factor are detected in both cortical and medullary epithelial cells, whereas thymosin B3 has been found only in the cortical population. Hall suggests that there are bidirectional circuits between the central nervous system and the immune system. Thymosins, interleukins, complement components, enkephalins, adreno-corticotropic hormones and thyroid stimulating hormones present in thymus cells produce neuroendocrine effects on both systems.

Thymic epithelial cells express class II MHC antigen, in three basic patterns. The subcapsular epithelial layer is class II negative. The remaining cortex, including the thymic nurse cells, is strongly class II positive in a reticular pattern that extends from the outer cortex to the corticomedullary border. The inner membrane of thymic nurse cells is strongly positive for class I as well. The medulla is densely class I and II positive but it is uncertain that class II is expressed on the spatulate epithelial cells because treatments that deplete bone marrow derived dendritic cells renders the medulla class II negative. Dendritic cells of the donor's class I and II allotype are present in the medulla by three weeks following bone marrow reconstitution while the cortical reticulum retains that of the host. All cells in the thymus express MHC class I antigens but the highest expression is on nonepithelial cells. Although self/non-self education and acquisition of haplotype restriction of developing T-cells requires intrathymic expressed MHC antigens, it is not known whether MHC+ epithelial cells or MHC+ mesenchymal cells contribute equally or not.

T-Cell maturation

Development and maturation of T-cells is under the influence of the epithelial and mesenchymal stroma. The epithelial cell stroma exhibits regional specialization with regard to structure and expression of antigenic markers. The mesenchymal stroma consists primarily of mononuclear phagocytes, Langerhans-like cells and unusual cell types like the rhabdomyocyte which appear to represent anatomic and antigenic stereotypes of normal host cells. Interaction of precursor T-cells with these various stromal elements presumably determines whether immunoreactivity or tolerance results.

The thymus provides a receptive microenvironment for bone marrow pre-T cells to enter, clonally expand, select antigen and MHC haplotype specificities, mature and exit to the periphery after elimination of rogue clones. The thymus releases fewer than 5% of its daily cell production, the remainder apparently dying with the organ. The route of traffic of pre-T cells to specific regions of the thymus and out again has long been debated and the controversy is still with us. The prevailing dogma has been that Pre-T cells multiply in the cortex, begin to mature and then migrate toward the medulla where recirculating cells exit via postcapillary venules near the corticomedullary junction and non-recirculating cells exit via efferent lymphatics in the capsule and interlobular septae. Proliferation of immature T-cells occurs in various discrete cortical microenvironments including inside cytoplasmic envelopes of nurse cells located in the outer cortex. Immature T-cells enclosed by nurse cells do not differentiate into cytotoxic T-cells and a minority express markers indicating helper cell commitment (Andrews et al 1985) but the ultimate fate of T-cells educated by nurse cells is unknown. Other T-cells divide and mature in association with distinctive epithelial cells and/or mononuclear cells of bone marrow origin.

Cell differentiation markers, are valuable in describing the development of mature, functional T-cells. In mice, some marrow stem cells stain faintly for the Thy-l (theta) antigen but do not express CD-4 or CD8 antigens. Double negative cells that express Thy-l and Jlld markers are apparently ready for differentiation because they can be induced to express markers of mature T-cells in vitro and in vivo. No all double negative thymocytes can be considered T-cell precursors, however. Class I MHC antigens and common lymphocyte antigen (CD-5) are expressed on precursor T-cells throughout intrathymic and extrathymic maturation. Expression of CD-2 (Tll, SRBC receptor) molecules by immature T-cells is believed to facilitate formation of T-cell aggregates and adhesive interactions with thymic epithelial cells through binding with lymphocyte-function associated antigen (LFA-3) expressed on cell membranes. As pre-T cells infiltrate the thymic epithelium they lose surface TL (thymic lymphoma) antigen, express more Thy-l copies and begin to express CD-4 and CD-8 antigens. Double positive (CD-4+, CD-8+) T-cells that express antigen receptor are regarded as immature "cortical" cells that will differentiate into specific accessory T-cells after exposure to antigens and/or help from other T-cells. A proportion of peripheral blood T-cells express both CD-4 and CD-8 indicating that simultaneous expression does not prevent these cells from exiting the thymus. Expression of a single functional phenotype such as CD-4 (helper) or CD-8 (suppressor, cytotoxic) is associated with entry into the thymic medulla, exit to the periphery and inclusion in the recirculating pool. The L3T4 marker (mouse equivalent of CD-4) identifies the helper T-cell and does not react with the Lyt-l antigen on it or any other lymphocyte. Some CD-4 T-cells have been identified that have cellular cytotoxicity as a principal function. The CD-8 (Lyt-2 in mouse) expressing T-cell functions as suppressor cells or are responsible for class I MHC restricted, antigen specific cellular cytotoxicity against virally infected target cells and tumour cells. CD-8 cells are also present in the recirculating pool but cannot be negatively selected from efferent lymph by antigen as are CD-4 helper cells. The CD-8 T-cell is also able to return to the thymic cortex after emigrating while the CD-4+ T-cell is barred from returning unless the thymic perivascular microenvironment is altered as in Myasthenia Gravis. Some CD-8+ and all double positive cells appear to be short-lived because they rapidly disappear from the circulation after adult thymectomy or low level total body irradiation which effects thymic cortical lymphocytes.

Acquisition of MHC haplotype restriction is one of the most intriguing aspects of T-cell development in the thymus. During maturation in the thymus a dramatic selection takes place. Only those T-cell clones that express receptors capable of recognizing antigen in conjunction with the allelic forms of the MHC proteins expressed by thymic epithelial cells (or marrow-derived mononuclear cells) can mature fully. Mature helper T-cells are "restricted" to responding to antigen associated with the class II MHC antigen they saw in the thymic environment regardless of their own genotype. The same is true for mature cytotoxic T-cells who must recognize the class I MHC antigens, they saw during development, simultaneously with the viral or tumour antigen in order to deliver the lethal hit. Specificity for the foreign antigen is not sufficient to activate the respective T-cell if the antigen is expressed on an antigen-presenting cell or target cell that has MHC antigens different from those that were seen during intrathymic development.

The T-cell antigen receptor is analogous to membrane immunoglobulin in B-cells. The T-cell receptor is composed of two polypeptides (alpha, beta or gamma, delta) associated in the membrane with CD3 and a number of accessory molecules such as CD4/CD8 and CD2. The T-cell receptor complex binds to immunogenic peptides only if they are presented within MHC antigens of antigen-presenting cells or target cells. "Processed and presented" antigens are bound in a cleft lined by the polymorphic regions of the MHC antigens (Bjorkman et al 1987) which forms them into "recognizable" 10 angstrom alpha helices. At the gene level, the alpha and beta chain genes are like immunoglobulin heavy chain genes where variable, diversity, joining and constant gene segments are rearranged prior to expression of the protein on the T-cell membrane. This genetic rearrangement and expression is believed to occur after the pre-T cell arrives in the thymus and before it leaves for the periphery. Unlike immunoglobulin, the T-cell receptor for antigen cannot effectively bind antigen except in the context of MHC gene products expressed on accessory target cells.

Bursa of Fabricius

It is a primary lymphoid organ of birds which is situated near the cloaca. It consists of lymphoid centers that contain epithelial cells and lymphocytes. These lymphocytes are antibody producing cells. The B-cells undergo maturation in this organ. Like thymus the bursa is largest at hatching and undergoes atrophy with maturation.

- Mammals do not have Bursa of Fabricius.
- A structure in mammals that functions equivalent to that of birds is the bone marrow.

Functions of Bursa of Fabricius

1. It produces B lymphocytes. These are responsible for antibody production and humoral immunity.
2. These B-cells mature and go to the peripheral lymphoid organs and are seeded there.
3. Following appropriate antigenic stimulation, B lymphocytes transform into plasma cells and produce antibodies.

Peyer's patches

These are quite large aggregates of lymphoid tissue found in the small intestine. The overlying 'dome' epithelium contains large numbers of intraepithelial lymphocytes. Some of the epithelial cells have complex microfolds in their surfaces. They are known as M cells and are believed to be important in the transfer of antigen from the gut lumen to Peyer's patches. Peyer's patches facilitate the generation of an immune response within the mucosa. B-cell precursors and memory cells are stimulated by antigen in Peyer's patches. Cells pass to the mesenteric lymph nodes where the immune response is amplified. Activated lymphocytes pass into the blood stream via the thoracic duct. These cells then home in the gut and carry out their final effector functions. HEVs are not present in Peyer's patches and the mechanism by which cells home in on mucosal sites is unknown. Cell surface molecules known as addressins may have a role.

SECONDARY LYMPHOID ORGANS

Secondary lymphatic tissues are a site of stimulation of mature lymphocytes. It also create sites for Ag-lymphocyte interaction. Secondary lymphatic tissues control the quality of immune responses. Differences among the various lymphatic tissues significantly affect the form of immunity and relate to how antigens are acquired by these organs. **Lymph nodes** are filters of lymph, the **spleen** is a filter of blood and mucosal associated lymphatic tissues (MALT) acquire antigens by transcytosis to lymphoid tissue from the "external" environment across specialized follicle-associated epithelial cells. Various type of organized lymphoid' tissue are located along the vessel of lymphatic system.

- Antigen is collected and brought to tissue and lymphocytes are exposed to antigen.
  - lymph nodes—antigen from intracellular tissue fluids
  - spleen-blood—borne antigen
  - MALT—antigen from mucosal surfaces. Includes Peyer's patches, tonsils, and adenoids.

Lymph

Lymph is a clear body fluid which clots like blood. Lymph forms when dissolved proteins and solutes filter out of venules and capillaries because of local differences in luminal hydrostatic and osmotic pressure. Whenever the epithelial barrier is broached, the neuropeptide "substance P" is released by axons and increases lymph production through hyperemic effects. Hyperemia increases fluid transudate. This aids fluid transport to the regional lymph node of cells, antigens and cytokines. Afferent lymphatic vessels have their origins in the reticular connective tissue beneath the epithelium of the skin, gut, and urogenital tract and within the mesenchymal tissues of all organs. The endothelial cells of lymphatics are anchored to the surrounding reticular fiber meshwork, causing the lymphatics to dilate rather than collapse when tissue hydrostatic pressure exceeds the pressure of fluid in the lymphatic (Figure 3.4). This opens valve-like junctions between endothelial cells and allows intravasation of tissue fluids. Lymph flows unidirectionally toward lymph nodes because valves prevent back flow under normal physiological conditions. Lymph capillaries merge into larger lymphatics which drain into lymph nodes. Efferent lymph from regional lymph nodes may drain into one or more additional nodes before flowing into major efferent lymphatics. The thoracic duct carries lymph draining from the gut and the lower half of the body.

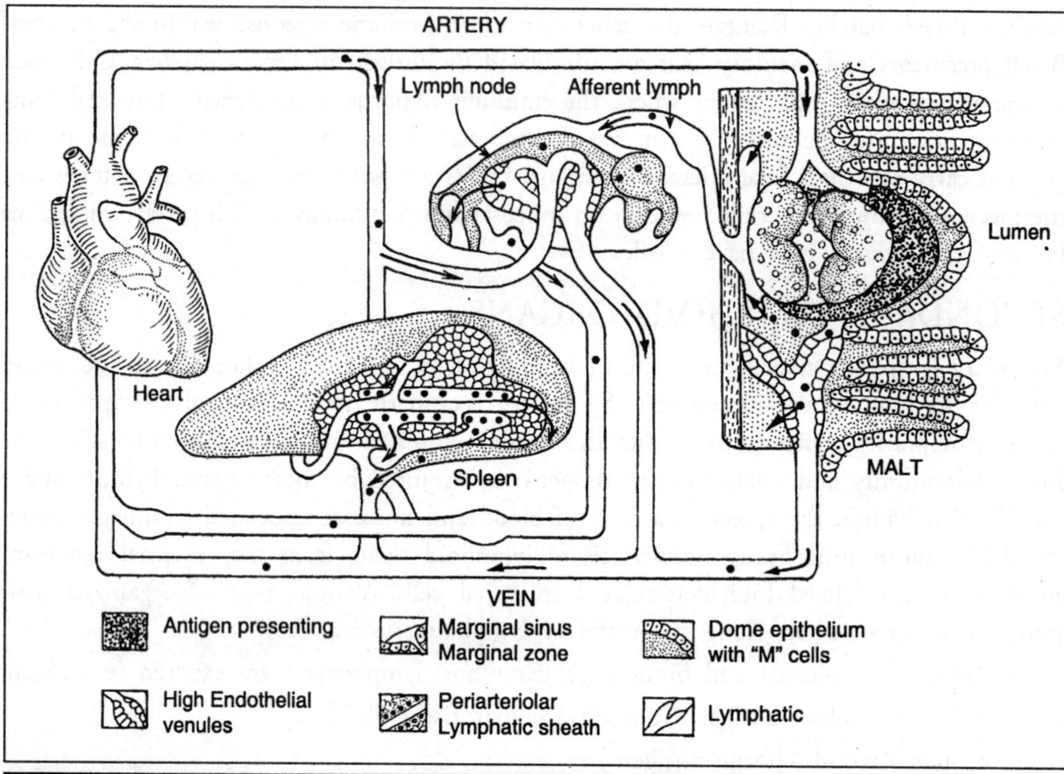

Fig. 3.4: Lymphatic system

Lymph nodes

Lymph nodes are small bean shaped structures lying along the course of lymphatics. They are aggregated in particular sites such as the neck, axillae, groins and para-aortic region. Knowledge of the sites of lymph nodes is important in physical examination of patients. Lymph nodes have two main functions. (1) Phagocytic cells act as filters for particulate matter and micro-organisms. (2) Antigen is presented to the immune system.

Structure

Lymph nodes have a fibrous capsule from which trabeculae extend towards the centre thus forming a framework (Figure 3.5). The Lymph node is made up of three components: A Lymph node is surrounded by a fibrous capsule. It has an outer cortex and inner medulla

- lymphatic sinuses
- blood vessels
- parenchyma (cortex, paracortex, medulla)

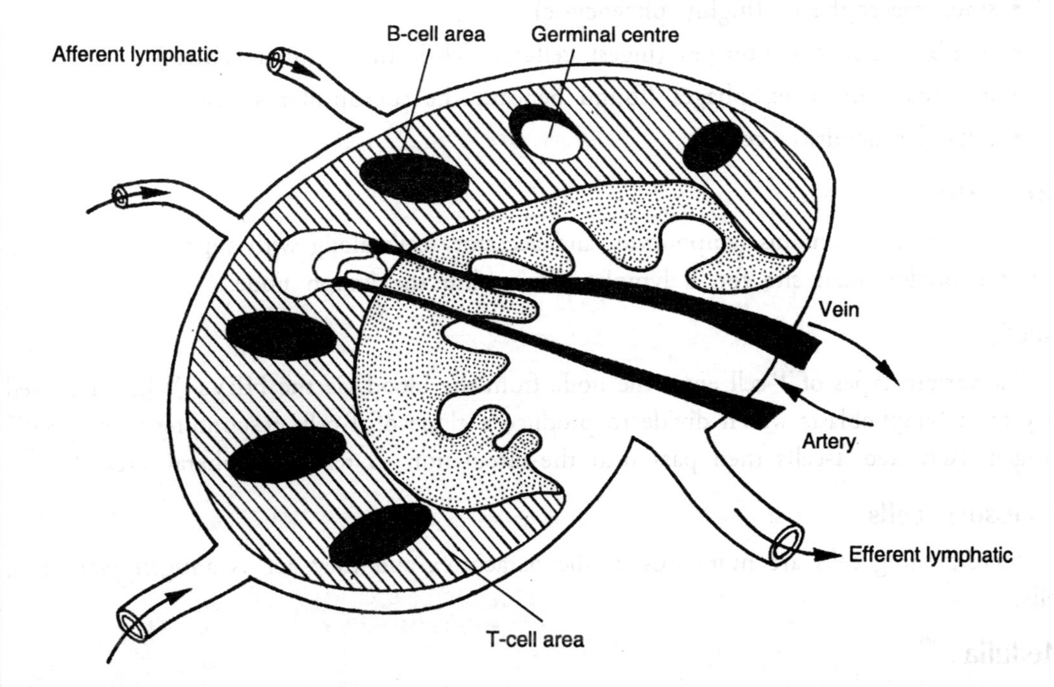

Fig. 3.5: Structure of Lymph node

Cortex

In the cortex the lymphocytes accumulate into primary and secondary follicles. These follicles contain lymphocytes and macrophages.

B-cells

These enter the lymph node via HEVs and pass to the follicles. If activated by antigenic stimulation they proliferate and remain in the node. Unstimulated B-cells, however, pass out rapidly from the node to return to the general circulation. Activated B-cells within the lymphoid follicles are known as follicle centre cells.

The pale staining central area of a secondary follicle is known as a germinal centre and this is surrounded by a mantle zone consisting of small, naive B-cells and a few T-cells. The follicle centre cells within the germinal centres consist of cells with cleaved nuclei (centrocytes) and cells with larger more open nuclei and several nucleoli (centroblasts). Stimulated mature B-cells responding to antigen change into centrocytes and then centroblasts. The centroblasts leave the follicle and pass to the paracortex and medullary sinuses, where they become immunoblasts. The immunoblasts divide to give rise to plasma cells or memory B-cells which are ready for their next encounter with specific antigen.

Accessory cells

Lymphocytes alone are not to make an effective immune response. They are assisted by so-called accessory cells. These may be grouped as follows:

- sinus macrophages (highly phagocytic)
- tingible body macrophages (ingest cellular debris in germinal centres)
- marginal zone macrophages (found beneath the subcapsular sinus)
- follicular dendritic cells

Paracortex

The paracortex contains lymphocytes and accessory cells along with supporting cells and it is the predominant site for T lymphocytes within the lymph node.

T-cells

The various types of T-cell enter the node from the blood via the HEVs. When activated they form lymphoblasts which divide to produce a clone of T-cells responding to a specific antigen. Activated T-cells then pass into the circulation to reach peripheral sites.

Accessory cells

Interdigitating cells are numerous in the paracortex and they act as antigen presenting cells.

Medulla

In the medulla the lymphocytes are arranged as elongated branching bands or medullary cords. The cortical follicles and medullary cords contain B lymphocytes and constitute bursa/bone marrow dependent areas. Between the cortical follicles and medullary cords, there is a broad, ill-defined intermediate zone called para cortical area. The para cortical area contains T lymphocytes and constitutes the thymus dependent area.

The medulla comprises:

- large blood vessels
- medullary cords
- medullary sinuses

The medullary cords are rich in plasma cells which produce antibodies that pass out of the node via the efferent lymphatic. Macrophages are also numerous within the medulla.

Passage of lymph

Lymph passes into the node through the afferent lymphatic into the marginal sinus, though the cortical sinuses to reach the medullary sinuses before leaving via the efferent lymphatic. Particulate matter in the lymph is removed by macrophages. Antigens are taken up by antigen presenting cells and these facilitate the specific immune response. Less than 10% of lymphocytes enter the node in the lymph, the large majority entering from the blood via the HEVs. Lymph nodes are connective tissue bags filled with mobile cells organized into functional compartments by a meshwork of reticulin fiber ensheathed by fibroblastic reticular cells; and, supplied by a system of specialized blood vessels and nerves. A "cortex" and "medulla" is distinguishable in histological preparations by the relative density of small lymphoid cells contained in the reticular matrix. The superficial cortex

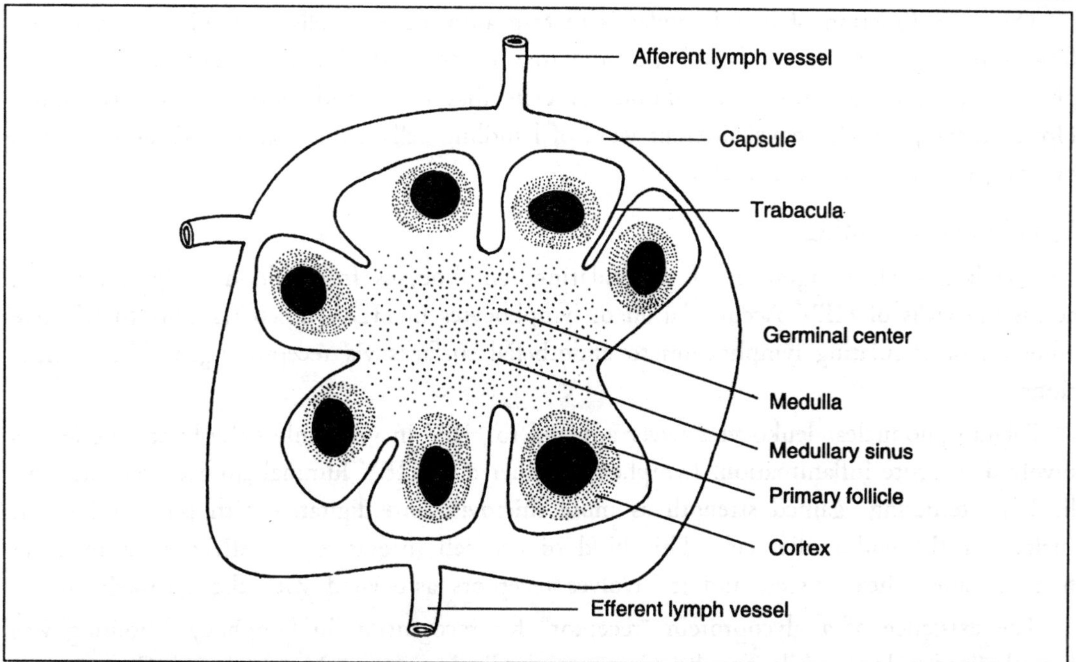

Fig. 3.6: Sectional view of Lymph node

contains a lymphatic sinus, macrophage rich zone and B-cell follicles. The deep cortex is a high traffic zone where migrant or recirculating T- and B-lymphocytes enter from the blood. Lymphocytes in the deep cortex are directed toward specific B- or T-cell microenvironments by fibroblastic reticular cell corridors where the lymphocytes encounter antigen presenting cells. If they are not activated by antigen displayed in Class II MHC along with costimulatory signals the T- or B-cells crawl into lymphatic channels and out of the lymph node. The medullary cords contain sessile B-lymphoblasts and plasma cells which accumulate there after immune reactions and the medullary lymph sinuses are the highways out of the lymph node. The cortex is organized into hemispheric lobules where the flat surfaces face the afferent lymph supply and the round central borders merge with stromal chords and sinuses to form the medulla. Cortical lobules in lymph nodes of all mammalian species have a constant vertical dimension. The lateral dimensions of cortical lobules are highly variable which accounts for the rapid 2-4 fold node enlargement that occurs 6-18 hours after antigen inoculation.

Function of lymph nodes

1. Lymph nodes act as filter for lymph.
2. Phagocytosis of foreign particles takes place in lymph node.
3. They help the proliferation and circulation of T- and B-cells.
4. They enlarge following local antigenic stimulus.
5. Lymph node is highly efficient in trapping antigen.
6. The plasma cells secrete antibodies.

The main function of lymph nodes is to trap antigens and cells containing antigen that flow into them via afferent lymphatics and to provide a site for clonal expansion of lymphoid cells recruited from the millions of cells that enter and leave via various routes. However, lymph nodes have different ways of handling cells, solutes and particles as sources of antigens that will be presented.

Lymphocytes homing

The large-scale emigration of recirculating lymphocytes from blood to lymph occurs across the walls of HEV. Accumulation of lymphocytes in HEV lumen was due to selective adhesion of circulating lymphocytes to endothelial surfaces via receptor ligand-like interactions.

Polymorphonuclear leukocytes never adhered to HEV in vivo unless the lymph node was involved in acute inflammation. Lymphocyte adhered to HEV luminal surfaces via microvilli. This "tethering" gained strength as more microvilli interdigitated with pits on luminal surface of the endothelial cell. This kind of cell-cell interaction is effective in resisting hydrodynamic shear forces, and it involves receptors associated with the cytoskeleton.

The existence of a glycoprotein "receptor" for recognition in lymphocyte homing was hypothesized a long while ago by Gesner and Ginsburg, and Marchesi and Gowans.

The specificity of organ specific homing could result from specific combinations of complimentary adhesive molecules and a cascade of cell interactions.

The original "Homing Receptors" that enabled "organ specific" lymphocyte homing are part of the consensus. Both MadCAM-1 (Mucosal Vascular Addressin) and PNad-1 (peripheral node addressin) are bound by alpha 4 beta 7 integrin and/or L-selectin because of specificity for sulfated carbohydrates. Specific oligosaccharides inhibited in vitro binding of lymphocytes to HEV thereby demonstrating that the lymphocyte receptor recognizes a carbohydrate moiety (Sialyl Lewis X) on the HEV.

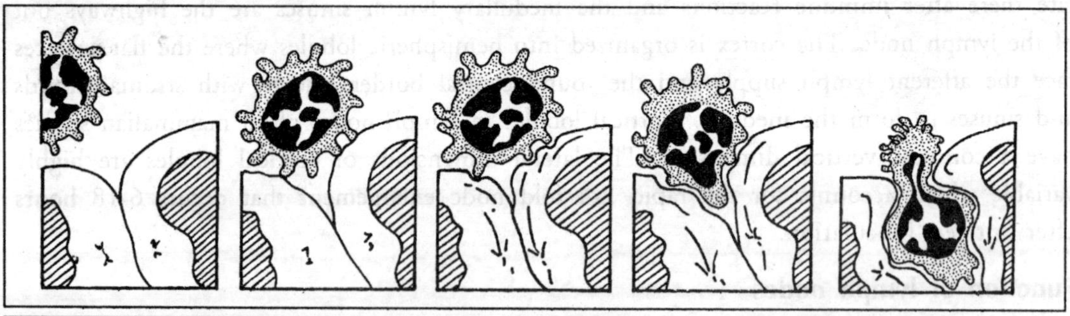

Fig. 3.7: Lymphocytes homing

The unique permeability of the HEV from lymphatic tissue toward the blood permits cytokines or other factors in adjacent lymphatic tissue to diffuse along spaces between HEV endothelial cells to yield chemotactic gradients. Fibroblastic reticular cells ensheathing reticular fibers serve as conduits that carry "chemokines" and other factors directly to the HEV perivenular channel.

This unique anatomic system provides for rapid and discrete communication between tissue and blood interface. When antigen inoculation causes altered blood flow, cytokine release, and increased tissue fluid pressures in regional lymph nodes, the resultant interendothelial gradient in HEV accelerates lymphocyte migration into the lymphatic tissue.

After squeezing out of perivenular channel into corridors and entering the deep cortical interstitium, lymphocytes crawl along cell processes and through a latticework of reticular fiber until they reach reactive foci in T- or B-cell areas.

They also cross lymphatic endothelium to enter intermediary sinuses that deliver them into the efferent lymph. Contact guidance along fibroblastic reticular cells was proposed as a mechanism for intranodal redistribution of migrants because lymphocytes bind to these cells and migrate rapidly on their surfaces in vitro.

Aggregation or repulsion of lymphocyte sets was also proposed to explain compartmentalization of lymphoid cells within lymphatic tissues.

THE LYMPHOCYTE TRAFFICKING

The deep cortex of lymph nodes, which appears in histological sections as a diffusely cellular area containing specialized vascular structures called high endothelial venules (HEV).

High endothelial venules (HEV), has also been called the T dependent cortex because of the adverse effect of neonatal thymectomy upon maturation of recirculating lymphocytes, and because a preponderance of T-cells are present in immunohistological preparations.

B-cells also migrate into the deep cortex. In "resting" lymph nodes the reticular connective tissue surrounding HEV contains numerous small lymphocytes in cords of reticular cell matrix between the HEV and lymphatic vessels of the cortical "intermediary" sinuses.

The size of the deep cortex depends upon the rate of lymphocyte immigration via HEV, the transit time of cells traversing the reticulum, and the rate of egress of immigrants via efferent lymphatic channels.

Indeed, the paracortex is capable of rapid 3 to 5 fold enlargement 6 to 24 hours after a strong stimulus induced by viral infection, injection or immunologic adjuvants, or inoculation of alloantigens.

Accelerated lymphocyte traffic usually results in logjams of small lymphocytes in cortical sinuses. While most of the acute enlargement of the deep cortex during the first 24 hours after inoculation is due to increased lymphocyte traffic into the node, some of it is due to sequestration and activation of antigen reactive cells causing them to remain to proliferate or release lymphokines. The outcome of this selection by antigen and cytokines is the subsequent outpouring of antigen reactive cells in the efferent lymph between 72 hours and 100 hours after inoculation.

The two alternating images are of lymph nodes from a normal rat and from a rat whose recirculating lymphocytes were depleted by continuous thoracic duct drainage for up to 14 days. Depletion of recirculating lymphocytes reduces a lymph node to its stromal and vascular components. The Paracortex is collapsed into a small mass of vessels and connective tissue. Germinal centers (follicle small lymphocyte mantle zones are absent) are all that remain of the B-cell compartment. Both images are shown at exactly the same magnification.

The deep cortex shrinks rapidly after chronic thoracic duct drainage or neonatal thymectomy. Both procedures are experimental manipulations that selectively deplete recirculating lymphocytes. Although the deep cortex is commonly known as the T-dependent cortex it is not comprised solely of T-cells. The lymphocytes populating the deep cortex are about 75% T-cells and 25% B-cells, 90% of them are small lymphocytes and virtually all belong to the recirculating pool. Small B-cells appear to migrate up to the superficial cortex after briefly accumulating around HEV. Small T-cells move out through the perivenular channels into corridors of reticulum comprising the paracortex.

There are slightly more CD8+ T-cells compared with CD4+ T-cells in resting lymph nodes but the ratio can vary considerably. This also applies to the ratio of T-cells to B-cells. Normal T- and B-cell recirculation through lymph nodes seems to be under control of constitutively expressed "Secondary Lymphoid Tissue Chemokine (SLC) and "B-Lymphocyte Chemokine" (BLC), shown in insitu hybridization preparations by Alexei Tumanov. After alloantigenic stimulation lymph nodes fill up with T-cells. In contrast, after inoculation of an antigen which selectively induces an antibody response, the deep cortex contains nearly 60% B-cells. Therefore, the deep cortex is a dynamic lymphatic compartment which responds to stimulation by cytokines, antigens, and other biological response modifiers.

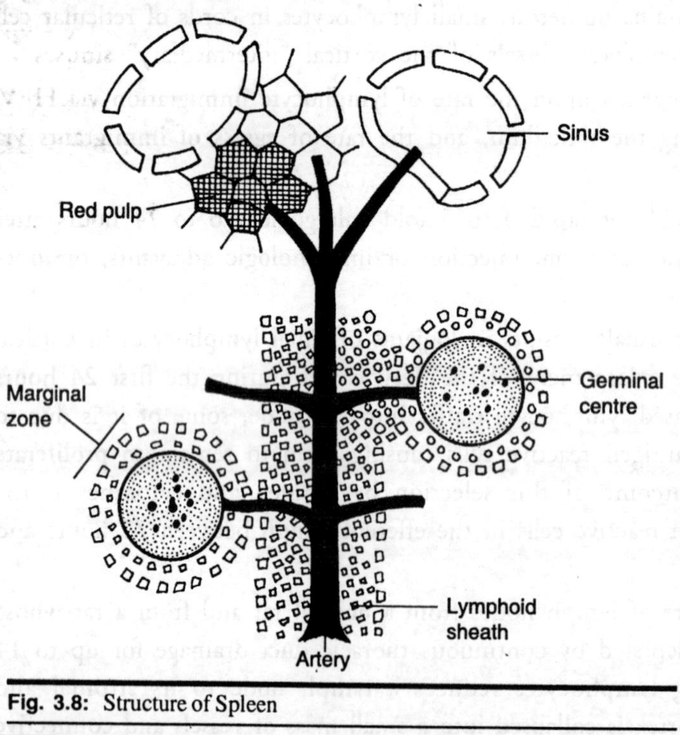

Fig. 3.8: Structure of Spleen

SPLEEN

The spleen is large, ovoid secondary lymphoid organ situated in the left abdominal cavity, while lymph node are specialized for trapping antigen from local tissue, the specialized for filtering blood and trapping blood born antigen. The spleen is located in the upper left quadrant of the abdomen. It has two main functions acting as part of the immune system and as a filter.

The spleen has a thin connective tissue capsule from which short septa extend inwards.

These septa are, in turn, connected to a complex reticulin framework. There are two distinct components of the spleen, the red pulp and the white pulp. The red pulp consists of large numbers of sinuses and sinusoids filled with blood and are responsible for the filtration function of the spleen. The white pulp consists of aggregates of lymphoid tissue and is responsible for the immunological function of the spleen:

Red pulp

There is a complex system of blood vessels within the red pulp arranged to facilitate removal of old or damaged red blood cells from the circulation. A small proportion of the splenic blood flow passes through more rapidly without undergoing this process of filtration.

White pulp

The white pulp contains T-cells, B-cells and accessory cells. There are many similarities with lymph node structure. The purpose of the white pulp is to mount an immunological response to antigens within the blood. The white pulp is present in the form of a periarteriolar lymphoid sheath. This sheath contains B-cell follicles and T-cells. At the edge of the T zone is a region known as the marginal zone where larger lymphocytes and antigen presenting dendritic cells are located.

The spleen is large ovoid secondary lymphoid organ. It is an organ weighing about 150 grams in adults and located in the left abdomen. The spleen filters blood and is the largest single lymphoid organ in the body. It has a dense fibrous capsule with muscular trabeculae extending inward to subdivide the spleen into lobules and provide anchors for the reticular meshwork supporting the cellular parenchyma.

The spleen is grossly divisible into white pulp, formed of cylindrical collections of lymphocytes around the arteries, and red pulp, which contains erythrocyte-rich blood in cords of reticulum. The splenic cords are a continuous sponge-like reticular tissue which crisscross between fenestrated walls of splenic venous sinuses. The cords contain erythrocytes, lymphocytes, macrophages, granulocytes, and plasma cells.

The splenic venous sinuses are lined by elongated endothelial cells that resemble barrel staves. The venous sinus wall is ringed by hoops of reticular fibers. All blood cells liberated into the splenic cords must squeeze through slits between these endothelial cells to enter the sinuses and return to the blood circulation. Splenic sinus lining cells express antigens associated with monocytes, macrophages, endothelial cells (Factor VIII, transferrin receptor), and T-lymphocytes. The spleen monitors for abnormal cells in blood by mechanical seiving in venous sinuses and exploration of surface antigens by mononuclear phagocytes.

The spleen is supplied by the splenic artery which enters through the hilus and branches to pass through trabeculae into medium sized arteries surrounded by a sheath of lymphatic tissue (periarteriolar lymphatic sheath, PALS).

These vessels terminate in small arterioles which pass through PALS and empty into the reticulum of the red pulp cords or pulp sinuses. Within the PALS narrow perpendicular branches of the central arteriole supply a ring of sinuses within the marginal zone of the PALS.

These sinuses fill readily with radio opaque perfusates and materials used to make vascular corrosion casts for scanning electron microscopy; otherwise they would not be noticed. Marginal sinuses are critically important to lymphocyte recirculation in the spleen.

Their supply vessels arise perpendicularly which may concentrate lymphoid cells through skimming the cells from the peripheral stream of the central arteriole. Blood from red pulp cords and marginal sinuses empties into the venous sinusoidal system which connects by gradual transition to pulp veins and splenic veins.

The spleen has not afferent lymphatic vessels, but efferent lymphatics arise within the white pulp near central arterioles. The efferent lymphatics cross perpendicularly between PALS areas to form an interconnecting network in cleared preparations of spleens whose lymphatics are labelled with carbon. The structure of the lymphatic network provides for cellular communication among the anatomically separated white pulp areas.

Lymphatic trunks pass through trabeculae, merge with each other, and exit at the hilus. Major efferent lymphatics of the spleen wind down the hilar blood vessels to join the thoracic duct near the pancreas. Parts of the intra-splenic lymphatic network have anatomical relationships resembling the "bridging zones" of Mitchell because they appear to link white pulp with red pulp in histological sections.

Noradrenergic fibers enter the hilus of the spleen with the vasculature, follow the trabeculae and branching vasculature, and ramify mainly within the white pulp along the central arterioles and periarteriolar lymphatic sheath. Fibers branch from a dense periarterial plexus to permeate and terminate among diffuse small lymphoid cells and accessory cells in the PALS.

Immunocytochemical data indicate that nerve fibers with neuropeptide-Y, met-enkephalin, cholecystokinin 8, and neurotensin immunoreactivity are present along the central artery with sparse fibers entering lymphoid tissue. Cholecystokinin-8 bearing fibers are more prevalent in the PALS, however.

FUNCTIONS

In principle, the function of the spleen and its response to antigen are much like those lymph node. The essential difference is that the spleen is major site for initiation of immune responses to blood born antigens, whereas lymph node are involved in responses antigen in lymph.

1. The spleen is the primary site for initiation of immune responses to antigens and pathogens that have invaded the blood stream in addition to being a partner in every other immune response in the body.
2. The spleen is important filter of blood, the filtration of the cells, debris and micro-organisms occurs primarily in the reticulum of the red pulp chords and in the sieve-like endothelium of venous sinuses.

3. The spleen receives a high proportion of the cardiac output which contributes to its effectiveness as a filter of blood and a site for lymphocyte recirculation and lodging.
4. The reticular cells provide anchoring sites for mononuclear phagocytes that perform reticuloendothelial functions in the red pulp and marginal zone.
5. The white pulp of the spleen contains a peripheral lymphatic tissue microenvironment for antigen trapping, cellular collaboration, lymphocyte proliferation and antibody production in addition to providing young bone marrow emigrant B-cells a place to complete maturation.
6. The antigen-laden mononuclear cells and lymphoblasts, released into efferent lymph from other lymphatic tissues 48 to 100 hours after antigen priming, lodge in the spleen and set up satellite zones of T- and B-cell proliferation.
7. During active immune responses lymphoblastic B-cells committed to plasma cell differentiation lodge in the red pulp cords and sinuses where they mature and commence secreting antibody.

LYMPHOCYTE TRAFFIC

Lymphocyte traffic in the spleen moves across a blood tissue interface into sites of antigen presentation in the marginal zone and periarteriolar lymphatic sheath. Antibodies, enzymes, and drugs that are known to block lymphocyte homing to lymph nodes do not interfere with entry of the PALS by recirculating lymphocytes.

Many of the lymphocytes entering the spleen in the arterial blood bypass the white pulp and flow into the reticulum of the red pulp cords and sinuses to exit via the splenic vein.

These cells are returned to the blood within 2 to 3 hours after entering the spleen. However, some small arterioles terminate in sinuses adjacent to the white pulp; others empty directly into the marginal zone.

Lymphocytes in the marginal sinuses enter the marginal zone, leaving other leukocytes in the lumen to be carried away in the blood.

Lymphocytes arriving in the marginal zone move across a vascular interface that is populated by antigen-binding macrophages and B-cells to enter the PALS.

Once in the PALS, lymphocytes may remain within the spleen for up to 12 hours. The PALS is analogous to the lymph node paracortex because it contains recirculating T- and B-cells in addition to interdigitating dendritic cells.

Segregation of T- and B-cells into discrete zones within the white pulp is not rigorous. Follicles located in the border between the PALS and the marginal zone contain predominantly B-cells, and more recirculating T- than B-cells are found in the PALS.

After variable periods of residence in the PALS, some recirculating cells exit through the efferent lymphatics, while others move into the red pulp sinuses. The magnitude of lymphocyte recirculation via efferent lymph versus venous return has not been determined, but traffic studies with viable lymphocytes whose motility has been arrested indicate that over

80% of the lymphocytes which leave the spleen in the venous blood never migrated into white pulp.

If immunocompetant cells encounter an appropriate antigenic stimulus, they proliferate in the same manner as that seen in peripheral nodes; B-cells form large germinal centers and antibody secreting cells can be found within the splenic cords after migration through the blood or direct movement out from the white pulp.

The marginal zone of the spleen contains a heterogenous assortment of mononuclear cells with specialized functions. A principal function of the marginal zone is antigen trapping.

Intravenously injected tracer antigens reproducibly accumulate in marginal zones regardless of the nature of the antigen. A subset of macrophages in the marginal zone appear to be important in presenting type-2 thymus independent antigens to B-cells.

The interfollicular areas are diffusely infiltrated with small and medium sized lymphocytes, usually of T-cell origin but numerous IgM+/IgD+ B-cells are also present. Recirculating lymphocytes emigrate from the blood into Peyer's patches via a network of HEV surrounding follicles. Lymphocyte recirculation across HEV in patches is very rapid. Accumulation of labelled lymphocytes peaks 90 minutes after infusion, which is the shortest transit time of any organized lymphatic tissue. Lymphocyte traffic to the patch is augmented by recirculation across venules which are not recognizable as HEV. Some lymphoid cells enter the lamina propria via thin-walled venules adjacent to intestinal crypts far away from the Peyer's patch and flow along submucosal lymphatics until they reach lymphatic plexuses in interfollicular areas.

T-Lymphocyte traffic patterns related to mucosal and peripheral lymphatic tissues are segregated with some crossing over occurring in mesenteric lymph node and the spleen.

MUCOSAL ASSOCIATED LYMPHATIC TISSUES (MALT)

In addition to lymph node, spleen and lymphocytes, there are some lymphatic tissue located beneath the gastrointestinal and respiratory tracts. The mucosa lining the alimentary, respiratory, genito-urinary and other surfaces is exposed to antigens. So these areas have rich collection of lymphoid cells. These lymphoid cells are called Peyer's patches or scattered isolated lymphoid follicles. The lymphatic tissues in theses site constitute mucosal immune system. These are called as mucosa associated lymphoid organs (MALT).

The mucosal lymphatic tissues are nonencapsulated submucosal lymphoid nodules and diffuse lymphocytic infiltrates in the submucosa of intestinal and respiratory tracts. These organs work collectively with regional lymph nodes and spleen to produce B- and T-effector cells which lodge in lamina propria and in intraepithelial locations wherever there is mucosa. Nasopharyngeal lymphatic tissues (i.e. tonsils and adenoids in man), broncus lymphatic tissues, Peyer's patches, appendix and isolated follicles in intestinal mucosae vary with regard to type of surface epithelium (stratified squamous, ciliated columnar, or absorptive columnar) and relative proportions of T- and B-cells (Tonsils have 60% T-cells compared to 25%-40% T-cells in respiratory or intestinal patches) but the similarity of these tissues to Peyer's patches is greater than the difference, especially since all have "M"cells in their follicle associate epithelium.

The prototypical mucosal lymphatic tissue is the Peyer's patch which has a unique dome epithelium that is specialized to sample environmental antigens. Peyer's patches contain lymphoid compartments that are analogous to the deep cortex and follicles of lymph nodes, but there are no afferent lymphatics and no medullary cords for local accumulation of plasma cells. Each Peyer's patch contains multiple individual B-cell follicles separated by diffuse lymphoid tissue in interfollicular areas.

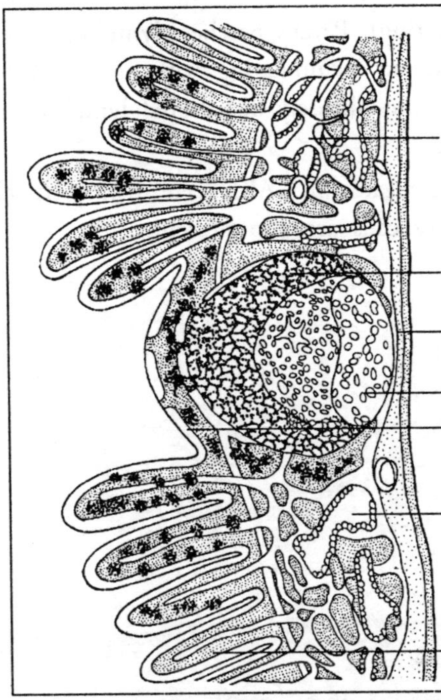

Fig. 3.9: Structure of Mucosa associated lymphoid organs

Functions of malt

1. They contain lymphoid and phagocytic cells and a special cell called Mcells, Both B- and T-cells are present.
2. Predominant antibody produced is IgA in the mucosal lining. Others such as IgG, IgM, and IgE are also produced.
3. IgA antibodies provide first line of defense against infectious agents.
4. The MALT appears to have two paradoxically opposite immune purposes while also participating in continued diversification of the immunoglobulin repertoire.
5. The MALT amplify development of committed B-cells for secretory IgA responses to enviromental antigens IgA responses to environmental antigens and programs certain environmental antigens for systemic tolerance induction.
6. The tolerance that results from mucosal immunization does not effect production of B-cells committed to IgA secretion. "Mucosal" tolerance is manifested by antigen-specific suppression of delayed cutaneous hypersensitivity and reduced IgG expression.
7. The MALT therefore exerts a Yin Yang effect on local versus systemic immunity initiated in unique mucosal and peripheral lymphoid microenvironments.

Gut associated lymphoid tissue (GALT)

The GALT system is exceptionally complex and is composed of several interactive sites. Lymphoid tissues in the gut, i.e. from the adenoids and tonsils up to the follicles of colon are called gut associated lymphoid tissues (GALT). This system comprises: tonsils, adenoids

(Waldeyer's ring), Peyer's patches, lymphoid aggregates in the appendix and large intestine, lymphoid tissue accumulating with age in the stomach, small lymphoid aggregates in the oesophagus and diffusely distributed lymphoid cells and plasma cells in the lamina propria of the gut.

Large aggregates of GALT have distinct B-cell follicles and T-cell areas. Antigen presenting accessory cells are also present.

CHAPTER 4

IMMUNOGLOBULIN

INTRODUCTION

Antibodies or immunoglobulins (Ig) are a group of glycoproteins present in the blood serum and tissue fluids of mammals. Serum glycoproteins can be separated according to their charge in an electric field and classified as albumin alpha-1 globulin, alpha-2 globulin; beta globulin, and gamma globulin. Notice that the gamma globulin band is wide because it contains a heterogeneous class of immunoglobulins; namely IgG, IgA, IgM and IgD. These differed from each other in molecular size, structure charge, amino acid composition, and carbohydrate content. Most antibodies in serum are in the IgG class. Has a similar mobility to IgD but cannot be qualitatively similar because of its low concentration in serum. Immunoglobulin are gamma globulin protein that are found in blood or other bodily fluids of vertebrates, and are used by the immune system to identify and neutralize foreign objects, such as bacteria and viruses. They are typically made of basic structural units—each with two large heavy chains and two small light chains—to form, for example, monomers with one unit, dimers with two units or pentamers with five units. Antibodies are produced by a kind of white blood cell called a B-cell. There are several different types of antibody heavy chain, and several different kinds of antibodies, which are grouped into different *isotypes* based on which heavy chain they possess. Five different antibody isotypes are known in mammals, which perform different roles, and help direct the appropriate immune response for each different type of foreign object they encounter.

Blood can be separated in a centrifuge into a fluid and a cellular fraction. The fluid fraction is the **plasma** and the cellular fraction contains red blood cells, leukocytes, and platelets. Plasma contains all of the soluble small molecules and macromolecules of blood, including fibrin and other proteins required for the formation of blood clots. If the blood or plasma is allowed to clot, the fluid phase that remains is called **serum.** It has been known since the turn of the century that antibodies reside in the serum. The first evidence that antibodies were contained in particular serum protein fractions came from a classic experiment by A. Tiselius and E.A. Kabat, in 1939. They immunized rabbits with the protein ovalbumin (the albumin of egg whites) and then divided the immunized rabbits' serum into two aliquots. Electrophoresis of one serum aliquot revealed four peaks corresponding to albumin and the alpha, beta, and gamma globulins. The other serum aliquot was reacted with ovalbumin, and the precipitate that formed was removed; the remaining serum proteins, which did not react with the antigen, were then electrophoresed. A comparison of the electrophoretic profiles of these two serum aliquots revealed that there was a significant drop in the gamma globulin.

The study of immunoglobulin or antibodies began in 1890 when Emil von Behring and Shibasaburo Kitasato described antibody activity against diphtheria and tetanus toxins. Behring and Kitasato put forward the theory of humoral immunity, proposing that a mediator in serum could react with a foreign antigen. Their idea prompted Paul Ehrlich to propose the side chain theory for antibody and antigen interaction in 1897, when he hypothesized that receptors (described as "side chains") on the surface of cells could bind specifically to toxins – in a "lock-and-key" interaction—and that this binding reaction was the trigger for the production of antibodies. Other researchers believed that antibodies existed freely in the blood and, in 1904, Almroth Wright suggested that soluble antibodies coated bacteria to label them for phagocytosis and killing; a process that he named opsoninization.

In the 1920s, Michael Heidelberger and Oswald Avery observed that antigens could be precipitated by antibodies and went on to show that antibodies were made of protein. The biochemical properties of antigen-antibody binding interactions were examined in more detail in the late 1930s by John Marrack. The next major advance was in the 1940s, when Linus Pauling confirmed the lock-and-key theory proposed by Ehrlich by showing that the interactions between antibodies and antigens depended more on their shape than their chemical composition. In 1948, Astrid Fagreaus discovered that B-cells, in the form of plasma cells, were responsible for generating antibodies.

A major advance in these structural studies was the discovery in the early 1960s by Gerald Edelman and Joseph Gally of the antibody light chain, and their realization that this protein was the same as the Bence-Jones protein described in 1845 by Henry Bence Jones. Edelman went on to discover that antibodies are composed of disulfide bond-linked heavy and light chains. Around the same time, antibody-binding (Fab) and antibody tail (Fc) regions of IgG were characterized by Rodney Porter. Together, these scientists deduced the structure and complete amino acid sequence of IgG, a feat for which they were jointly awarded the 1972 Nobel prize in Physiology or Medicine. While most of these early studies focused on IgM and IgG, other immunoglobulin isotypes were identified in the 1960s: Thomas Tomasi discovered secretory antibody (IgA) and David Rowe and John Fahey identified IgD, and IgE was identified by Kikishige Ishizaka and Teruki Ishizaka as a class of antibodies involved in allergic reactions.

Although the general structure of all immunoglobulin is very similar, a small region at the tip of the protein is extremely variable, allowing millions of antibodies with slightly different tip structures to exist. This region is known as the hypervariable region. Each of these variants can bind to a different target, known as an antigen. This huge diversity of antibodies allows the immune system to recognize an equally wide diversity of antigens. The unique part of the antigen recognized by an antibody is called an epitope. These epitopes bind with their antibody in a highly specific interaction, called induced fit, that allows antibodies to identify and bind only their unique antigen in the midest of the millions of different molecules that make up an organism. Recognition of an antigen by an antibody *tags* it for attack by other parts of the immune system. Antibodies can also neutralize targets directly by, for example, binding to a part of a pathogen that it needs to cause an infection.

The large and diverse population of antibodies is generated by random combinations of a set of gene segments that encode different antigen binding sites (or *paratopes*), followed by random mutations in this area of the antibody gene, which create further diversity. Antibody genes also re-organize in a process called class switching that changes the base of the heavy chain to another, creating a different isotype of the antibody that retains the antigen specific variable region. This allows a single antibody to be used by several different parts of the immune system. Production of antibodies is the main function of the humoral immune system.

IMMUNOGLOBULIN STRUCTURE

All immunoglobulin molecules have a basic structure composed of four polypeptide chains connected to each other by disulfide bond. The antibody molecule is made up of two identical heavy and two identical light chains held together by interchain disulfide bonds. These chains can be separated by reduction of the S–S bonds and acidification. Each light chain usually consists of about 220 amino acids and has a molecular weight of 25000. The heavy chains are structurally distinct for each immunoglobin class or subclass regions. Each heavy chain usually consists of about 440 amino acids and has a molecular weight of 50000. In the most abundant type of antibody, **immunoglobulin G (IgG)**, the exposed hinge region is extended in structure due to the high proline content and is therefore vulnerable to proteolytic attack. Therefore the molecule is split by papain to yield two identical **Fab** fragments, each with a single combining site for antigen, and a third fragment, Fc, which lacks the ability to bind antigen. Pepsin strikes at a different point and cleaves the Fc from the remainder of the molecule to leave a large fragment, which is formulated as F(ab)2 since it is still divalent with respect to antigen binding just like the parent antibody.

Antibodies have move than one antigen-combining site for example, most human antibodies have two combining sites or are bivalent. Some bivalent antibody molecules can combine to form multimedia antibodies that have up to ten combining sites.

The four chains are arranged in the form of a flexible Y with a hinge region. This hinge region the entire antibody molecule to assume a T shape. The stalk the Y is termed as the crystallizable fragment (Fc) a contains the site at which the antibody molecule cabind to a cell. The top of the Y consists of a two antigen-binding fragments (Fab) that bind with compatible antigenic determinant sites. The fabric mints have both constant and variable regions, where as Fc region composed only of constant region. Intrachain disulfide bonds create loops within each antibody chain (Figure 4.1).

More specially, the light chain exists in two distinct forms called kappa (k) and lambda (Y). These can be distinguished by the amino acid sequence. The light chain variable domain is hyper variable region that differ in amino acid sequence more frequently than the rest of the variable domain.

In the heavy chain the NH_2–terminal domain has a pattern of variability similar to that of the V_k and V_y domains and is termed the Vh domain. The other domains of the heavy chains are termed constant region and are numbered CH_1, CH_2, CH_3 and sometimes CH_4 starting with the domain next to the variable domain. The constant domains of the heavy chain form the constant (CH) region.

IMMUNOGLOBULIN CLASSES

Types or classes of antibodies

In human there are five classes of heavy chain called gamma, alpha, mu, delta and epsilon. The properties of these five heavy chains determine, respectively, the five-immunoglobulin classes- IgG, IgA, IgM, IgD, and IgE. Each immunoglobulin class differs in its general properties, half-life, distribution in the body, and interaction with other components of the host's defensive systems. They are each named with an "Ig" prefix that stands for immunoglobulin, another name for antibody, and differ in their biological properties, functional locations and ability to deal with different antigens.

Within two of the major immunoglobulin classes there are variants or subclasses. For e.g. IgG can be grouped into four subclasses (IgG_1, IgG_2, IgG_3 and IgG_4) and IgA into two subclasses (IgA1 & IgA2). Each subclass has differences in the amino acid composition of the heavy chain.

Immunoglobulin G (IgG)

IgG is the major immunoglobulin in the human serum, accounting for 70 to 75% of the immunoglobulin pool. IgG is present in blood plasma and tissue fluids. IgG class act against bacteria and viruses by opsonising the invaders and neutralizing toxins. It is also one of the two immunoglobulin classes that activate complement by classical pathways. IgG is the only immunoglobulin molecule able to cross placenta and provide naturally acquired passive immunity for newborn. There are about four IgG subclasses (IgG_1, IgG_2, IgG_3 and IgG_4) that vary chemically in their chain composition and the number and arrangement of interchain disulphide bond. About 65% of serum IgG is IgG1 and 23% is IgG2.

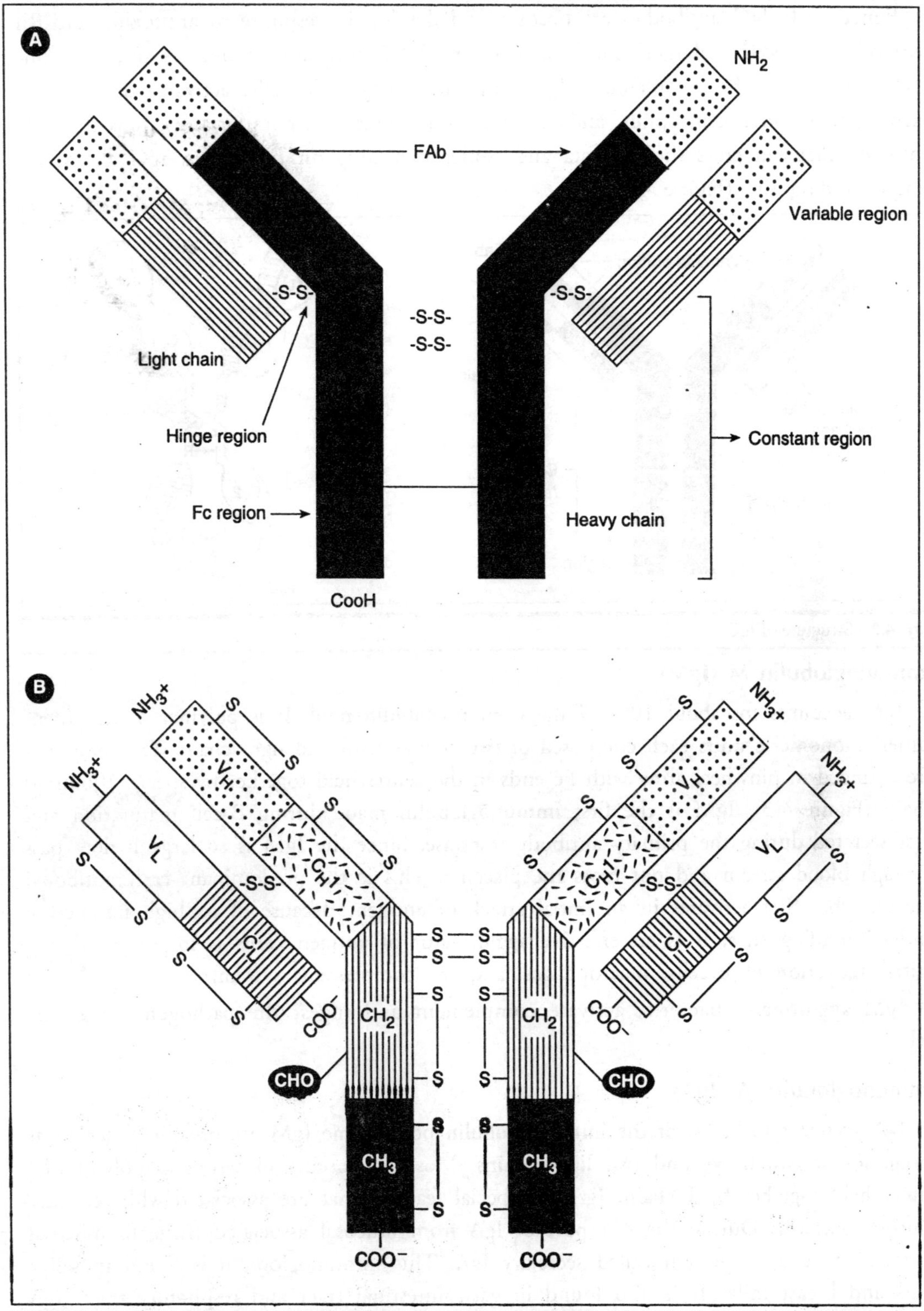

Fig. 4.1: General structure of Immunoglobulin

Functionally IgG antibodies are opsonic and develop in response to antitoxin. Anti-Rh antibodies are IgG1 and IgG3 subclass. IgG1 and IgG3 also bind monocytes and macrophages and activate the complement. IgG stimulates phagocytic cells, activates the complement system, binds neutrophils, and can neutralise toxins. Most importantly, it is the only antibody that can cross the placenta and confer immunity on the foetus. IgG also has a single binding site (Figure 4.2).

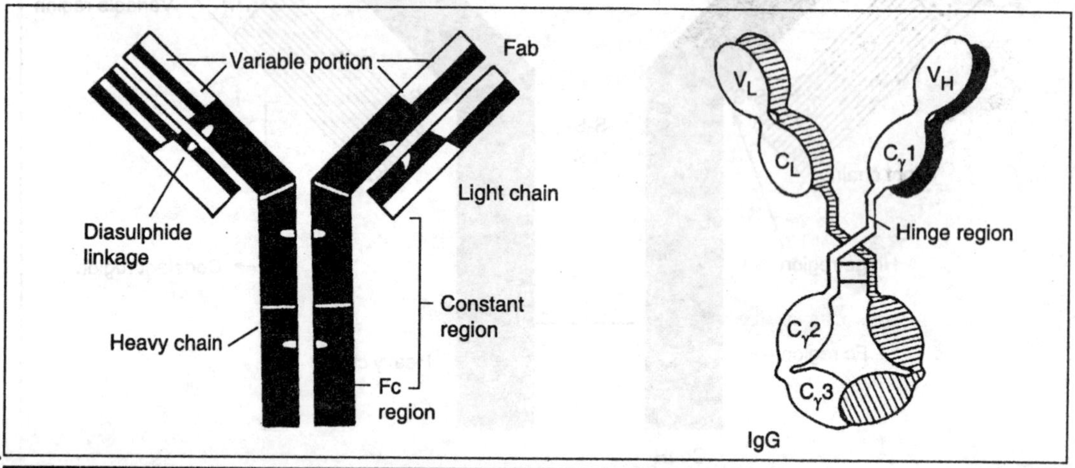

Fig. 4.2: Structure of IgG

Immunoglobulin M (IgM)

IgM accounts for about 10% of the immunoglobulin pool. It is polymer of five (pentamer) monomeric units each composed of two heavy chain and light chain. The monomer are arranged in pinwheel array with Fc ends in the centre held together by special J(joining) chain (Figure 4.3). IgM is the first immunoglobulin made during B-cell maturation and first secreted during the primary antibody response. Since the IgM is so large it does pass through blood stream and not cross the placenta. This is the predominant early antibody; the one that first activates in an initial attack of antigen. Because of its high number of antigen binding sites, it is an effective agglutinator of antigen. This is important in the initial activation of B-cells, macrophages, and the complement system.

IgM agglutinates bacteria, activate complement and ingest the pathogen phagocytic cells.

Immunoglobulin A (IgA)

IgA accounts for 15% of the immunoglobulin pool. Some IgM are present in serum as monomer of two heavy and two light chains. Most IgA occurs in serum as polymerized dimer held together by J chain. IgA has special features that are associated with secretary mucosal surfaces. During the transport of IgA from mucosal associated tissue to mucosal surfaces, it acquires protein called secretary IgA. This immunoglobulin is found in saliva tears and breast milk. It is also found in gastrointestinal tract and respiratory tract. IgA found in mucous secretions of the respiratory tract and the upper part of the digestive tract

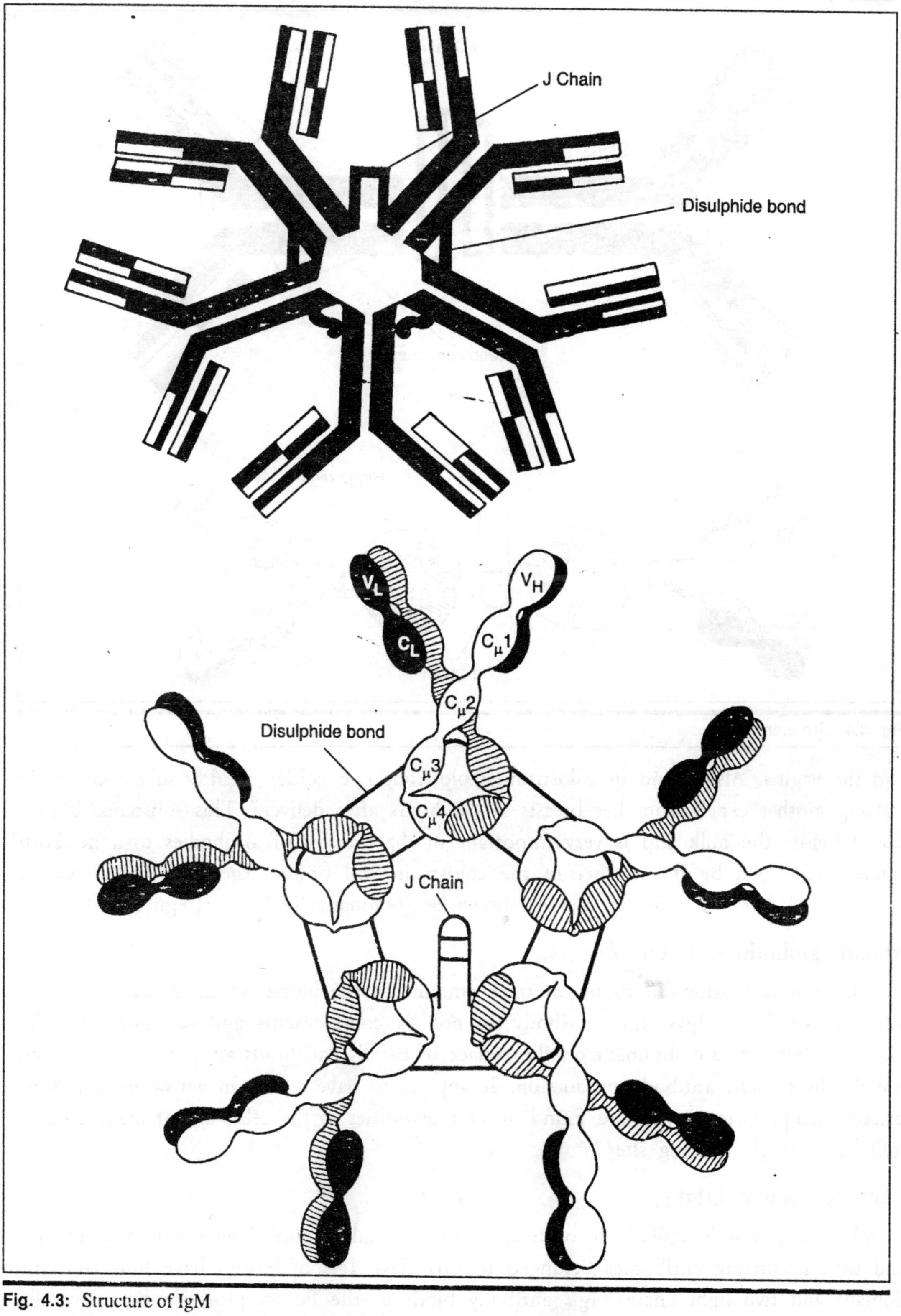

Fig. 4.3: Structure of IgM

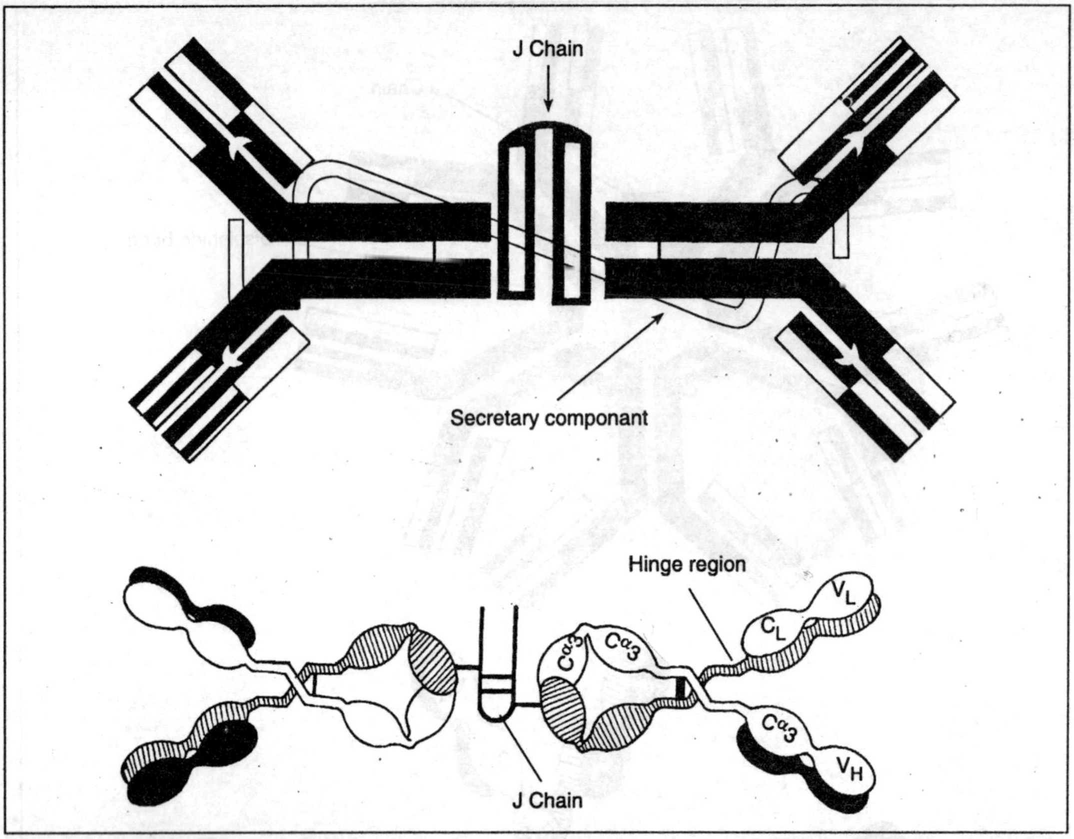

Fig. 4.4: Structure of IgA

and the vagina. Also found in colostrum. Colostrum is a golden liquidy substance that a nursing mother expels from her breasts 24-48 hours after delivery. This substance is produced before the milk and is very important in the transfer of antibodies to a newborn infant. IgA given by the mother in the colostrum will protect the baby for about six months. IgA has two binding sites as shown diagrammatically below (Figure 4.4).

Immunoglobulin D (IgD)

IgD is immunoglobulin found in trace amount in the blood serum. It has monomer structure similar to IgG. IgD antibody do not fix complements and can not cross the placenta, .but they are abundant on the surface of B-cells and binds antigen, thus signaling the B-cells to start antibody production. It appears to have a role in activating and suppressing lymphocyte activity and found in large quantities in the cell walls of many B-cells. IgD has a single binding site.

Immunoglobulin E(IgE)

IgE makesup only 0.00005% of total immunoglobulins pool. The classic anaphylactic and skin sensitizing antibodies belonged to this class. IgE molecules have four constant domain and two light chains. IgE antibody binds to the Fc receptor on the mast cells

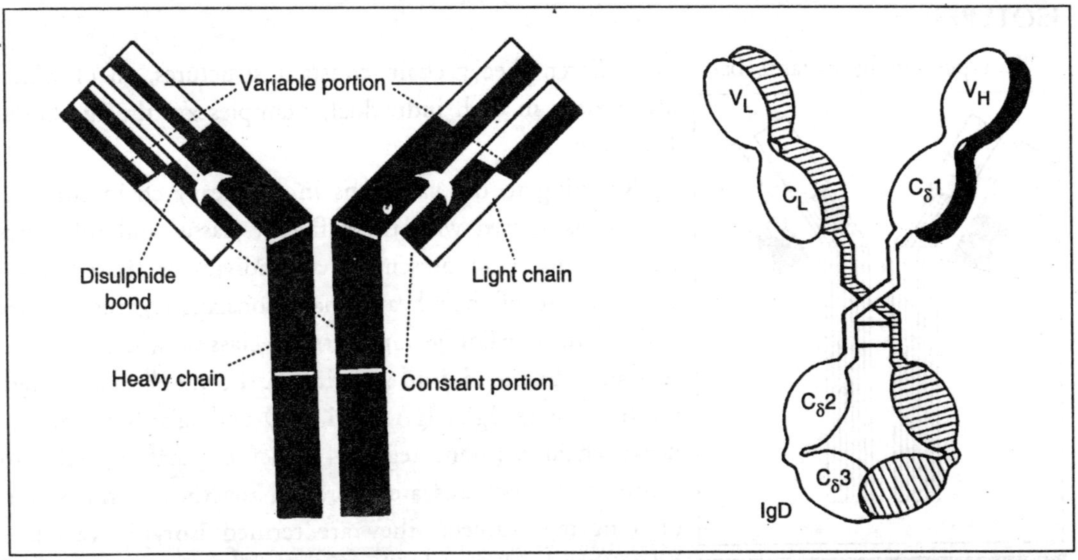

Fig. 4.5: Structure of IgD

surface (Figure 4.6). When antigen enters two IgE molecule cross linked the antigen and activate the mast cell that leads to degranulation of mast cells. Most importantly activates histamine secreting cells. Also appears to play a role in parasitic infection. IgE has a single binding site.

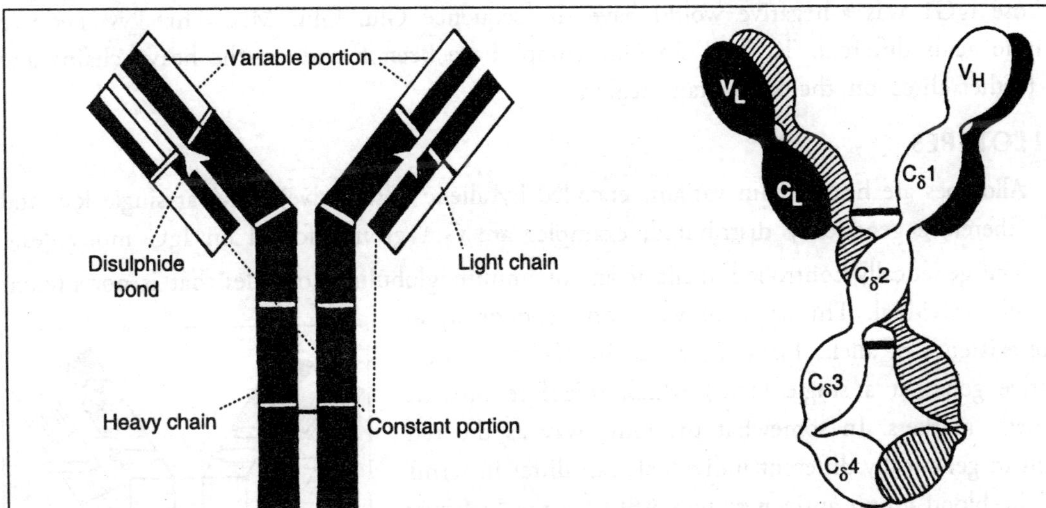

Fig. 4.6: Structure of IgE

STRUCTURAL VARIANTS

Within two of immunoglobulin classes, there are variants or subclasses. For example IgG can be grouped into four subclasses (IgG1, IgG2, IgG3, IgG4) and IgA into two subclasses IgA1 IgA2. Each sub classes has differences in amino acid sequence composition of heavy chain. These variations can be classified as:

ISOTYPES

Isotypes are Ig variants based on different heavy chain constant structures, all of which are present in each individual; examples are the Ig·classes, IgG, IgA, etc.

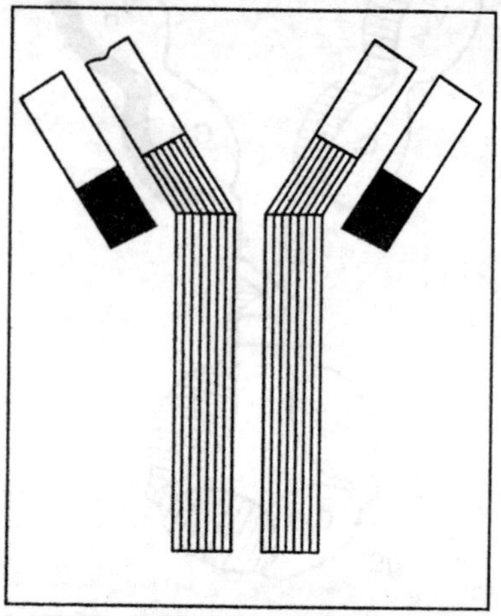

Fig. 4.7: Structure of Isotype variant

Referring to the variations in the heavy chain constant regions associated with the different classes and subclasses that are normally present in all individuals. Based upon the structure of their heavy chain constant regions, Igs are divided into major groups termed **classes**, which may be further subdivided into **subclasses**. In the human, there are five classes: IgG, IgA, IgM, IgD and IgE. Since all the heavy chain constant region, structures that give rise to classes and subclasses are expressed together in the serum of a normal subject, they are termed **isotypic variants.** Likewise, the light chain constant regions (C_L) exist in isotypic forms known as k and l which are associated with all heavy chain isotypes. Because the light chains in a given antibody are identical, Igs are either k or l but never mixed (unless specially engineered in the peptide chain. Take, for example, the G1m(a) locus on IgG1). An individual with this allotype would have the peptide sequence Asp. Glu. Leu. Thr. Lys on each of their IgG1 molecules. Another person whose IgG1 was a-negative would have the sequence Glu. Glu. Met. Thr. Lys, i.e. two amino acids different. To date, 25 Gm groups have been found on the heavy chains and a further three on the k constant region.

ALLOTYPES

Allotypes are heavy chain variants encoded by allelic (alternative) genes at single loci and are therefore genetically distributed; examples are Gm groups found on IgG molecules.

The genetically controlled allelic forms of immunoglobulin molecules that is not present in all individual. This type of variation depends upon the existence of allelic forms (encoded by alleles or alternative genes at a single locus) which therefore provide genetic markers. In somewhat the same way as the red cells in genetically different individuals can differ in terms of the blood group antigen system ABO, so the Ig heavy chains differ in the expression of their allotypic groups. Typical allotypes are the **Gm specificities** on IgG (Gm = *marker* on IgG). Allotypic differences at a given Gm locus usually involve one or two amino acids in the one end of the variable domain where they form parts of the b-turn loops and are clustered close to each other in space.

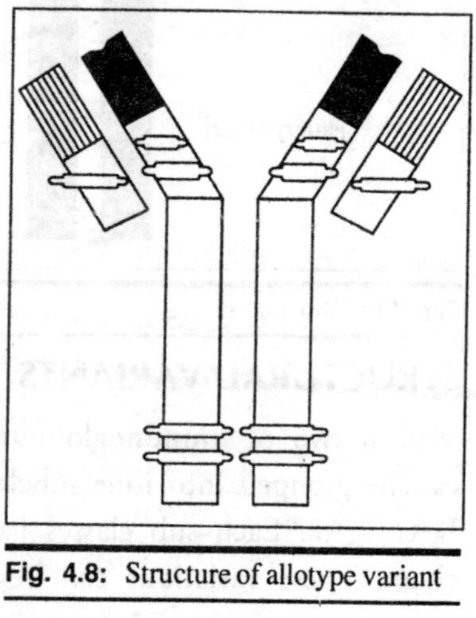

Fig. 4.8: Structure of allotype variant

IDIOTYPES

An idiotype is the collection of antigenic determinants on an antibody, usually associated with the hypervariable regions, recognized by other antigen-specific receptors, either antibody (the anti-idiotype) or T-cell receptors.

Referring to individual specific immunoglobulin molecules that differed in the hyper variable region of the Fab portion.The many variations of immunoglobulin structure reflect the diversity of antibody-generated by the immune response.

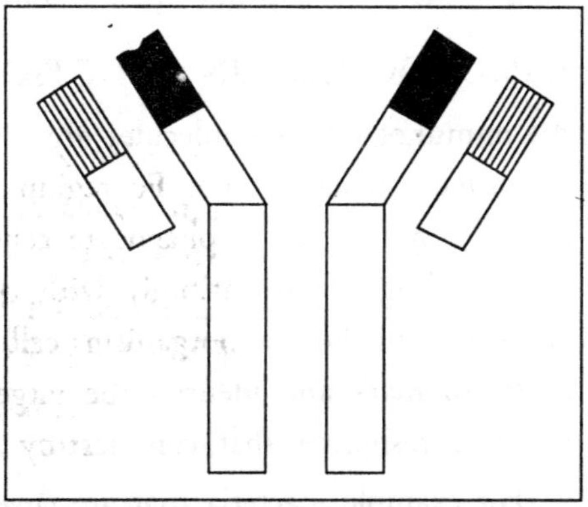

Fig. 4.9: Structure of Idiotype

It is possible to obtain antibodies that recognize the isotypic and allotypic variants described above. Similarly one can also raise antisera that are specific for individual antibody molecules and discriminate between one monoclonal antibody and another independently of isotypic or allotypic structures. Such antisera define the individual determinants characteristic of each antibody, collectively termed the **idiotype**.

Not surprisingly, it turns out that the idiotypic determinants are located in the variable part of the antibody associated with the hypervariable regions. The reader will (or should) be startled to learn that it is possible to raise autoanti-idiotypic sera since this means that individuals can make antibodies to their own idiotypes. Some T-cell receptors are also capable of recognizing idiotypes.

CLASS SWITCHING

Isotype or class switching is a biological process occurring after activation of the B-cell, which allows the cell to produce different classes of antibody (IgA, IgE, or IgG). The different classes of antibody, and thus effector functions, are defined by the constant (C) regions of the immunoglobulin heavy chain. Initially, naïve B-cells express only cell-surface IgM and IgD with identical antigen binding regions. Each isotype is adapted for a distinct function, therefore, after activation, an antibody with a IgG, IgA, or IgE effector function might be required to effectively eliminate an antigen. Class switching allows different daughter cells from the same activated B-cell to produce antibodies of different isotypes. Only the constant region of the antibody heavy chain changes during class switching; the variable regions, and therefore antigen specificity, remain unchanged. Thus the progeny of a single B cell can produce antibodies, all specific for the same antigen, but with the ability to produce the effector function appropriate for each antigenic challenge. Class switching is triggered by cytokines; the isotype generated depends on which cytokines are present in the B-cell environment.

Class switching occurs in the heavy chain gene locus by a mechanism called class switch recombination (CSR). This mechanism relies on conserved nucleotide motifs, called *switch (S) regions*, found in DNA upstream of each constant region gene (except in the δ-chain).

The DNA strand is broken by the activity of a series of enzymes at two selected S-regions. The variable domain exon is rejoined through a process called non-homologous end joining (NHEJ) to the desired constant region (γ, α or ε). This process results in an immunoglobulin gene that encodes an antibody of a different isotype.

IMMUNOGLOBULIN FUNCTION

All immunoglobulin molecules are bifuncitonal. The Fab region is concerned with binding to antigen, where as the Fc region mediates binding ro host tissue, various cells of the immune system, some phagocyte cells, or the first component of the complement system. The binding of an antibody with on antigen usually does not cause destruction of the antigen or of the microorganism, cell, or agent to which it is attached. Rather, the antibody serves to mark and identify the target for immunologic attack and to activate nonspecific immune responses that can destroy the target.

For example, bacteria that are covered with antibodies are better targets for phagocytosis by neutrophils and macrophages. This ability of an antibody to stimulate phagocytosis termed as opsonization. Immune destruction is also promoted by antibody-induced activation of the complement system. Both opsonization and the complement system are discussed and in phagocytes have an intrinsic ability to bind directly to microorganisms by non specific cell surface receptors, form phagosomes, and digest the process in which microorganisms or other particles are coated by antibody or complement, there by being prepared for "recognition" and ingestion by phagocytic cells. Opsonizing antipolice can be directed against surface components of bacteria. (e.g. Capsular polysaccharides M protein of group A streptococci and peptidoglycuns of staphylococci).

The inflammatory response can be triggered by non-immune and immune factors. An antigen triggers inflammation by two different routes. The first is mediated by IgE's attachment to the surface of a mast cells and basophils, which results in the release of histamine. Histamine causes most of the characteristic events of he inflammatory response. The second route involves the complement pathway in which the C_3a and C_5a fragments bind to mast cells and platelets, triggering the release of histamine. Immunity to disease like diarrhea depends on the production of specific antibodies that in active the toxins produced by the bacteria. This process is termed toxin neutralization. Once neutralized the toxin antibody complex is either unable to attach to receptor sites on host target cells or unable to enter the cell.

For example diphtheria toxin inhibits protein synthesis after binding the cell surface by the B fragment and subsequent passage of the active. A fragment into the cytosol of the target cell. Thus, the antibody blocks the toxic effect by inhibiting the entry of the fragment or the binding of the B fragment. Antiserum containing neutralizing antibody against a toxin is called antitoxin.

IgG, IgM and IgA antibodies can bind to some viruses during their extracellular phase and inactive them. This antibody mediated viral inactivating is called viral neutralization.

The capacity of bacteria to colonize the muscle surfaces of mammalian hosts is dependent impart on their algidity to adhere to musical epithelial cells. Recent studies have demonstrated that secretary IgA antibodies inhibit certain bacterial adherence promoting factors. Thus IgA has a unique rote in protecting the host against infection from certain pathogenic bacteria and perhaps from the other microorganism on musical surfaces.

When cells are infected with viruses, changes often occur in the plasma membrane proteins that led to the development of an antibody response against the cells. Cytolysis of the infected cells may be achieved by the membrane attack and complement or by the binding .of natural killer cells to the attuned cells after attachment to the Fc region of the antibody. Antibody provides the bridging mechanism for an intimate association between the effectors and target cells. After attachment, the natural killer cells destroys the target cell by releasing catatonic mediators or directly causing cytolysis. The process is called antibody dependent cell mediated cytotoxicity.

Immune reactions against protozoan and helminthes parasites are only partially understood. Parasites that have a tissue invasive phase in their life cycle are often associated with both eosinophilia (an excessive number of eosinophils in the blood) and elevated IgE levels. Recent evidence shows that in the presence of elevated IgE, eosinophils can bind in the parasites and discharge their lysosomal granules. Degranulation releases lytic and inflammatory mediators that destroy the parasites.

DIVERSITY OF ANTIBODIES

One unique property of antibodies is their remarkable diversity. According to current estimates, each human or mouse can synthesize more than 10 million different kinds of antibodies. How is this diversity generated? The answer is rearrangement of antibody gene segments, somatic mutations and generation of different codons during antibody gene splicing.

Immunoglobulin genes are split or interrupted genes with many exons. Embryonic B cells contain a small number of exons, close together on the same chromosome, that determine the constant (c) region of the light chains (Figure 4.10). Separated from them, but on the same chromosome, is a larger cluster of exons that determines the variable region of the light chains. During B-cell differentiation one exon for the constant region is spliced on to one exon for the variable region. It is the splicing than produces a complete light-chain antibody gene. A similar splicing mechanism also occurs to join the constant and variable exons of the heavy chains.

Because of light-chain genes usually actually consists of three parts, and the heavy chain consists of four to formation of a finished antibody molecule is slightly more complicated than previously outlined (Figure 4.11). The germ line DNA for the light chain gene contains multiple coding sequences called V and J (joining) chain. This DNA joining process is termed combinatorial joining science it can create many combinations of the V and J regions. When the light chain gene is transcribed, transcription continuous through portion combinatorial joining in the formulation of heavy chain genes occurs by means of

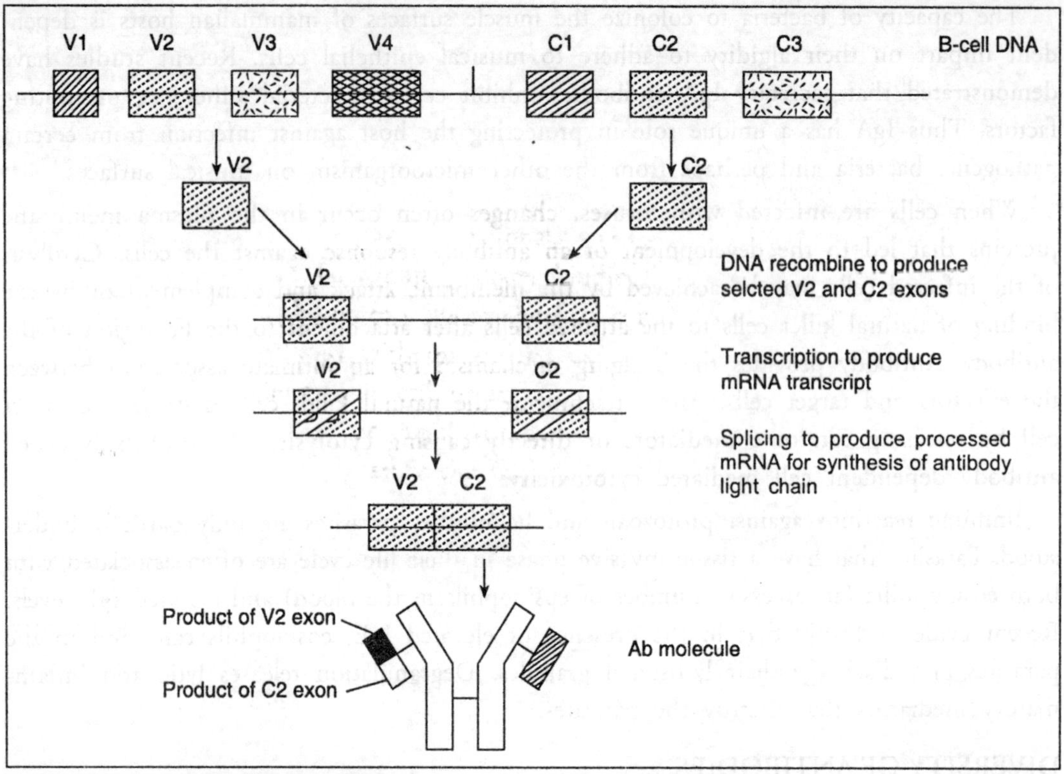

Fig. 4.10: Gene shuffling and antibody diversity

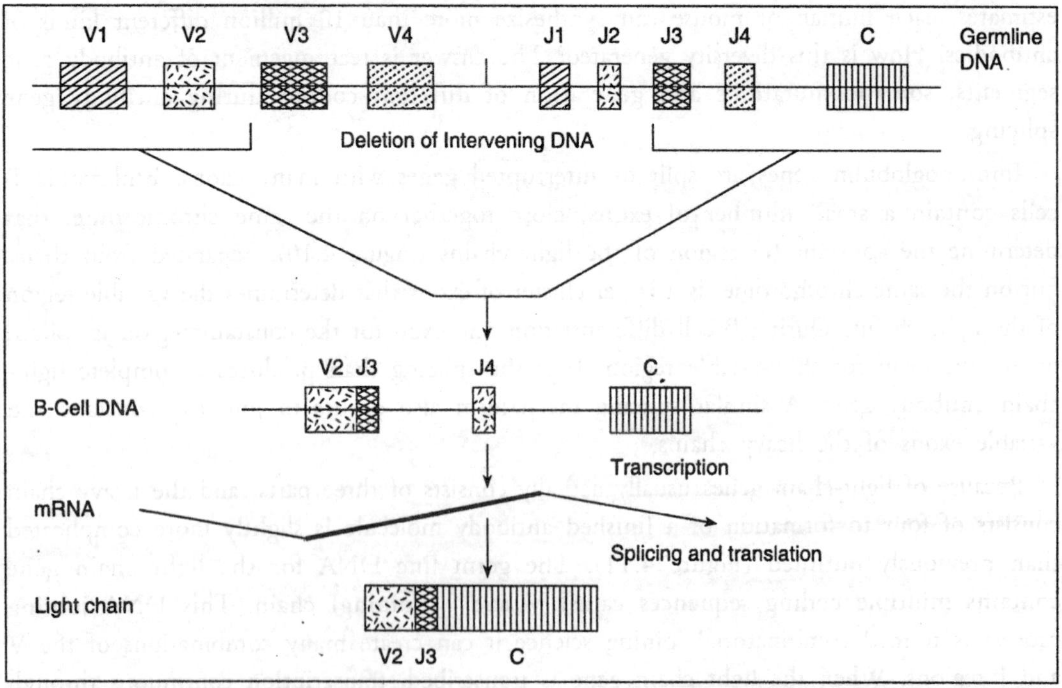

Fig. 4.11: Light chain production in mouse

DNA splicing of the heavy chain counterparts of V and J along with a thing set of D sequences. Initially, all heavy chains have this type of constant region. This corresponds to class IgM. Another DNA splice joins the VDII region with a different constant region that can subsequently change the class of antibody produced by the B-cell.

The amount of antibody diversity in the mouse that can be generated by combitorial joining. In this animal the K light chains are formed from combination of about 250Vk and 4Ik regions giving approximated 250 × 4 = 1,000 different K chains. The y chains have their k counter parts.

The heavy chain approximately 250 V_H, 10D, and $4J_H$ regions giving 250 × 10 × 4 = 10,000 different combinations. Because any light chain can combine with any heavy chain, there will be at least 1000 × 10,000 = 10^7 possible antibody types.

The value of 10^7 different antibodies is actually an underestimate science antibody diversity is further animated by two processes:

1. The V regions of germ line DNA are susceptible to a high rate of somatic mutation during B-cell. Development in the bone marrow. These mutation allow B-cell clones to produce different polypeptide sequences.
2. The junction of either VJ or VDJ splicing in combinatorial joining can occur between different nucleotides and thus generated different codons in the spliced gene. E.g. On VJ splicing event can join the V sequence CCTCCC with the J sequence TGGTGG in two ways. CCTCCC + TGGTGG = CCGTGG, which codes for the amino acids proline and tryptophan and CCTCCC + TGGTGG = CCTCGG, which codes for proline and arginine. Thus the same VJ joining could produce polypeptides differing in a single amino acid.

CHAPTER 5

ANTIGEN

NATURE OF ANTIGEN

An antigen is an agent which recognized by the body as non-self and can provoke an immune response. It may be chemical like protein or polysaccharide or it may be biological entity like microbe virus and parasites. A substances capable of inducing a specific immune response are called antigens; it is more appropriately called an immunogens. The immune system distinguishes between "self" and "non-self" through an elaborate recognition process. Prior to birth the body somehow makes an inventory of the proteins and various other large molecules present and removes most T-cells specific for self-determinants. Subsequently self-substances and be distinguished from non-self substances, and lymphocytes can produce specific immunologic reactions against the latter leading to their removal.

The foreign substances such as proteins, nucleoproteins, polysaccharides and some glycolipids, to which lymphocyte respond are called antigens or immunogens. Most antigens are large, complex molecules with a molecular weight generally greater than about 10,000. The ability of a molecule to function as an antigen depends on its size and the complexity of its structure.

Antigens are macromolecules that elicit an immune response in the body. Antigens can be proteins, polysaccharides, conjugates of lipids with proteins (lipoproteins) and polysaccharides (glycolipids). Antigens that enter the body from the environment; these would include inhaled macromolecules (e.g. proteins on cat hairs that can trigger an attack of asthma in susceptible people), ingested macromolecules (e.g., shellfish proteins that trigger an allergic response in susceptible people), molecules that are introduced beneath the skin (e.g. on a splinter or in an injected antigens) that are generated within the cells of the body; these would include proteins encoded by the genes of viruses that have infected a cell aberrant proteins that are encoded by mutant genes; such as mutated genes in cancer cells.

FACTORS INFLUENCING IMMUNOGENICITY

To protect against infectious disease, the immune system must be able to recognize bacteria, bacterial products, fungi, parasites, and viruses as immunogens. In fact, the immune system actually recognizes particular macromolecules of an infectious agent, generally either proteins or polysaccharides. Proteins are the most potent immunogens, with polysaccharides ranking second. In contrast, lipids and nucleic acids of an infectious agent generally do not serve as immunogens unless they are complexed with proteins or polysaccharides. Immunologists tend to use proteins or polysaccharides as immunogens in most experimental studies of humoral immunity. For cell-mediated immunity, only proteins and some lipids and glycolipids serve as immunogens. These molecules are not recognized directly. Proteins must first be processed into small peptides and then presented together with MHC molecules on the membrane of a cell before they can be recognized as immunogens. Immunogenicity is not an intrinsic property of an antigen but rather depends on a number of properties of the particular biological system that the antigen encounters.

A. CONTRIBUTION OF THE IMMUNOGENS

1. Foreignness

The immune system *normally* discriminates between self and non-self such that only foreign molecules are immunogenic. When an antigen is introduced into an organism, the degree of its immunogenicity depends on the degree of its foreignness. Generally, the greater the phylogenetic distance between two species, the greater the structural (and therefore the antigenic) disparity between them. For example, the common experimental antigen bovine serum albumin (BSA) is not immunogenic when injected into a cow but is strongly immunogenic when injected into a rabbit. Moreover, BSA would be expected to exhibit greater immunogenicity in a chicken than in a goat, which is more closely related to bovines.

2. Size

There is not absolute size above which a substance will be immunogenic. However, in general, the larger the molecule the more immunogenic it is likely to be. There is a correlation between the size of a macromolecule and its immunogenicity. The most active immunogens tend to have a molecular mass of 100,000 daltons (Da). Generally, substances with a molecular mass less than 5000–10,000 Da are poor immunogens, although a few substances with a molecular mass less than 1000 Da have proven to be immunogenic.

3. Chemical composition

In general, the more complex the substance is chemically the more immunogenic it will be. The antigenic determinants are created by the primary sequence of residues in the polymer and/or by the secondary, tertiary or quaternary structure of the molecule.

4. Physical form

In general particulate antigens are more immunogenic than soluble ones and denatured antigens more immunogenic than the native form.

5. Degradability

Antigens that are easily phagocytosed are generally more immunogenic. This is because for most antigens, the development of an immune response requires that the antigen be phagocytosed, processed and presented to helper T-cells by an antigen presenting cell (APC).

B. GENETIC FACTORS

Some substances are immunogenic in one species but not in another. Similarly, some substances are immunogenic in one individual but not in others (i.e. responders and non-responders). The species or individuals may lack or have altered genes that code for the receptors for antigen on B-cells and T-cells or they may not have the appropriate genes needed for the APC to present antigen to the helper T-cells.

1. Age

Age can also influence immunogenicity. Usually the very young and the very old have a diminished ability to mount and immune response in response to an immunogen.

C. METHOD OF ADMINISTRATION

The **dose** of administration of an immunogens can influence its immunogenicity. There is a dose of antigen above or below which the immune response will not be optimal. Generally the subcutaneous route is better than the intravenous or intragastric routes. The **route** of antigen administration can also alter the nature of the response. Substances that can enhance the immune response to an immunogens are called adjuvants. The use of **adjuvants**, however, is often hampered by undesirable side effects such as fever and inflammation.

CHEMICAL NATURE OF IMMUNOGENS

The vast majority of immunogens are **proteins**. These may be pure proteins or they may be glycoproteins or lipoproteins. In general, proteins are usually very good immunogens. Pure **polysaccharides** and lipopolysaccharides are good immunogens. **Nucleic acids** are usually poorly immunogenic. However, they may become immunogenic when single stranded or when complexed with proteins. In general **lipids** are non-immunogenic, although they may be haptens.

TYPES OF ANTIGENS

There are two different types of antigens: T-independent antigens and T-dependent antigens.

T-independent Antigens

T-independent antigens are antigens which can directly stimulate the B-cells to produce antibody without the requirement for T-cell help. In general, polysaccharides are T-independent antigens. The responses to these antigens differ from the responses to other antigens.

Properties of T-independent antigens

- **Polymeric structure:** These antigens are characterized by the same antigenic determinant repeated many times.
- **Polyclonal activation of B-cells:** Many of these antigens can activate B-cell clones specific for other antigens (polyclonal activation). T-independent antigens can be subdivided into Type 1 and Type 2 based on their ability to polyclonally activate B-cells. Type 1 T-independent antigens are polyclonal activators while Type 2 are not.
- **Resistance to degradation:** T-independent antigens are generally more resistant to degradation and thus they persist for longer periods of time and continue to stimulate the immune system.

T-dependent antigens

T-dependent antigens are those that do not directly stimulate the production of antibody without the help of T-cells. Proteins are T-dependent antigens. Structurally these antigens are characterized by a few copies of many different antigenic determinants.

Haptens

Many small organic molecules are not antigenic by themselves but can become antigenic if they bond to a larger carrier molecule such as a protein. They can not stimulate antibody formation by themselves, but can react with antibodies once formed. Such a small molecules are termed haptens [*Latin hatpin-to grasp*]. When lymphocytes are stimulated by the combined molecule, they can react to pair either the happen or the larger carrier molecule. One example of a hapten is penicillin. By itself, penicillin is not antigenic. However when it combines with certain serum proteins of sensitive individuals the resulting molecule does initiate a severe and sometimes fatal allergic immune reaction. In these instances the hapten is acting as a antigenic determinant on the carrier molecule.

A substance that is non-immunogenic but which can react with the products of a specific immune response. Haptens are small molecules which could never induce an immune response when administered by themselves but which can when coupled to a carrier molecule. Free haptens, however, can react with products of the immune response after such products have been elicited. Haptens have the property of antigenicity but not immunogenicity.

HAPTEN-CARRIER CONJUGATES

Hapten-carrier conjugates are immunogenic molecules to which haptens have been covalently attached. The immunogenic molecule is called the carrier. Structurally these conjugates are characterized by having native antigenic determinants of the carrier as well as new determinants created by the hapten (haptenic determinants). The actual determinant created by the hapten consists of the hapten and a few of the adjacent residues, although the antibody produced to the determinant will also react with free hapten. In such conjugates the type of carrier determines whether the response will be T-independent or T-dependent.

ANTIGENIC DETERMINANTS OR EPITOPES

Each antigen can have several antigenic determinant sites or epitope. These areas of the molecule stimulate production and combine with specific antibodies. Antibodies are formed most readily in response to determinants that project from the foreign molecule or to terminal residues of a specific polymer chain.

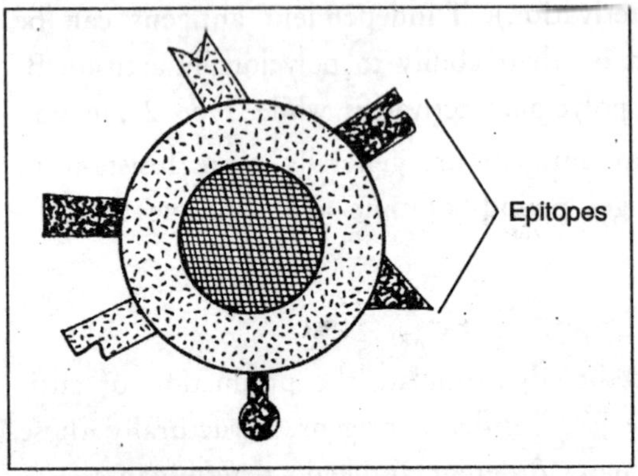

Fig. 5.1: Epitopes on antigen

Antigen molecules each have a set of antigenic determinants or epitope. Each antibody only recognizes one epitope rather than the whole multivalent antigen. The surface of an antigen is its valence. If one determinant site is present, the antigen is monovalent. Most antigens, however, have more than one determinant site and are termed multivalent. Multivalent antigens generally illicit a stronger immuneresponse than do monovalent antigen sometimes a multivalent antigen, termed a heterophile of heterologous antigen, can react with antibodies produced response to a different antigen.

Epitope recognized by B-cells

Antigenic determinants recognized by B-cells and the antibodies secreted by B-cells are created by the primary sequence of residues in the polymer (linear or sequence epitopes or determinants) and/or by the secondary, tertiary or quaternary structure of the molecule (conformational epitopes or determinants). In general antigenic determinants are small and are limited to approximately 4-8 residues, (amino acids and/or sugars). The combining site of an antibody will accommodate an antigenic determinant of approximately 4-8 residues. Although, in theory, each 4-8 residues can constitute a separate antigenic determinant, in practice, the number of antigenic determinants per antigen is much lower than what would theoretically be possible. Usually the antigenic determinants are limited to those portions of the antigen that are accessible to antibodies.

Epitopes recognized by T-cells

Antigenic determinants recognized by T-cells are created by the primary sequence of amino acids in proteins. T-cells do not recognize polysaccharide or nucleic acid antigens. This is why polysaccharides are generally T-independent antigens and proteins are generally T-dependent antigens. The determinants need not be located on the exposed surface of the antigen since recognition of the determinant by T-cells requires that the antigen be proteolytically degraded into smaller peptides. Free peptides are not recognized by T-cells, rather the peptides associate with molecules coded for by the major histocompatibility complex (MHC) and it is the complex of MHC molecules + peptide that is recognized by T-cells. In general antigenic determinants are small and are limited to approximately 8-15 amino acids. Although, in theory, each 8-15 residues can constitute a separate antigenic determinant, in practice, the number of antigenic determinants per antigen is much less than what would theoretically be possible. The antigenic determinants are limited to those portions of the antigen that can bind to MHC molecules. This is why there can by differences in the responses of different individuals.

Superantigens

When the immune system encounters a conventional T-dependent antigen, only a small fraction (1 in 10^4–10^5) of the T-cell population is able to recognize the antigen and become activated (monoclonal/oligoclonal response). However, there are some antigens which polyclonally activate a large fraction of the T-cells (up to 25%). These antigens are called **superantigens.**

Examples of superantigens include: Staphylococcal enterotoxins (food poisoning), Staphylococcal toxic shock toxin (toxic shock syndrome), Staphylococcal exfoliating toxins (scalded skin syndrome) and Streptococcal pyrogenic exotoxins (shock). Although the bacterial superantigens are the best studied there are superantigens associated with viruses and other microorganisms as well. The diseases associated with exposure to superantigens are, in part, due to hyper activation of the immune system and subsequent release of biologically active cytokines by activated T-cells.

NON-PEPTIDE BACTERIAL ANTIGENS

It is well known that nonprotein antigens are recognized by the immune system. It has been shown that antigenic lipid and derived from pathogenic bacteria such as *Mycobacterium tuberculosis* are presented by members of the CD1 family of nonclassical class I molecules.

The CD1 family of molecules associates with B2–microglobulin and has general structural similarity to class I MHC molecules CD1 gene products are found on antigen—presenting cells. There are five genes encoding human CD1 molecules (CD1a-e), which are not located with on the MHC but on a different chromosomes. These genes have been identified in all mammalian species studied. Sequence identity of CD1 with classical class I in considerably lower than that among the class I molecules.

Antigens presented by CD1 molecules reversed them to be lipid components (mycalic acid) of the *M. tuberalosis* cell wall further studies of CD1 presentation indicated that a glycolipid (lipoarabinomanan) from *Mycobacterium leprous* could also be printed by there molecules.

ANTIGEN-ANTIBODY REACTIONS

NATURE OF Ag-Ab ACTIONS

A. Lock and Key Concept: The combining site of an antibody is located in the Fab portion of the molecule and is constructed from the hypervariable regions of the heavy and light chains. X-Ray crystallography studies of antigen-antibody interactions show that the antigenic determinant nestles in a cleft formed by the combining site of the antibody as illustrated in Figure 5.2. Thus, our concept of antigen-antibody reactions is one of a key (i.e. the antigen) which fits into a lock (i.e. the antibody).

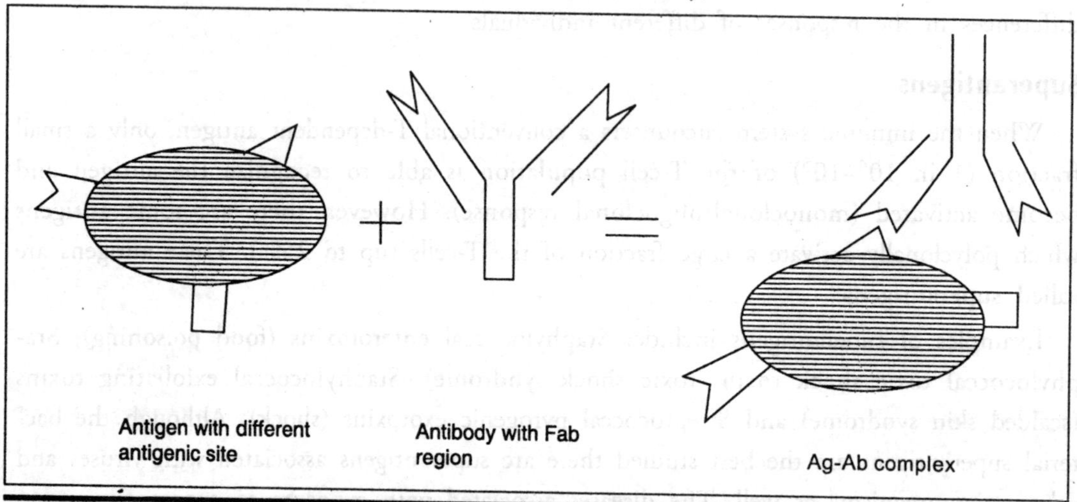

Fig. 5.2: Shows the antigen-antibody complex

B. Non-covalent Bonds: The bonds that hold the antigen to the antibody combining site are all non-covalent in nature. These include hydrogen bonds, electrostatic bonds, Van der Waals forces and hydrophobic bonds. Multiple bonding between the antigen and the antibody ensures that the antigen will be bound tightly to the antibody. Since antigen-antibody reactions occur via non-covalent bonds, they are by their nature reversible.

AFFINITY AND AVIDITY

Affinity: Antibody affinity is the strength of the reaction between a single antigenic determinant and a single combining site on the antibody. It is the sum of the attractive and repulsive forces operating between the antigenic determinant and the combining site of the antibody as illustrated in Figure 5.2. Affinity is the equilibrium constant that describes the antigen-antibody reaction. Most antibodies have a high affinity for their antigens.

Avidity: Avidity is a measure of the overall strength of binding of an antigen with many antigenic determinants and multivalent antibodies. Avidity is influenced by both the valence of the antibody and the valence of the antigen. Avidity is more than the sum of the individual affinities. To repeat, affinity refers to the strength of binding between a single antigenic determinant and an individual antibody combining site where as avidity refers to the overall strength of binding between multivalent antigens and antibodies.

SPECIFICITY AND CROSS REACTIVITY

Specificity

Specificity refers to the ability of an individual antibody combining site to react with only one antigenic determinant or the ability of a population of antibody molecules to react with only one antigen. In general, there is a high degree of specificity in antigen-antibody reactions. Antibodies can distinguish in following way:1) the primary structure of an antigen, 2) isomeric forms of an antigen, and 3) secondary and tertiary structure of an antigen.

Cross reactivity

Cross reactivity refers to the ability of an individual antibody combining site to react with more than one antigenic determinant or the ability of a population of antibody molecules to react with more than one antigen. Cross reactions arise because the cross reacting antigen shares an epitope in common with the immunizing antigen or because it has an epitope which is structurally similar to one on the immunizing antigen (multispecificity).

Factors affecting measurement of antigen-antibody reactions

The only way that one knows that an antigen-antibody reaction has occurred is to have some means of directly or indirectly detecting the complexes formed between the antigen and antibody. The ease with which one can detect antigen-antibody reactions will depend on a number of factors.

Affinity: The higher the affinity of the antibody for the antigen, the more stable will be the interaction. Thus, the ease with which one can detect the interaction is enhanced.

Avidity: Reactions between multivalent antigens and multivalent antibodies are more stable and thus easier to detect.

Antigen to antibody ratio: The ratio between the antigen and antibody influences the detection of antigen-antibody complexes because the size of the complexes formed is related to the concentration of the antigen and antibody.

Physical form of the antigen: The physical form of the antigen influences how one detects its reaction with an antibody. If the antigen is a particulate, one generally looks for agglutination of the antigen by the antibody. If the antigen is soluble one generally looks for the precipitation of the antigen after the production of large insoluble antigen-antibody complexes.

Agglutination Tests

Agglutination/Hemagglutination—When the antigen is particulate, the reaction of an antibody with the antigen can be detected by agglutination (clumping) of the antigen. The general term agglutinin is used to describe antibodies that agglutinate particulate antigens. When the antigen is an erythrocyte the term hemagglutination is used. All antibodies can theoretically agglutinate particulate antigens but IgM, due to its high valence, is particularly good agglutinin and one sometimes infers that an antibody may be of the IgM class if it is a good agglutinating antibody.

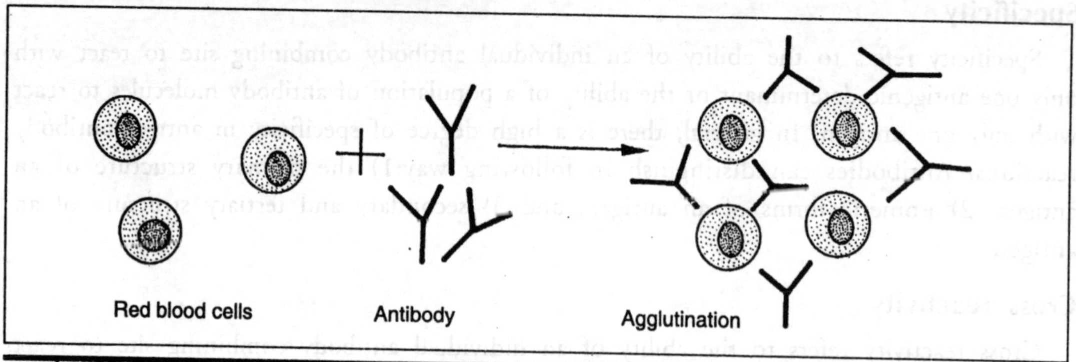

Fig. 5.3: Hemagglutination reaction

Qualitative agglutination test

Agglutination tests can be used in a qualitative manner to assay for the presence of an antigen or an antibody. The antibody is mixed with the particulate antigen and a positive test is indicated by the agglutination of the particulate antigen (Figure 5.3).

For example, a patient's red blood cells can be mixed with antibody to a blood group antigen to determine a person's blood type. In a second example, a patient's serum is mixed with red blood cells of a known blood type to assay for the presence of antibodies to that blood type in the patient's serum.

Quantitative agglutination test

Agglutination tests can also be used to measure the level of antibodies to particulate antigens. In this test, serial dilutions are made of a sample to be tested for antibody and then a fixed number of red blood cells or bacteria or other such particulate antigen is added. Then the maximum dilution that gives agglutination is determined. The maximum dilution that gives visible agglutination is called the titer. The results are reported as the reciprocal of the maximal dilution that gives visible agglutination.

Prozone effect—Occasionally, it is observed that when the concentration of antibody is high (i.e. lower dilutions), there is no agglutination and then, as the sample is diluted, agglutination occurs (See Figure 5.4). The lack of agglutination at high concentrations of antibodies is called the prozone effect. Lack of agglutination in the prozone is due to antibody excess resulting in very small complexes that do not clump to form visible agglutination.

Applications of agglutination tests

i. Determination of blood types or antibodies to blood group antigens.
ii. To assess bacterial infections. e.g. A rise in titer of an antibody to a particular bacterium indicates an infection with that bacterial type.
iii. Although the test is easy to perform, it is only semi-quantitative.

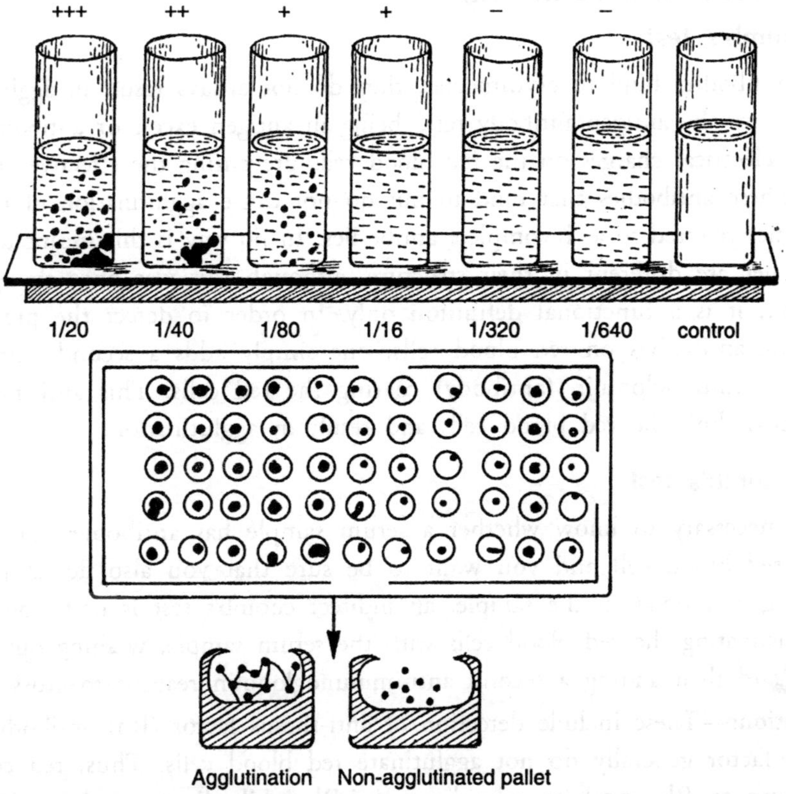

Fig. 5.4: Tube agglutination test for determining titer

Passive hemagglutination

The agglutination test only works with particulate antigens. However, it is possible to coat erythrocytes with a soluble antigen (e.g. viral antigen, a polysaccharide or a hapten) and use the coated red blood cells in an agglutination test for antibody to the soluble antigen (Figure 5.5). This is called passive hemagglutination. The test is performed just like the agglutination test. Applications include detection of antibodies to soluble antigens and detection of antibodies to viral antigens.

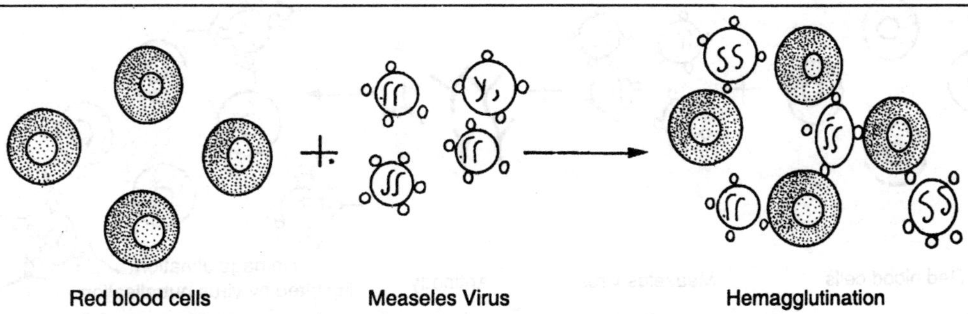

Fig. 5.5: Passive hemagglutination

Coomb's Test (Antiglobulin Test)

Direct coomb's test

When antibodies bind to erythrocytes, they do not always result in agglutination. This can result from the antigen/antibody ratio being in antigen excess or antibody excess or in some cases electrical charges on the red blood cells preventing the effective cross linking of the cells. These antibodies that bind to but do not cause agglutination of red blood cells are sometimes referred to as incomplete antibodies. In no way is this meant to indicate that the antibodies are different in their structure, although this was once thought to be the case. Rather, it is a functional definition only. In order to detect the presence of non-agglutinating antibodies on red blood cells, one simply adds a second antibody directed against the immunoglobulin (antibody) coating the red cells. This anti-immunoglobulin can now cross link the red blood cells and result in agglutination.

Indirect coomb's test

If it is necessary to know whether a serum sample has antibodies directed against a particular red blood cell and you want to be sure that you also detect potential non-agglutinating antibodies in the sample, an indirect coomb's test is performed. This test is done by incubating the red blood cells with the serum sample, washing out any unbound antibodies and then adding a second anti-immunoglobulin reagent to cross link the cells.

Applications—These include detection of anti-rhesus factor (Rh) antibodies. Antibodies to the Rh factor generally do not agglutinate red blood cells. Thus, red cells from Rh^+ children born to Rh^- mothers, who have anti-Rh antibodies, may be coated with these antibodies. To check for this, a direct coombs test is performed.

Hemagglutination Inhibition—The agglutination test can be modified to be used for the measurement of soluble antigens. This test is called hemagglutination inhibition. It is called hemagglutination inhibition because one measures the ability of soluble antigen to inhibit the agglutination of antigen-coated red blood cells by antibodies. In this test, a fixed amount of antibodies to the antigen in question is mixed with a fixed amount of red blood cells coated with the antigen (see passive hemagglutination above). Also included in

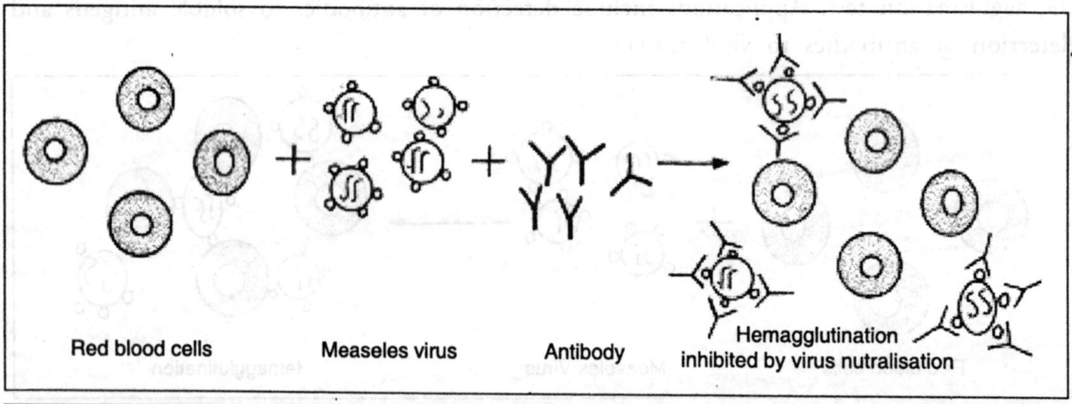

Fig. 5.6: Hemagglutination inhibition

the mixture are different amounts of the sample to be analyzed for the presence of the antigen. If the sample contains the antigen, the soluble antigen will compete with the antigen coated on the red blood cells for binding to the antibodies, thereby inhibiting the agglutination of the red blood cells.

By serially diluting the sample, you can quantitate the amount of antigen in your unknown sample by its titer. This test is generally used to quantitate soluble antigens and is subject to the same practical considerations as the agglutination test of the red blood cells. By serially diluting the sample, you can quantitate the amount of antigen in your unknown sample by its titer. This test is generally used to quantitate soluble antigens and is subject to the same practical considerations as the agglutination test.

PRECIPITATION TESTS

Radial immunodiffusion (Mancini): In radial immunodiffusion antibody is incorporated into the agar gel as it is poured and different dilutions of the antigen are placed in holes punched into the agar. As the antigen diffuses into the gel, it reacts with the antibody and when the equivalence point is reached a ring of precipitation is formed as illustrated in (Figure 5.7).

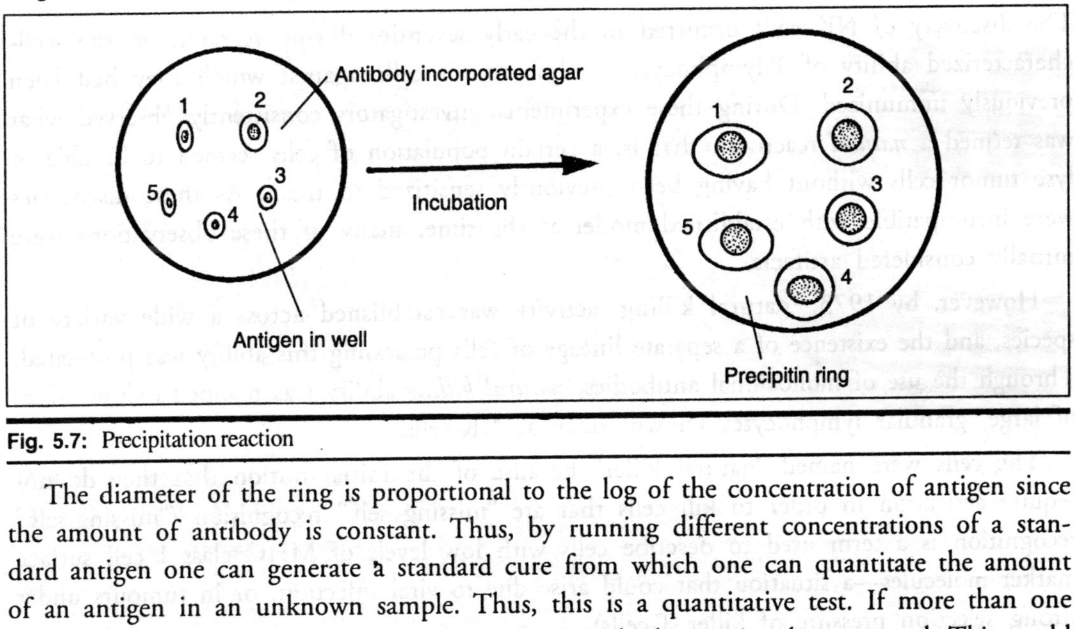

Fig. 5.7: Precipitation reaction

The diameter of the ring is proportional to the log of the concentration of antigen since the amount of antibody is constant. Thus, by running different concentrations of a standard antigen one can generate a standard cure from which one can quantitate the amount of an antigen in an unknown sample. Thus, this is a quantitative test. If more than one ring appears in the test, more than one antigen/antibody reaction has occurred. This could be due to a mixture of antigens or antibodies. This test is commonly used in the clinical laboratory for the determination of immunoglobulin levels in patient samples.

CHAPTER

6

NATURAL KILLER (NK) CELLS

HISTORY AND DISCOVERY

The discovery of NK cells occurred in the early seventies during research on the well-characterized ability of T-lymphocytes to lyse tumor cells against which they had been previously immunized. During these experiments, investigators consistently observed what was termed a *natural* reactivity, that is, a certain population of cells seemed to be able to lyse tumor cells without having been previously sensitized to them. As these discoveries were incompatible with established model at the time, many of these observations were initially considered artifacts.

However, by 1973, 'natural killing' activity was established across a wide variety of species, and the existence of a separate lineage of cells possessing this ability was postulated. Through the use of monoclonal antibodies, *natural killing* ability was mapped to the subset of large, granular lymphocytes known today as NK-cells.

The cells were named "natural killer" because of the initial notion that they do not require activation in order to kill cells that are "missing self" recognition ("missing-self" recognition is a term used to describe cells with low levels of MHC class I cell surface marker molecules—a situation that could arise due to viral infection, or in tumours under strong selection pressure of killer T-cells).

With the discovery of activating receptors almost two decades after the discovery of the inhibitory receptors these cells continue to be called by the same name, though "natural" no longer means the same thing. The term "natural killer" continues to be justified by:

- a morphology characteristic of activated cytotoxic lymphocytes, e.g. large size, high protein synthesis activity in the abundant endoplasmic reticulum (ER), and preformed granules.

- the mature state (does not require much new protein synthesis and remodelling before starting to kill).
- the rapid killing activity observed in freshly-isolated NK cells.

Natural killer cells (or NK cells) are a type of cytotoxic lymphocyte that constitute a major component of the innate immune system. NK cells play a major role in the rejection of tumors and cells infected by viruses. The cells kill by releasing small cytoplasmic granules of proteins called perforin and granzyme that cause the target cell to die by apoptosis.

NK cells are defined as large granular lymphocytes that do not express T-cell antigen receptors (TCR) or Pan T marker CD3 or surface immunoglobulins (Ig) B-cell receptor but that usually express the surface markers CD16 (FcγRIII) and CD56 in humans, and NK1.1/NK1.2 in certain strains of mice.

They were named "natural killers" because of the initial notion that they do not require activation in order to kill cells that are "missing self" markers of major histocompatibility complex (MHC) class I. However, it is now known that the cells are activated.

The natural killer cell was first described in 1976, when it was shown that the body contains a small population of large, granular lymphocytes that display cytotoxic activity against a wide range of tumour cells in the absence of any previous immunization with the tumour. NK cells were subsequently shown to play an important role in host defense both against tumor cells and against cells infected with some, though not all, viruses. These cells, which constitute 5%–10% of lymphocytes in human peripheral blood, do not express the membrane molecules and receptors that distinguish T- and B-cell lineages. Although NK cells do not have T-cell receptors or immunoglobulin incorporated in their plasma membranes, they can recognize potential target cells in two different ways. In some cases, an NK cell employs NK cell receptors to distinguish abnormalities, notably a reduction inthe display of class I MHC molecules and the unusual profile of surface antigens displayed by some tumour cells and cells infected by some viruses. Another way in which NK cells recognize potential target cells depends upon the fact that some tumor cells and cells infected by certain viruses display antigens against which the immune system has made an antibody response, so that antitumor or antiviral antibodies are bound to their surfaces. Because NK cells express CD16, a membrane receptor for the carboxyl-terminal end of the IgG molecule, called the Fc region, they can attach to these antibodies and subsequently destroy the targeted cells. This is an example of a process known as antibody-dependent cell mediated cytotoxicity (ADCC).

It has since become evident that NK cells represent a third class of lymphocytes distinct from B- and T-cells. The most striking distinction is that NK cells do not express a clonally distributed antigen receptor that is subject to somatic diversification. One clear indication of this is that NK cells do not rearrange B- and T-cell receptor genes, and, unlike B and T-cells, arise normally in animals with deficiencies in genes that encode the V(D)J recombinase. The NK cells are now known to employ several types of stimulatory receptors specific for molecules associated with infected, transformed, or stressed cells. NK cells are also regulated by inhibitory receptors specific for class I MHC molecules. The distinctive specificity of the cells, along with their fast response and lack of a memory stage, is the basis for considering the cells part of the innate (as opposed to adaptive) immune system. NK cells play several roles in immune responses. Once activated, NK cells are potent cytotoxic cells capable of lysing certain tumor cell lines, cells infected with various intracellular pathogens, and antibody-coated cells. Activated NK cells also produce cytokines, notably the inflammatory cytokines interferon- (IFN-) and tumour necrosis factor α (TNF-α). These cytokines can directly induce antipathogen activity by macrophages and neutrophils and can also help to regulate the adaptive immune response to an antigen. Given their strong cytolytic activity and the potential for auto-reactivity, natural killer cell activity is tightly regulated. Natural killer cells must receive an activating signal, which can come in a variety of forms, the most important of which are listed below.

1. 'Cytokines'

   The cytokines interferon play a crucial role in NK cell activation. As these are stress-molecules, released by cells upon viral infection, they serve to signal to the NK-cell the presence of viral pathogens.

2. 'Fc Receptor'

   NK cells, along with macrophages and several other cell types, express the FcR molecule, an activating biochemical receptor that binds the Fc portion of antibodies. This allows natural killer cells to target cells against which a humoral response has been mobilized and to lyse cells through antibody-dependant cellular cytotoxicity (ADCC).

3. 'Activating and inhibitory receptors'

   Aside from the Fc receptor, natural killer cells express a variety of receptors that serve to either activate or suppress their cytolytic activity. These receptors bind to various ligands on target cells, both endogenous and exogenous, and have an important role in regulating the NK cell response.

Mechanism

NK cells are cytotoxic; small granules in their cytoplasm contain proteins such as perforin and proteases known as granzymes. Upon release in close proximity to a cell slated for killing, perforin forms pores in the cell membrane of the target cell through which the granzymes and associated molecules can enter, inducing apoptosis. The distinction between apoptosis and cell lysis is important in immunology: lysing a virus-infected cell would only release the virions, whereas apoptosis leads to destruction of the virus inside.

NK cells are activated in response to interferons or macrophage-derived cytokines. They serve to contain viral infections while the adaptive immune response is generating antigen-specific cytotoxic T-cells that can clear the infection. Patients deficient in NK cells prove to be highly susceptible to early phases of herpes virus infection.

In order for NK cells to defend the body against viruses and other pathogens, they require mechanisms that enable the determination of whether a cell is infected or not. The exact mechanisms remain the subject of current investigation, but recognition of an "altered self" state is thought to be involved. To control their cytotoxic activity, NK cells possess two types of surface receptors: *activating receptors* and *inhibitory receptors.* Most of these receptors are not unique to NK cells and can be present in other T-cell subsets as well. These inhibitory receptors recognize MHC class I alleles, which could explain why NK cells kill cells possessing low levels of MHC class I molecules. This inhibition is crucial to the role played by NK cells. MHC class I molecules consist of the main mechanism by which cells display viral or tumour antigens to cytotoxic T-cells. A common evolutionary adaption to this seen in both intracellular microbes and tumours is a chronic down-regulation of these MHC I molecules, rendering the cell impervious to T-cell mediated immunity. It is believed that NK cells, in turn, evolved as an evolutionary response to this adaption, as the loss of the MHC would deprive these cells of the inhibitory effect of MHC and render these cells vulnerable to NK cell mediated lysis.

Receptor types

NK cell receptor types (with inhibitory as well as some activating members) are differentiated by structure:

1. **CD94 : NKG2** (heterodimers)—a C-type lectin family receptor, conserved in both rodents and primates and identifies non-classical (also non-polymorphic) MHC I molecules like HLA-E. Though indirect, this is a way to survey the levels of classical (polymorphic) HLA molecules, however, because expression of HLA-E at the cell surface is dependent upon the presence of classical MHC class I leader peptides.
2. **Ly49** (homodimers)—a relatively ancient, C-type lectin family receptor; are of multigenic presence in mice, while humans have only one pseudogenic Ly49; the receptor for classical (polymorphic) MHC I molecules.
3. **KIR** (killer cell immunoglobulin-like receptors)—belong to a multigene family of more recently-evolved Ig-like extracellular domain receptors; are present in non-rodent primates; and are the main receptors for both classical MHC I (HLA-A, HLA-B, HLA-C) and also non-classical HLA-G in primates. Some KIRs are specific for certain HLA subtypes.
4. **ILT** or **LIR** (leucocyte inhibitory receptors)—are recently-discovered members of the Ig receptor family.

The three major families of NK cell receptors specific for class I MHC proteins are designated KIR (killer cell immunoglobulin-like receptors), Ly49 receptors, and CD94/NKG2 receptors. Each family is composed of several related isoforms. As their name implies, KIR are members of the immunoglobulin superfamily, whereas Ly49 and CD94/NKG2 receptors are structurally similar to C-type lectins. Ly49 genes were initially identified in mice. Although Ly49 genes may be functionally expressed in many species, including some primates, the human genome reportedly contains only a single nonfunctional Ly49 pseudogene. Conversely, KIR were discovered in humans and are found in various primates but have not been well studied in mice and other rodents. Despite their dissimilar structures,

KIR and Ly49 proteins function very similarly in NK cell recognition. CD94/NKG2 receptors are the only MHC-specific receptors that are known to be functionally expressed in both humans and rodents and presumably in most other animal species.

Unlike KIR and Ly49 receptors, CD94/NKG2 receptors recognize classical class I MHC molecules in an indirect manner. All three of these gene families were initially discovered as a result of the inhibitory function of some family members, each of which contain ITIMs in their cytoplasmic domains. Unexpectedly, each family also contains stimulatory isoforms, which are characterized by the presence of a charged amino acid residue in the transmembrane domain that enables the receptor to associate with the KARAP/DAP12 signalling molecule. Like the inhibitory isoforms, many if not all of the stimulatory isoforms recognize class I MHC molecules.

Functions

The existence of stimulatory MHC-specific receptors expressed by NK cells was not initially anticipated, and their biological function is not well understood. One role, as proposed previously, may be to aid in attacking cells that selectively down-regulate class I MHC molecules. However, several other possibilities may be considered. One idea is that stimulatory MHC-specific receptors evolved to counter the evasion tactics that viruses employ to inhibit NK cells. If the stimulatory receptors bind to the class I "decoys" that some viruses encode, the balance of signalling could shift back in the direction of NK cell activation. Alternatively, stimulatory receptors may preferentially bind MHC molecules that have complexed with specific pathogen-derived peptides. Another possibility is suggested by the observation that stimulatory MHC-specific receptors are generally of lower affinity than are their inhibitory counterparts. In this scheme, stimulatory receptors are saturated only when class I MHC molecules on a target cell are overexpressed, as occurs in response to locally produced interferons at the site of infections. Saturation of the stimulatory receptors may overcome inhibitory signalling, leading to lysis of the class I–overexpressing cells.

Finally, it is possible that some of the stimulatory receptors cross-react with non-MHC ligands associated with infection, such as pathogen-encoded molecules or host-encoded molecules whose expression is induced by infection. Direct evidence demonstrates that at least one stimulatory receptor, called Ly49H, recognizes a virally encoded MHC "decoy." Although a member of the Ly49 family, Ly49H exhibits no known MHC reactivity. This receptor plays an important role in the NK cell response to MCMV. NK cells are known to play an important role in limiting cytomegalovirus replication early in the infectious cycle. The effectiveness of this response is under genetic control, in such a way that certain mouse strains, such as C57BL/6, are more resistant to the virus than others, such as DBA/2 or BALB/c. Genetic segregation studies demonstrated the existence of a key dominant gene, called *Cmv1 r*, that confers resistance to MCMV, especially affecting viral replication in the spleen. Genetic evidence suggests that *Cmv1 r* encodes Ly49H, and that virus-sensitive strains lack a functioning Ly49h gene. When virus-resistant C57BL/6 mice were injected with an antibody that blocks the Ly49H receptor, they became susceptible to the virus. In addition, virus infection leads to preferential accumulation of the subset of NK cells that express Ly49H. Apparently, MCMV infection leads to expression in target cells of a ligand that interacts with Ly49H, which triggers a strong NK cell response against the infected cells. The ligand was identified as a viral product called m157, which is structurally related to class I MHC proteins. Of significance is that the m157 protein also binds to certain inhibitory Ly49 receptors, which suggests that, in some circumstances, it can function as a "decoy" to aid the virus in evading NK cells. A plausible evolutionary scenario to explain these findings is that the inhibitory function of m157 evolved first as a viral adaptation to evade the immune system. Natural selection may then have given rise to Ly49H in the host as an evolutionary "countermeasure."

DEVELOPMENT AND DIFFERENTIATION

NK cells, like other lymphocytes, are derived from pluripotent hematopoietic stem cells. Intermediate stages are thought to include a lymphocyte-restricted progenitor (restricted to T, B, and NK lineages), followed by a T/NK restricted intermediate. NK cell development but not T-cell development is impaired in mice treated with strontium 89 and other agents that disrupt the bone marrow, which suggests that critical phases of NK cell development occur primarily in the bone marrow. In contrast, although small numbers of NK cells do develop in the fetal thymus, NK cell development is not impaired in athymic mice, which lack conventional T-cells. Thus, NK cell development does not require a maturation step in the thymus. Indeed, NK cell activity is often elevated in athymic mice in comparison with normal mice, perhaps because of compensatory mechanisms or a higher incidence of infections in such mice. Analysis of mice with mutations in cytokine genes or their receptors has demonstrated that IL-15, a cytokine produced by bone marrow stromal cells, plays a central role in NK cell development or survival or both. Other cytokines are believed to be necessary at an earlier stage, in order to induce expression of IL-15 receptors on progenitor cells. Cell culture models of NK cell development suggest that stromal cells also supply additional undefined signals that enhance NK cell development and that cannot be replaced by any known cytokine.

NK cells are derived from an early lymphoid precursor cell type that lacks lineage markers (Lin -) and expresses flt3, the receptor for flt3 ligand. Exposure of the cell to flt3 ligand and other cytokines results in the expression of the receptor for interleukin (IL)–15. IL-15 from stromal cells induces NK cell differentiation, proliferation, or survival, or a combination of these, but is not sufficient to induce expression of most inhibitory NK cell receptors. Undefined stromal cell signals in conjunction with IL-15 stimulate expression Ly49 inhibitory receptors. The various Ly49 receptors are expressed more or less randomly in a given NK cell but are up-regulated over time in a cumulative manner. Initial expression of Ly49 receptors that do not interact with self class Ia major histocompatibility complex (MHC) molecules (lightly-shaded triangles) has no effect on subsequent expression of other Ly49 receptors.

Expression of a Ly49 receptor that strongly interacts with self class I MHC molecules (black triangle) results in inhibition of new receptor expression. Some NK cells (such as the one depicted on the bottom row) may never succeed in expressing Ly49 receptors specific for self class I MHC proteins. Such cells exhibit a hyporesponsive functional phenotype.

As NK cells develop, they acquire various markers and receptors progressively. An early marker is the IL-2/IL-15 receptor ß chain, expression of which is necessary for responsiveness to IL-15 and subsequent stages in development. Expression of at least some of the stimulatory receptors, such as NKG2D, appears to also be a relatively early event. In mice, expression of the NK1.1 antigen first occurs at a later stage, along with expression of many of the known inhibitory receptors. Little is known about NK cell ontogeny in humans.

Self-tolerance

A key function of inhibitory NK receptors is to prevent the lysis of normal cells. However, because of the variegated expression of the receptors and the fact that they differ in specificity for MHC alleles, some NK cells in every animal are expected to lack inhibitory receptors specific for a given MHC molecule or set of MHC molecules. This largely explains the capacity of NK cells to attack foreign cells, as discussed previously. However, this line of reasoning also suggests that some NK cells that fail to express inhibitory receptors specific for *self* MHC molecules will arise. Nonetheless, it is known that NK cells exhibit self-tolerance. What mechanisms prevent NK cell-mediated autoreactivity. As one explanation for self-tolerance of NK cells, it has been proposed that all peripheral NK cells in fact do express at least one inhibitory receptor specific for self class I MHC molecules, presumably as the result of a selective or developmental process. Direct evidence for this model is sparse, however. To the contrary, there is some evidence that there exists a subset of mouse NK cells that lack inhibitory receptors specific for *self* MHC molecules.

EFFECTOR FUNCTIONS

The primary effector functions of NK cells are cytolysis and cytokine release. In several scenarios, a key biologically important function of these cells is production of proinflammatory cytokines, especially IFN- and TNF-a. IFN- is also produced by Th1 cells and CTLs, but in the early stages of many immune responses, the cytokine is produced principally by NK cells. IFN-, like other interferons, induces an antiviral state in uninfected cells, but it also has numerous activities distinct from those of the other interferons (162). Important among these is its capacity to activate macrophages.

Activated macrophages are more effective in fusing phagosomes with lysozymes and produce potent antimicrobial substances such as oxygen radicals, nitric oxide, antimicrobial peptides, and proteases. Therefore, activated macrophages exhibit greatly enhanced capacity to destroy both intracellular and extracellular pathogens, including bacteria, parasites, and viruses. Activation of macrophages by IFN- is enhanced by TNF-a, which is also produced by stimulated NK cells.

CHAPTER 7

ANTIGEN-PRESENTING CELLS

INTRODUCTION

Antigen presentation is a process in the body's immune system by which macrophages, dendritic cells and other cell types capture antigens and then enable their recognition by T-cells. The basis of adaptive immunity lies in the capacity of immune cells to distinguish between the body's own cells, and infectious pathogens. The host's cells express "self" antigens that identify them as such. These antigens are different from those in bacteria ("non-self" antigens) or in virally-infected host cells ("missing-self"). The ability of the adaptive immune system to survey for infection requires specialized pathways of enabling recognition of pathogen-derived antigens by T-cells.

ANTIGEN RECOGNITION

Unlike B-cells, T-cells fail to recognize antigen in the absence of antigen-presentation, with the important exception of the superantigens. The T-cell receptor is restricted to recognizing antigenic peptides only when bound to appropriate molecules of the major histocompatibility complex (MHC), also known in humans as human leukocyte antigen (HLA).

With the exception of some cell types, all cells are capable of presenting antigen and activating the adaptive response. Some cells, however, are specially equipped to acquire and present antigen, and to prime naive T-cells. Dendritic cells, B-cells, and macrophages play a major role in the innate response, and also act as professional antigen-presenting cells (APC).

These professional APCs are equipped with special immunostimulatory receptors that allow for enhanced activation of T-cells (Figure 7.1). Several different types of T-cell can be activated by professional APCs, and each type of T-cell is specially equipped to deal with different pathogens, whether the pathogen is bacterial, viral or a toxin. The type of T-cell activated, and therefore the type of response generated, depends, in part, on the context in which the antigen was first encountered by the APC.

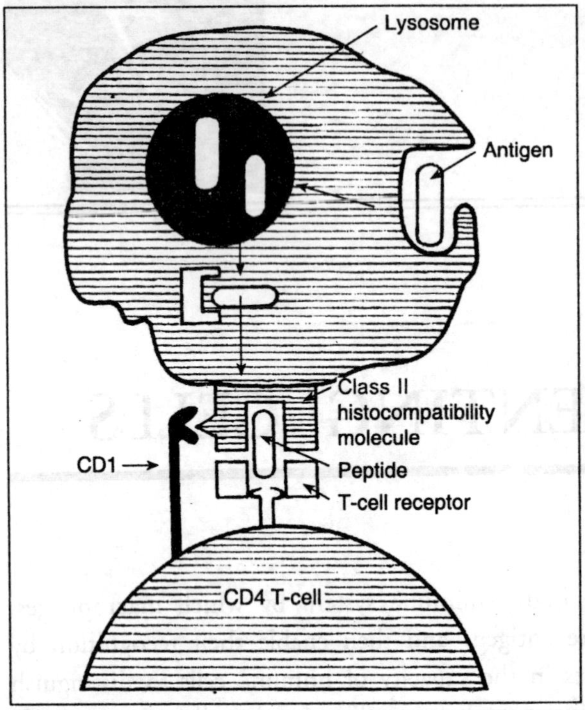

Fig. 7.1: Antigen recognition

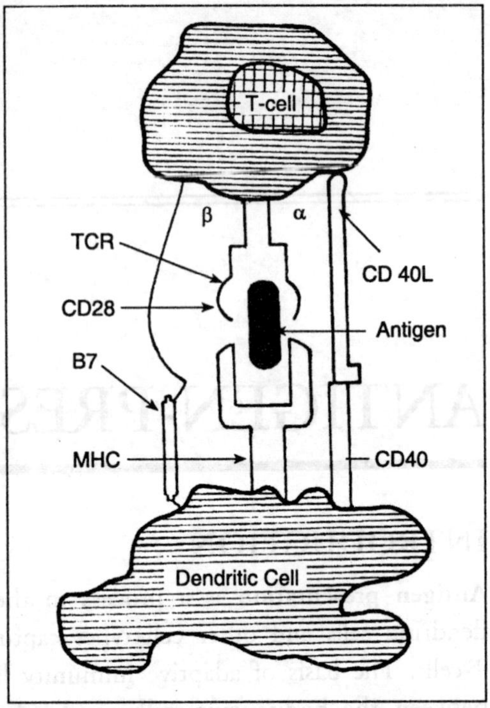

Fig. 7.2: Mechanism of co-stimulation

CO-STIMULATION

The immune system contains of three types of antigen-presenting cells. It will be explained below that these three types of APC's present different sets of antigens and also may serve to activate T-cells at different points during the immune response. The most relevant property of APC's is that, in addition to antigen presentation, they provide co-stimulatory signals via B7.1 and B7.2. It is important to note that APC's do not constitutively express these co-stimulatory molecules (Figure 7.2). Since these cells potentially phagocytose both self and infectious materials, there has to be some mechanism for the recognition of infection (non-self). Upon this recognition, the APC's will up regulate their co-stimulatory molecules (namely B7's), and only then will they activate T-cells, by interactiong with the B7 receptor called CD28.

Antigen-Presenting Cells

Although some Ags can stimulate the immune response directly, T-cell–dependent acquired immune responses typically require antigen-presenting cells (APCs) to present Ag-derived peptides within major histocompatibility complex (MHC) molecules. Intracellular Ag (e.g. viruses) can be processed and presented to CD8 cytotoxic T-cells by any nucleated cell because all nucleated cells express class I MHC molecules. However, extracellular Ag must be processed into peptides and complexed with surface class II MHC molecules on professional APCs to be recognized by CD4 helper T (T_H) cells. The following cells constitutively express class II MHC molecules and therefore act as professional APCs:

- B-cells
- Monocytes
- Macrophages
- Dendritic cells

Monocytes in the circulation are precursors to tissue macrophages. Monocytes migrate into tissues, where over about 8 hr they develop into macrophages under the influence of macrophage colony-stimulating factor (M-CSF), secreted by various cell types (e.g. endothelial cells, fibroblasts). At infection sites, activated T-cells secrete cytokines (e.g. interferon-γ [IFN-γ]) that induce production of macrophage migration inhibitory factor, preventing macrophages from leaving.

Macrophages are activated by IFN-γ and granulocyte-macrophage colony-stimulating factor (GM-CSF). Activated macrophages kill intracellular organisms and secrete IL-1 and tumor necrosis factor-α (TNF-α). These cytokines potentiate the secretion of IFN-γ and GM-CSF and increase the expression of adhesion molecules on endothelial cells, facilitating leukocyte influx and destruction of pathogens.

Dendritic cells are present in the skin (as Langerhans' cells), lymph nodes, and tissues throughout the body. Dendritic cells in the skin act as sentinel APCs, taking up Ag, then travel to local lymph nodes where they can activate T-cells. Follicular dendritic cells are a distinct lineage, do not express class II MHC molecules, and therefore do not present Ag to T_H cells. However, they have receptors for the crystallizable fragment (Fc) region of IgG and for complement, which enable them to bind with immune complexes and present the complex to B-cells in germinal centers of secondary lymphoid organs.

The types of antigen-presenting cells

A variety of cells can function as antigen-presenting cells. Their distinguishing feature is their ability to express class II MHC molecules and to deliver a co-stimulatory signal. Three cell types are classified as *professional* antigen-presenting cells: dendritic cells, macrophages, and B lymphocytes. These cells differ from each other in their mechanisms of antigen uptake, in whether they constitutively express class II MHC molecules, and in their co-stimulatory activity. Several other cell types, classified as *nonprofessional* antigen-presenting cells, can be induced to express class II MHC molecules or a co-stimulatory signal. Many of these cells function in antigen presentation only for short periods of time during a sustained inflammatory response.

Because nearly all nucleated cells express class I MHC molecules, virtually any nucleated cell is able to function as a target cell presenting endogenous antigens to TC cells. Most often, target cells are cells that have been infected by a virus or some other intracellular microorganism. However, altered self-cells such as cancer cells, aging body cells, or allogeneic cells from a graft can also serve as targets.

There are main three types of APCs

1. Macrophages
2. Dendritic cells
3. B-cells

MACROPHAGES

Macrophages must be activated by phagocytosis of particulate antigens before they express class II MHC molecules or the co-stimulatory B7 membrane molecule. These cells are part of the innate response. Unlike T- and B-cells, they do not contain any specific receptors. Macrophages continuously phagocytose self-proteins and cells in their vicinity, during normal tissue repair and aging (e.g. old red blood cells). All of these proteins are degraded and presented on MHC-II. These self-proteins however, do not activate T-cells, because in the absence of infection, macrophages express low levels of MHC-II, and almost no co-stimulator (B7). Further, T-cells with high affinity receptor for self-peptides have been deleted during T-cell development in the thymus.

In the case of infection, however, macrophages posses certain types of receptors that recognize differential carbohydrate patterns on foreign cells. They also have receptors for specific bacterial products such as lipopolysaccharide (LPS) (endotoxin). When these molecules bind their bacterial ligands, they stimulate the macrophages to up regulate MHC-II and B7, providing these cells with strong antigen presentation properties. They also start to secrete cytokines that aid in their functions. It is at this point that antigen presentation by MHC II will activate Th cells. Since viral proteins are internally synthesized by host cells, they possess the same carbohydrate pattern as the host. Thus macrophages do ingest them as part of the general tissue debris, present the antigen as peptides, but do not become activated.

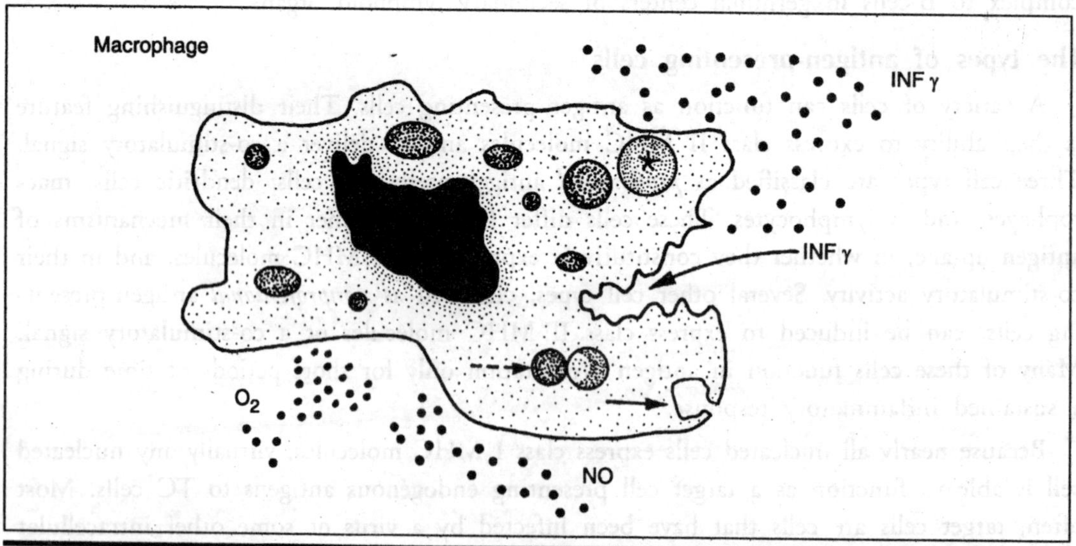

Fig. 7.3: Macrophages vacuole containing ingested extracellular material, open arrow indicates fusion of lysosome with phagocytosed vacuole, phagolysosome formation

DENDRITIC CELLS

Dendritic cells are the most effective of the antigen-presenting cells. Because these cells constitutively express a high level of class II MHC molecules and costimulatory activity, they can activate naive Th cells. Dendritic cells are also very numerous in the thymus, where they act in positive and negative selection during T-cell development. Dendritic cells present viral particles, they should also activate CD8 cells, the main effector cell to fight viral infections (intracellular). Unlike macrophages however, dendritic cells can also recognize viral particles as non-self. In addition, they can present antigen via **both** MHC I and MHC II. Thus they can activate both CD8 and CD4 T-cells, directly. Once the T cells are activated, they will leave the lymph nodes and travel to the sites of inflammation. These cells are mostly found in the skin and mucosal epithelium, where they are referred to as Langerhan's cells. These cells continuously express high levels of co-stimulatory B7. During ingestion of self proteins, they remain in the epithelium, inaccessible to "naive" T-cells. Naive T-cells circulate through blood and the lymphatic system. Only activated and/or memory T-cells can migrate to tissues. Upon recognition of infectious particles, these cells migrate through the lymphatics to the nearest lymph node (Dendritic cells also possess non-specific receptors that can recognize infectious particles). In the follicles of the lymph node they come into close contact with naive T-cells.

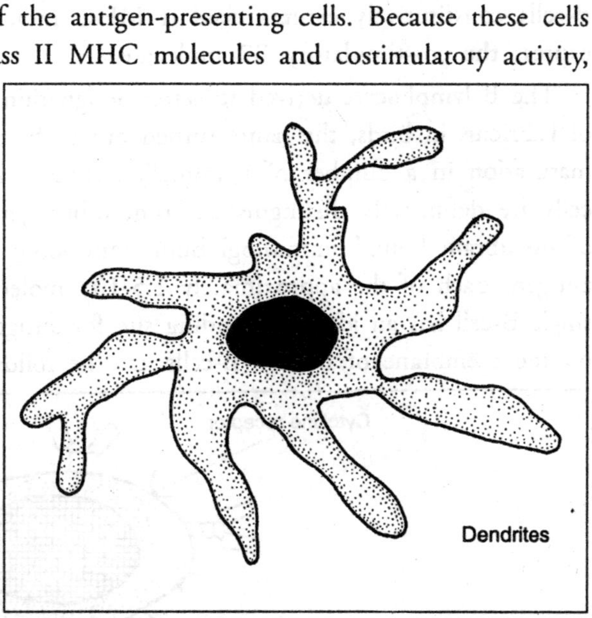

Fig. 7.4: Dendritic cells

The **dendritic cell** (DC) acquired its name because it is covered with long membrane extensions that resemble the dendrites of nerve cells. Dendritic cells can be difficult to isolate because the conventional procedures for cell isolation tend to damage their long extensions. The development of isolation techniques that employ enzymes and gentler dispersion has facilitated isolation of these cells for study in vitro. There are many types of dendritic cells, although most mature dendritic cells have the same major function, the presentation of antigen to Th cells. Four types of dendritic cells are known: Langerhans cells, interstitial dendritic cells, myeloid cells, and lymphoid dendritic cells. Each arises from hematopoietic stem cells via different pathways and in different locations. Despite their differences, they all constitutively express high levels of both class II MHC molecules and members of the co-stimulatory B7 family. For this reason, they are more potent antigen-presenting cells than macrophages and B-cells, both of which need to be activated before they can function as antigen-presenting cells (APCs).

B-CELLS

B-cells constitutively express class II MHC molecules but must be activated before they express the co-stimulatory B7 molecule.

The B lymphocyte derived its letter designation from its site of maturation, in the bursa of Fabricius in birds; the name turned out to be apt, for bone marrow is its major site of maturation in a number of mammalian species, including humans and mice. Mature B-cells are definitively distinguished from other lymphocytes by their synthesis and display of membrane-bound immunoglobulin (antibody) molecules, which serve as receptors for antigen. Each of the approximately 1.5-105 molecules of antibody on the membrane of a single B-cell has an identical binding site for antigen. Among the other molecules expressed on the membrane of mature B-cells are the following:

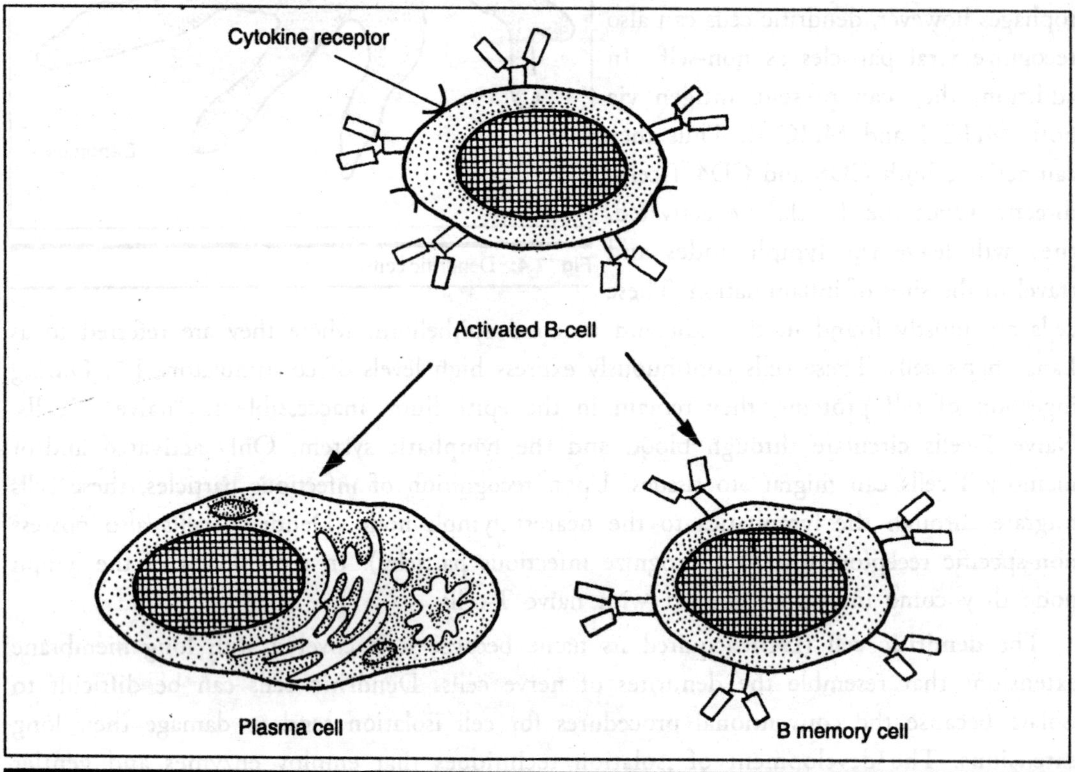

Fig. 7.5: B-cells

B220 (a form of CD45) is frequently used as a marker for B-cells and their precursors. However, unlike antibody, it is not expressed uniquely by B-lineage cells. **Class II MHC molecules** permit the B-cell to function as an antigen-presenting cell (APC). **CR1** (CD35) and **CR2** (CD21) are receptors for certain complement products. **Fc RII** (CD32) is a receptor for IgG, a type of antibody. **B7-1** (CD80) and **B7-2** (CD86) are molecules that interact with CD28 and CTLA-4, important regulatory molecules on the surface of different types of T-cells, including TH cells. **CD40** is a molecule that interacts with CD40 ligand on the surface of helper T-cells. In most cases this interaction is critical for the survival of antigen stimulated B-cells and for their development into antibody-secreting plasma cells or memory B-cells. B-cells are the least efficient antigen-presenting cells. Unlike the other two APC's, they possess specific antigen receptors, surface immunoglobulins. B-cells ingest soluble proteins by pinocytosis. They also posses specific uptake receptors in surface immunoglobulins. B-cells present antigen via MHC-II. But these cells do not express co-stimulatory molecules. In order to do so, they need to be activated by Th cells. The role of B-cells as APC's in vivo is not very well understood. The activated B-cell are of two types plasma cell and memory cells.

Table 7.1
Antigen-Presenting Cells and their location

	MHC-II Expression	Antigen type and presentation by MHC	Co-Stimulation (B7 expression	Location
Macrophage	Low levels. Induced by Bacteria and/or cytokines	Extracellular Antigens: presentation via MHC-II	Low levels. Induced by Bacteria and/or Cytokines	Lymphoid tissue Connective tissue Body cavities
Dendritic Cell	Always Expressed	Intracellular & Extracellular Antigens: presentation via MHC-I & II	Always expressed at high Levels	Lymphoid tissue Connective tissue epithelium
B-cell	Always Expressed Inducible upon activation	Extracellular Antigen binds to specific Ig receptors: presentation via MHC-II	Low levels. Inducible upon activation	Lymphoid tissues. Blood

CHAPTER 8

CELL ADHESION MOLECULES

INTRODUCTION

Cellular adhesion is the binding of a cell to another cell or to a surface or matrix. Cellular adhesion is regulated by specific cell adhesion molecules that interact with molecules on the opposing cell or surface. Such adhesion molecules are also termed "receptors" and the molecules they recognize are termed "ligands" (and sometimes "counterreceptors"). Since cells are not often found in isolation, rather they tend to stick to other cells or non-cellular components of their environment, a fundamental question is: what makes cells sticky? Cell adhesion generally involves protein molecules at the surface of cells, so the study of cell adhesion involves cell adhesion proteins and the molecules that they bind to.

For a cell adhesion protein like the one shown in the diagram, the intracellular domain binds to protein components of the cell's cytoskeleton. This allows for very tight adhesion. Without attachment to the cytoskeleton, a cell adhesion protein that is tightly bound to a ligand would be in danger of being hydrolyzed by extracellular hydrolytic enzymes. This will rip out the adhesion protein from the fragile cell membrane. For example, cadherin cell adhesion proteins are typically coupled to the cytoskeleton by way of special linking proteins called "catenins". Prokaryotes have adhesion molecules usually termed "adhesins". Adhesins may occur on pili (fimbriae), flagellae, or the cell surface. Adhesion of bacteria is the first step in colonization and regulates tropism (tissue- or cell-specific interactions).

Viruses also have adhesion molecules required for viral binding to host cells. For example, influenza virus has a hemagglutinin on its surface that is required for recognition of the sugar sialic acid on host cell surface molecules. HIV has an adhesion molecule termed gp120 that binds to its ligand CD4, which is expressed on lymphocytes.

Cell adhesion molecules (CAMs) are proteins located on the cell surface involved with the binding with other cells or with the extracellular matrix (ECM) in the process called cell adhesion. These proteins are typically transmembrane receptors and are composed of three domains: an intracellular domain that interacts with the cytoskeleton, a transmembrane domain and an extracellular domain that interacts either with other CAMs of the same kind (**homophilic** binding) or with other CAMs or the extracellular matrix (**heterophilic** binding). The vascular endothelium serves as an important "gatekeeper"regulating the movement of blood-borne molecules and leukocytes into the tissues. In order for circulating leukocytes to enter inflamed tissue or peripheral lymphoid organs, the cells must adhere to and pass between the Endothelial cells lining the walls of blood vessels a process called **extravasation.** Endothelial cells express leukocyte-specific **cell adhesion molecules (CAMs).** Some of these membrane proteins are expressed constitutively; others are expressed only in response to local concentrations of cytokines produced during an inflammatory response. Recirculating lymphocytes, monocytes, and granulocytes bear receptors that bind to CAMs on the vascular endothelium, enabling these cells to extravasate into the tissues.

FAMILIES OF CAMs

A number of endothelial and leukocyte CAMs have been cloned and characterized, providing new details about the extravasation process. Most of these CAMs belong to four families of proteins: the selectin family, the mucin-like family, the integrin family, and the immunoglobulin (Ig) superfamily.

Ig superfamily

Several adhesion molecules contain a variable number of immunoglobulin-like domains and thus are classified in the **immunoglobulin superfamily.** Included in this group are ICAM-1, ICAM-2, ICAM-3, and VCAM, which are expressed on vascular endothelial cells and bind to various integrin molecules. An important cell-adhesion molecule called MAdCAM-1 has both Ig-like domains and mucin-like domains. This molecule is expressed on mucosal endothelium and directs lymphocyte entry into mucosa. It binds to integrins by its immunoglobulin-like domain and to selectins by its mucin-like domain.

Integrins

The **integrins** are a family of heterophilic CAMs that bind IgSF CAMs or the extracellular matrix. They are heterodimers, consisting in two non-covalently linked subunits, called alpha and beta. The twenty four different alpha subunits are known that can link in many different combinations.with the nine different beta subunits, however not all combinations are observed. Some sources (for example, MeSH) don't consider integrins to be cell adhesion molecules. The integrins are heterodimeric proteins (consisting of an alpha and a beta chain) that are expressed by leukocytes and facilitate both adherence to the vascular endothelium and other cell-to-cell interactions. The integrins are grouped into categories according to which alpha subunit they contain. Different integrins are expressed by different populations of leukocytes, allowing these cells to bind to different CAMs that belong to the immunoglobulin superfamily expressed along the vascular endothelium. As described later, some integrins must be activated before they can bind with high affinity to their ligands. The importance of integrin molecules in leukocyte extravasation is demonstrated by leukocyte-adhesion deficiency (LAD), an autosomal recessive disease. It is characterized by recurrent bacterial infections and impaired healing of wounds. Integrin molecules migrate to the cell surface and congregate in high-avidity patches. Intracellular integrin domains associate with the leukocyte cytoskeleton, via mediation with cytosolic factors such as talin, α-actinin and vinculin. This association causes a conformational shift in the integrin's tertiary structure, allowing ligand access to the binding site. Divalent cations (e.g. Mg^{2+}) are also required for integrin-ligand binding. Integrin ligands ICAM-1 and VCAM-1 are activated by inflammatory cytokines, while ICAM-2 is constitutively expressed by some endothelial cells but downregulated by inflammatory cytokines. ICAM-1 and ICAM-2 share two homologous N-terminal domains; both can bind LFA-1.

During chemotaxis, cell movement is facilitated by the binding of β1 integrins to components of the extracellular matrix: VLA-3, VLA-4 and VLA-5 to fibronectin and VLA-2 and VLA-3 to collagen and other extracellular matrix components.

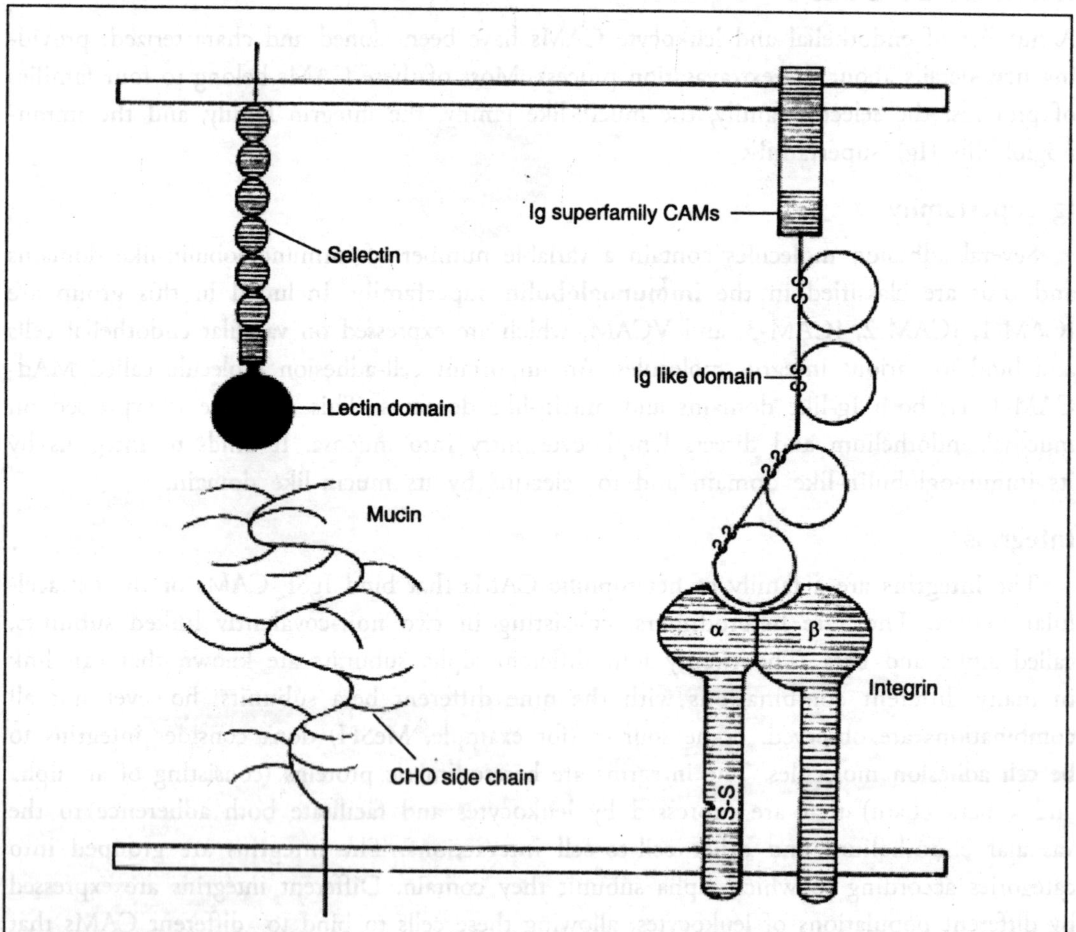

Fig. 8.1: General structure of CAMs

MUCIN

Mucin are the group of serine and threonine rich protein that are heavily glycosylated. Its extended structure allow to bind with selectin by using sialylated carbohydrate ligands. PSGL1 is mucin like molecule present on neutrophill interact with P-selectin and L-selectin.

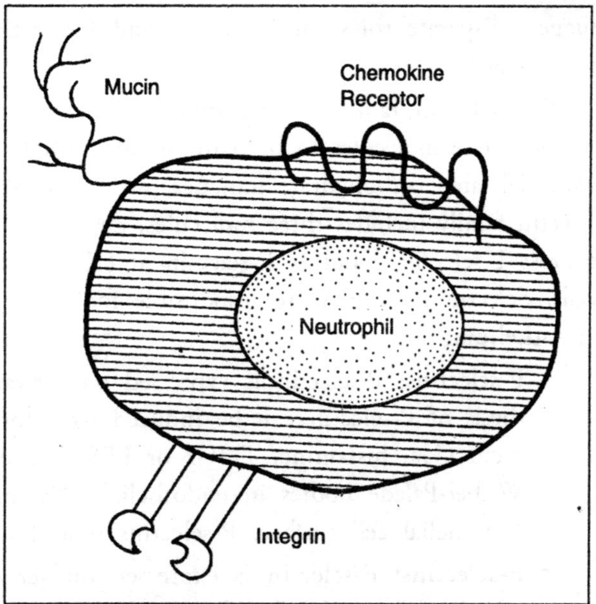

Fig. 8.2: Integrin, mucin-like receptor and chemokines express on neutrophil

The **mucins** are a group of serine- and threonine rich proteins that are heavily glycosylated. Their extended structure allows them to present sialylated carbohydrate ligands to selectins. For example, L-selectin on leukocytes recognizes sialylated carbohydrates on two mucin-like molecules (CD34 and GlyCAM-1) expressed on certain endothelial cells of lymph nodes. Another mucin-like molecule (PSGL-1) found on neutrophils interacts with E- and P-selectin expressed on inflamed endothelium.

Cadherins

The **cadherins** are a family of homophilic CAMs, Ca^{2+} dependent. The most important members of this family are E-cadherins (epithelial), P-cadherins (placental) and N-cadherins (neural).

Selectins

The selectin family includes three molecule designated as P, L, and E. The most circulating leukocytes express L-selectin whereas P-selectin and selectin are express on the vascular endothelial cells. Selectins are expressed shortly after cytokine activation of endothelial cells by tissue macrophages.

The **selectins** are a family of heterophilic CAMs that bind fucosylated carbohydrates, e.g. mucins. They are calcium-dependent. The three family members are E-selectin (endothelial), L-selectin (leukocyte) and P-selectin (platelet). The best-characterized ligand for the three selectins is P-selectin glycoprotein ligand-1 (PSGL-1), which is a mucin-type glycoprotein expressed on all white blood cells. Neutrophils and eosinophils bind to E-selectin. One of the reported ligands for E-selectin is the sialylated Lewis X Ag (sLe(x)). Eosinophils, like neutrophils, use sialylated, protease-resistant structures to bind to E-selectin, although the eosinophil expresses much lower levels of these structures on its surface. Ligands for P-selectin on eosinophils and neutrophils are similar sialylated, protease-sensitive, endo-beta-galactosidase-resistant structures, clearly different than those reported for E-selectin, and suggest disparate roles for P-selectin and E-selectin during recruitment during inflammatory responses.

The **selectin** family of membrane glycoproteins has a distal lectin-like domain that enables these molecules to bind to specific carbohydrate groups. Selectins interact primarily with sialylated carbohydrate moieties, which are often linked to mucin-like molecules. The selectin family includes three molecules, designated L, E, and P. Most circulating leukocytes express L-selectin, whereas E-selectin and P-selectin are expressed on vascular endothelial cells. Selectin molecules are responsible for the initial stickiness of leukocytes to vascular endothelium.

- **P-selectins:** P-selectin is expressed on activated endothelial cells and platelets. Synthesis of P-selectin can be induced by thrombin, leukotriene B4, complement fragment C5a, histamine, TNFα or LPS. These cytokines induce the externalisation of Weibel-Palade bodies in endothelial cells, presenting pre-formed P-selectins on the endothelial cell surface. P-selectins bind PSGL-1 as a ligand.
- **E-selectins:** E-selectin is expressed on activated endothelial cells. Synthesis of E-selectin follows shortly after P-selectin synthesis induced by cytokines such as IL-1 and TNFα E-selectins bind PSGL-1 and ESL-1.
- **L-selectins:** L-selectins are constitutively expressed on some leukocytes and are known to bind GlyCAM-1, MadCAM-1 and CD34 as ligands.

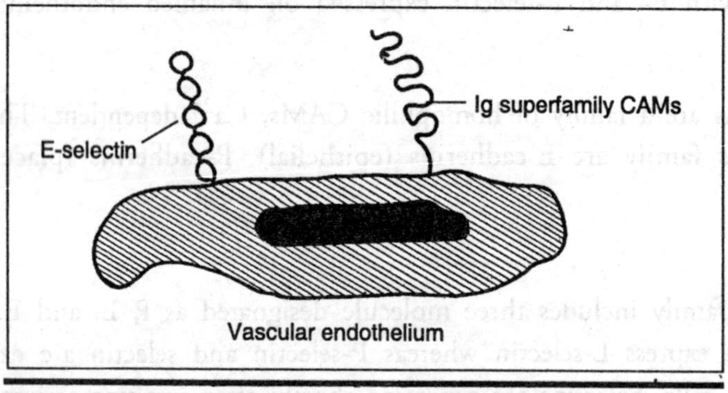

Fig. 8.3: Selectin molecule express on endothelial cells

Suppressed expression of some selectins results in a slower immune response. If L-selectin is not produced, the immune response may be ten times slower, as P-selectins (which can also be produced by leukocytes) bind to each other. P-selectins can bind each other with high affinity, but occur less frequently because the receptor-site density is lower than with the smaller E-selectin molecules. This increases the initial leukocyte rolling speed, prolonging the slow rolling phase.

Functions of CAM

The following are the functions of CAM

- Embryonic development
- Formation of nervous system

- Holding tissues together in adults
- Inflammation and wound healing
- Metastasis of tumours
- Transmits signal into and out of the cell.

CHAPTER 9

B-CELLS AND HUMORAL IMMUNITY

INTRODUCTION

B-lymphocytes constitute one of the major arms of the immune system, being responsible for humoral immunity. B-cells in humans and mice are produced throughout life, primarily in the fetal liver before birth and in the bone marrow afterwards. Their development from hematopoietic stem cells has been extensively characterized in mice, and the generation of numerous gene-targeted and transgenic lines in many cases has provided crucial information on the role of transcription factors, cellular receptors, and interactions that are critical in their generation.

B-cells are produced in the bone marrow and are distributed through the body in the lymph nodes. B-cells respond to the 'foreign' antigens of a pathogen by producing specific antibodies. Antibodies are complex proteins that are released into the blood and carried to the site of infection. B-cells do not fight pathogens directly.

B-cells also serve as antigen-presenting cells in T-cell activation. Resting B-cells express class II MHC molecules but fail to express co-stimulatory B7 molecules. Consequently, resting B-cells cannot activate naive T-cells, although they can activate the effector and memory T-cell populations. Upon activation, B-cells up-regulate their expression of class II MHC molecules and begin expressing B7. These activated B-cells can now activate naive T-cells as well as the memory and effector populations (Fig 9.1).

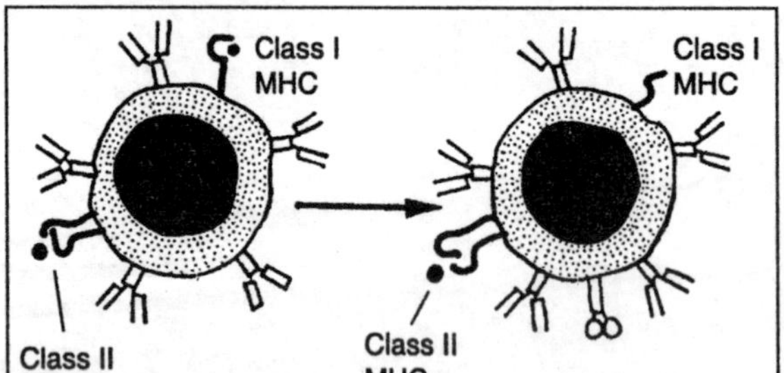

Fig. 9.1: Resting and activated B-cells

An antibody, or immunoglobulin, is a y-shaped protein molecule that is made by a B-lymphocyte in response to a particular antigen. Antibodies interact with the antigen and render it harmless. When a pathogen tries to invade the body for the first time, each of its antigens activates B-cell, which divides rapidly to produce a large population of cells. All the new cells are identical: we say they are clones, and they all secrete antibodies specific for the invading pathogen. When the infection is over, most of the newlymade B-cells die. This sequence of events is the primary immune response.

So that the body can respond more quickly next time, some of the activated B-cells persist in the body for several years. These memory cells 'remember' what the pathogen is like and, if it tries to invade again, they all divide rapidly to produce an even greater number of active B-cells, all capable of secreting specific antibody. This response is called a secondary immune response and is very much quicker and more effective than the primary response.

B-CELL DEVELOPMENT

B-cell development in mammals takes place in the bone marrow. When B-cells develop from their precursors they initiate a programme of differentiation which leads to rearrangement of the heavy chain gene segments. In order to allow the cell to test the functional integrity of the rearranged gene (only 1/3 are "in-frame"), Pre-B-cells express 2 single domain Ig-like protein of invariant sequence which substitute for the light chain. The formation of the complex of the mu heavy chain with these surrogate light chains somehow tells the cell to stop rearranging the heavy chain locus and start rearranging the kappa locus. Similarly if a successful light chain rearrangement is achieved and the light and heavy chains form a complete antibody then this complex tells the cell to stop rearranging light chains, thus ensuring that only a single specificity is produced (**allelic exclusion**).

Correspondingly those developing B-cell clones which fail to make a productive rearrangement at both one of their heavy chain alleles and a light chain locus will die. There are also mechanisms for ensuring the death of any newly produced B-cells whose antibody reacts strongly with self proteins on the surface of host cells.

B-CELL DEVELOPMENT IN MICE

In mice, B-cells are produced from hematopoietic stem cells through a complex process of differentiation that is gradually becoming understood. One of the goals of classical hematology is the delineation of differentiation pathways for different lineages of blood cells, and this is rapidly being achieved. Over the past decade there has been considerable progress in utilizing the ordered expression of a diverse set of cell surface and internal proteins, some with known functions and others whose roles are only suspected, to construct a description of the intermediate stages that cells transit as they develop into B-lymphocytes. Thus, uncommitted hematopoietic stem cells generate more restricted progeny, recognizable (in this example) by expression of IL-7Ra and these in turn produce B-lineage–restricted cells identified by expression of CD45R/B220 (and, importantly, by absence of CD19).

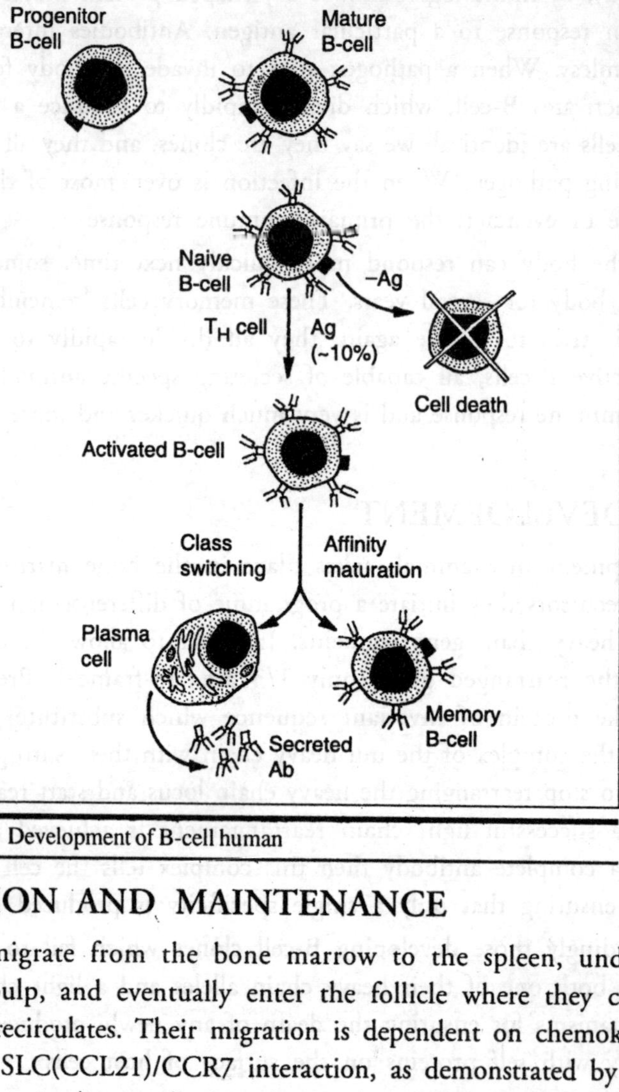

Fig. 9.2: Development of B-cell human

B-CELL MIGRATION AND MAINTENANCE

Newly formed B-cells migrate from the bone marrow to the spleen, undergo further maturation in the red pulp, and eventually enter the follicle where they constitute the mature B-cell pool that recirculates. Their migration is dependent on chemokines/receptor interactions, notably the SLC(CCL21)/CCR7 interaction, as demonstrated by the inability of mature B-cells to be retained normally in spleens of CCR7-null mice. The role of the CXCR5 receptor on B-cells in homing to the lymphoid follicle due a gradient of the B-lymphocyte chemoattractant CXCL13 is also well known; moreover, ÇXCL13 can directly induce Lta1ß2 on the recruited cells. Finally, it is also possible that the SDF1(CXCL12)/CXCR4 interaction, critical for normal B-lymphopoiesis, may also be important at this later stage, although investigation of this issue is complicated by the early defect. This is an ongoing area of investigation and may eventually be clarified by developmentally regulated gene-targeting studies

MEMORY B-CELLS

Memory B-cells were initially defined functionally as cells that could respond rapidly by production of high-affinity antibody when challenged in a host reconstituted with B-cells and T-cells from a primed animal. Subsequently, such cells have been purified based on their antigen-binding properties and shown to consist primarily of isotype-switched (IgG +) B-cells that continue to express CD45R/B220 and have distinctively lower levels of cell-surface BCR (317). They arise during the T-cell—dependent immune response, probably only from follicular B-cells, in the germinal center. They are very long-lived or self-regenerating, as cell transfer assays have shown that memory responses can be detected for long periods after the primary immunization. It is not clear whether all B-cells are capable of giving rise to memory B-cells. Memory or "secondary" B-cells were originally described as expressing distinctively low levels of CD24/HSA, recognized by the monoclonal antibody. Some years later, fractionation of naive spleen B-cell precursors into J11d low and J11d high subsets in a spleen-focus assay system showed that while rapid antibody secretion derived from J11d high cells, memory came largely from the J11d low subset. Subsequent experiments demonstrated that germinal centers (the site where most memory B-cells are generated) were only produced in cell transfers of J11d low B-cells and not with J11d high or CD5 + B-cells. Considering the rapid Ig secretory response of B-1 B-cells and MZ B-cells, both contained in the CD24/HSA high fraction, and the fact that most other CD24/HSA high cells are immature (transitional) B-cells, likely to be highly susceptible to apoptosis, it seems quite reasonable that memory B-cells would not be a major product of this fraction. Rather, the most likely candidate for the memory B-cell precursor is the follicular B-cell subset (Fo), which has a variable but lower expression of CD24/HSA. Whether there is heterogeneity for GC formation or memory B-cell generation within the Fo population remains to be determined. Whether the maintenance of memory requires periodic re-stimulation by antigen has been a long-standing controversial issue.

BCR

Humoral immunity provides immediate and long-term protection against a vast array of infectious agents as one vital component of the adaptive immune system. Expression of a B-cell receptor (BCR) signalling complex guides the development of immature B-cells in the bone marrow and is required for the survival of mature B-cells at homeostasis in the periphery. Infection disturbs homeostasis to signal and recruit antigen-specific B-cells into the adaptive immune response. B-Cell Receptors bind soluble antigens (like diphtheria toxoid, the protein introduced into your body in the DTP vaccine). The bound antigen molecules are engulfed into the B-cell by receptor-mediated endocytosis.The antigen is digested into fragments which are

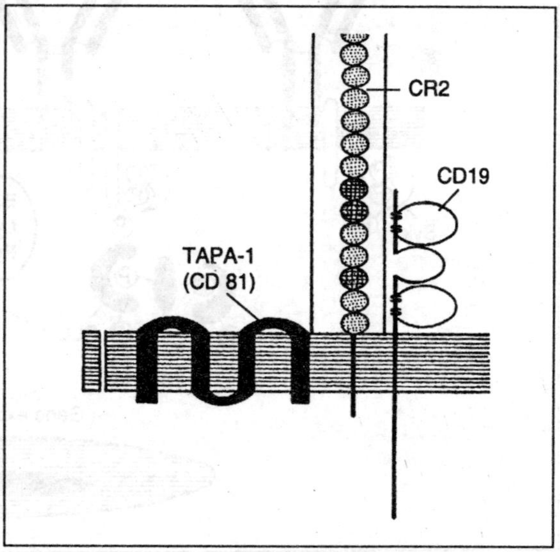

Fig. 9.3: Structure of B-cell receptor

then displayed at the cell surface nestled inside a class II histocompatibility molecule Helper T-cells specific for this structure (i.e. with complementary TCRs) bind the B-cell and secrete lymphokines that stimulate the B-cell to enter the cell cycle and develop, by repeated mitosis, into a clone of cells with identical BCRs switch from synthesizing their BCRs as integral membrane proteins to a soluble version differentiate into plasma cells that secrete these soluble BCRs, which we now call antibodies.

The B-cell coreceptor is a complex of three proteins: CD19, CR2 (CD21), and TAPA-1 (CD81) (Figure 9.3). CD19, a member of the immunoglobulin superfamily, has a long cytoplasmic tail and three extracellular domains. The CR2 component is a receptor of C3d, a breakdown product of the complement system, which is an important effector mechanism for destroying invaders; note that the involvement of C3d in the pathway for coreceptor activity reveals different arms of the immune system interacting with each other. CR2 also functions as a receptor for a membrane molecule and the transmembrane protein TAPA-1. In addition to the stimulatory coreceptor, another molecule, CD22, which is constitutively associated with the B-cell receptor in resting B-cells, delivers a negative signal that makes B-cells more difficult to activate. The CR2 component of the coreceptor complex binds to complement-coated antigen that has been captured by the mIg on the B-cell.

This crosslinks the coreceptor to the BCR and allows the CD19 component of the coreceptor to interact with the Ig component of the BCR. CD19 contains six tyrosine residues in its long cytoplasmic tail and is a major substrate of the protein tyrosine kinase activity that is mediated by crosslinkage of the BCR. Phosphorylation of CD19 permits it to bind a number of signalling molecules, including the protein tyrosine kinase Lyn.

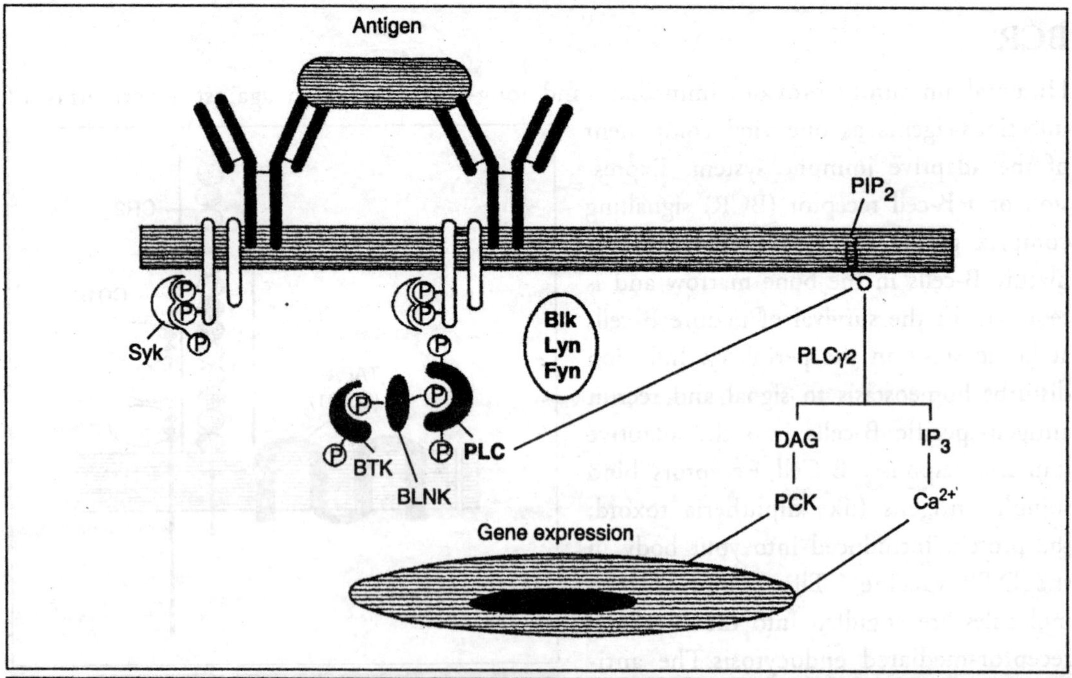

Fig. 9.4: Signal transduction by BCR.

CHAPTER 10

T-CELLS AND TCR

T-CELLS

T-cells belong to a group of white blood cells known as lymphocytes, and play a central role in cell-mediated immunity. They can be distinguished from other lymphocyte types, such as B-cells and NK cells by the presence of a special receptor on their cell surface called the T-cell receptor (TCR). The abbreviation T in T-cell stands for thymus, since it is the principal organ in the T-cell's development. T lymphocytes constitute the "cellular" arm of acquired/specific immunity. T-cells see fragmented, linear antigenic peptides when these are bound to proteins encoded by MHC genes. T lymphocytes play a central role in controlling the acquired immune response and furthermore serve as crucial effector cells through antigen specific cytotoxic activity and the production of soluble mediators called lymphokines (the cytokines produced by lymphocytes).

There are two major subsets of T lymphocytes that differ in effector function, MHC restriction and accessory molecule usage.

T-CELL SUBSETS

Several different subsets of T-cells have been described, each with a distinct function. Molecular association of CD8+ T-cells with MHC class I and CD4+ T-cells with MHC class II.

- **Helper T-cells** (T_H cells) are the "middlemen" of the adaptive immune system. Once activated, they divide rapidly and secrete small proteins called cytokines that regulate or "help" the immune response. Depending on the cytokine signals received, these cells differentiate into T_H1, T_H2, T_H17, or one of other subsets, which secrete different cytokines.

- **Cytotoxic T-cells** (T_C cells, or CTLs) destroy virally infected cells and tumour cells, and are also implicated in transplant rejection. These cells are also known as $CD8^+$ T-cells, since they express the CD8 glycoprotein at their surface. Through interaction with helper T-cells, these cells can be transformed into regulatory T-cells, which prevent autoimmune diseases such as experimental autoimmune encephalomyelitis.
- **Memory T-cells** are a subset of antigen-specific T-cells that persist long-term after an infection has resolved. They quickly expand to large numbers of effector T-cells upon re-exposure to their cognate antigen, thus providing the immune system with "memory" against past infections. Memory T-cells comprise two subtypes: central memory T-cells (T_{CM} cells) and effector memory T-cells (T_{EM} cells). Memory cells may be either CD4+ or CD8+.
- **Regulatory T-cells** (T_{reg} cells), formerly known as **suppressor T-cells,** are crucial for the maintenance of immunological tolerance. Their major role is to shut down T-cell-mediated immunity toward the end of an immune reaction and to suppress auto-reactive T-cells that escaped the process of negative selection in the thymus. Two major classes of CD4+ regulatory T-cells have been described, including the naturally-occurring T_{reg} cells and the adaptive T_{reg} cells. Naturally-occurring T_{reg} cells (also known as $CD4^+CD25^+FoxP3^+$ T_{reg} cells) arise in the thymus, whereas the adaptive T_{reg} cells (also known as Tr1 cells or Th3 cells) may originate during a normal immune response. Naturally-occurring T_{reg} cells can be distinguished from other T-cells by the presence of an intracellular molecule called FoxP3. Mutations of the FOXP3 gene can prevent regulatory T-cell development, causing the fatal autoimmune disease IPEX.
- **Natural killer T-cells** (NKT cells) are a special kind of lymphocyte that bridges the adaptive immune system with the innate immune system. Unlike conventional T-cells that recognize peptide antigen presented by major histocompatibility complex (MHC) molecules, NKT cells recognize glycolipid antigen presented by a molecule called CD1d. Once activated, these cells can perform functions ascribed to both T_h and T_c cells (i.e. cytokine production and release of cytolytic/cell killing molecules).
- **γδ T cells** represent a small subset of T-cells that possess a distinct TCR on their surface. A majority of T-cells δhave a TCR composed of two glycoprotein chains called α- and β- TCR chains. However, in γδ T-cells, the TCR is made up of one γ-chain and one δ-chain. This group of T-cells is much less common (5% of total T-cells) than the αβ T-cells, but are found at their highest abundance in the gut mucosa, within a population of lymphocytes known as intraepithelial lymphocytes (IELs). The antigenic molecules that activate γδ T-cells are still widely unknown. However, γδ T-cells are not MHC restricted and seem to be able to recognise whole proteins rather than requiring peptides to be presented by MHC molecules on antigen-presenting cells. Some recognize MHC class IB molecules though. Human Vγ9/Vδ2 T-cells, which constitute the major γδ T-cell population in peripheral blood, are unique in that they specifically and rapidly respond to a small non-peptidic microbial metabolite, HMB-PP, an isopentenyl pyrophosphate precursor.

- **Autoaggressive T-cells** are a unique T-cell subset that are characterized by the expression of CD40. CD40 typically is associated with antigen-presenting cells, but is also expressed on a subset of T-helper cells. **Th40** cells are found in all individuals but occur at drastically-expanded percentages in autoimmune subjects. This is true of autoimmune humans and mice. Th40 cells from type 1 diabetic subjects respond to known self-antigens, whereas Th40 cells from non-autoimmune subjects do not respond to those antigens. A crucial role of CD40 on T-cells is to induce RAG1 and RAG2, the recombinase proteins responsible for altering the T-cell receptor. The TCR is the means by which T-cells are able to recognize antigens. Dogma requires that RAG1 and RAG2 be expressed only in the thymus, during T-cell development. However RAGs are re-expressed in peripheral T-cells and CD40 engagement on Th40 cells induces RAGs expression. Following RAG expression changes in TCR occur. This means that Th40 cells are capable of adapting throughout an individual's lifetime. This process of altering TCR expression in the periphery is called **TCR revision.** Revision can be responsible for expanding the T-cell repertoire, but also could result in the generation of autoaggressive T-cells. TCR revision is therefore another means of T-cell tolerance. Alteration of autoaggressive TCR would necessarily tolerize those T.

T-cell development in the thymus

All T-cells originate from hematopoietic stem cells in the bone marrow. Hematopoietic progenitors derived from hematopoietic stem cells populate the thymus and expand by cell division to generate a large population of immature thymocytes. The earliest thymocytes express neither CD4 nor CD8, and are therefore classed as double-negative ($CD4^-CD8^-$) cells. As they progress through their development they become double-positive thymocytes ($CD4^+CD8^+$), and finally mature to single-positive ($CD4^+CD8^-$ or $CD4^-CD8^+$) thymocytes that are then released from the thymus to peripheral tissues. About 98% of thymocytes die during the development processes in the thymus by failing either **positive selection** or **negative selection,** whereas the other 2% survive and leave the thymus to become mature immunocompetent T-cells.

Positive selection

Double-positive thymocytes move deep into the thymic cortex where they are presented with self-antigens (i.e. antigens that are derived from molecules belonging to the host of the T-cell) complexed with MHC molecules on the surface of cortical epithelial cells. Only those thymocytes that bind the MHC/antigen complex with adequate affinity will receive a vital "survival signal." Developing thymocytes that do not have adequate affinity cannot serve useful functions in the body; the cells must be able to interact with MHC and peptide complexes in order to effect immune responses. Therefore, the other thymocytes with low affinity die by apoptosis (programmed cell death), and their remains are engulfed by macrophages. This process is called **positive selection.** Whether a thymocyte becomes a CD4+ cell or a CD8+ cell is also determined during positive selection. Double-positive cells that are positively selected on MHC class II molecules will become CD4+ cells, and cells positively selected on MHC class I molecules become CD8+ cells.

Negative selection

Thymocytes that survive positive selection migrate towards the boundary of the thymic cortex and thymic medulla. While in the medulla, they are again presented with self-antigen in complex with MHC molecules on antigen-presenting cells (APCs) such as dendritic cells and macrophages. Thymocytes that interact too strongly with the antigen receive an apoptosis signal that causes their death; the vast majority of all thymocytes initially produced end up dying during thymic selection. A small minority of the surviving cells is selected to become regulatory T-cells. The remaining cells will then exit the thymus as mature naive T-cells. This process is called **negative selection**, an important mechanism of immunological tolerance that prevents the formation of self-reactive T-cells capable of generating autoimmune disease in the host.

T-cell activation

Although the specific mechanisms of activation vary slightly between different types of T-cells, the "two-signal model" in CD4+ T-cells holds true for most. Activation of CD4+ T-cells occurs through the engagement of both the T-cell receptor and CD28 on the T-cell by the major histocompatibility complex peptide and B7 family members on the APC, respectively. Both are required for production of an effective immune response; in the absence of CD28 co-stimulation, T-cell receptor signalling alone results in anergy. The signalling pathways downstream from both CD28 and the T-cell receptor involve many proteins.

The first signal is provided by binding of the T-cell receptor to a short peptide presented by the major histocompatibility complex (MHC) on another cell. This ensures that only a T-cell with a TCR specific to that peptide is activated. The partner cell is usually a professional antigen-presenting cell (APC), usually a dendritic cell in the case of naïve responses, although B-cells and macrophages can be important APCs. The peptides presented to CD8+ T-cells by MHC class I molecules are 8-9 amino acids in length; the peptides presented to CD4+ cells by MHC class II molecules are longer, as the ends of the binding cleft of the MHC class II molecule are open.

The second signal comes from co-stimulation, in which surface receptors on the APC are induced by a relatively small number of stimuli, usually products of pathogens, but sometimes breakdown products of cells, such as necrotic-bodies or heat-shock proteins. The only co-stimulatory receptor expressed constitutively by naive T-cells is CD28, so co-stimulation for these cells comes from the CD80 and CD86 proteins on the APC. Other receptors are expressed upon activation of the T-cell, such as OX40 and ICOS, but these largely depend upon CD28 for their expression. The second signal licenses the T-cell to respond to an antigen. Without it, the T-cell becomes anergic, and it becomes more difficult for it to activate in future. This mechanism prevents inappropriate responses to self, as self-peptides will not usually be presented with suitable co-stimulation.

The T-cell receptor exists as a complex of several proteins. The actual T-cell receptor is composed of two separate peptide chains, which are produced from the independent T-cell receptor alpha and beta (TCRα and TXPβ) genes. The other proteins in the complex are the CD3 proteins: CD3εδ heterodimers and, most important, a CD3ζ homodimer, which has a total of six ITAM motifs. The ITAM motifs on the CD3ζ can be phosphorylated by Lck and in turn recruit ZAP-70. Lck and/or ZAP-70 can also phosphorylate the tyrosines on many other molecules, not least CD28, Trim, LAT and SLP-76, which allows the aggregation of signalling complexes around these proteins.

Phosphorylated LAT recruits SLP-76 to the membrane, where it can then bring in PLCγ, VAV1, ltk and potentially PI3K. Both PLCγ and PI3K act on PI(4, 5) P2 on the inner leaflet of the membrane to create the active intermediaries di-acyl glycerol (DAG), inositol-1, 4, 5-trisphosphate (IP3), and phosphatidylinositol-3, 4, 5-trisphosphate (PIP3). DAG binds and activates some PKCs, most important, in T-cells PKCθ, a process important for activating the transcription factors NF-κB and AP-1, IP3 is released from the membrane by PLCγ and diffuses rapidly to activate receptors on the ER, which induce the release of calcium. The released calcium then activates calcineurin activates NFAT, which then translocates to the nucleus. NFAT is a transcription factor, which activates the transcription of a pleiotropic set of genes, most notable, IL-2, a cytokine that promotes long term proliferation of activated T cells.

T-CELL MATURATION

Maturation of T-Cells in the Thymus

The first step is the rearrangement of the variable, joining and constant region genes of the chain of the T cell antigen receptor in a way very similar to that of heavy chain rearrangement needed for immunoglobulin synthesis. In fact, the same enzymes are used for both.

Production of a functional TCR chain, signals expression of both CD4 and CD8 on the cell surface. This induces the genetic rearrangements needed to produce a functional TCR chain and an increase in TCR membrane expression. CD3 is then expressed, which produces a functional TCR complex.

At this point, some of the T-cells stop making CD8, so only CD4 remains on their cell membrane. The others undergo the reverse process, so they express only CD8. T-cells then learn to not attack self tissues and to respond to antigen only if it is associated with a self histocompatibility antigen.

***This requires two steps*:**

- First, the immature, but CD4 or CD8 positive T-cells are exposed to cells in the thymus, which have class I and class II histocompatibility antigens on them. T-cells which are able to bind to one or the other of these antigens are protected, whereas the others die.
- Second, the cells that survive the above selection process are exposed to self antigens that have been taken up and associated with either class I or class II MHC antigen. Those that bind at this stage die (actually they commit suicide, called apoptosis).

The cells that survive are those that recognize non-self antigens associated with MHC antigens. After a little more maturation, they exit the thymus to perform their role in immune responses.

TABLE 10.1

Types of subset of T-cells and their functions

T cell type	Co-receptor	MHC restriction	Function
Helper T-cell	CD4	Class II	B ooll help macrophage activation help for CD8 T-cells cytokine secretion
Cytotoxic T-cell	Cd8	Class I	Killing virus infected cells Killing tumour cells?

T-CELL RECEPTORS

Structure of T-Cell receptor

The **T-cell receptor** or TCR is a molecule found on the surface of T lymphocytes (or T cells) that is generally responsible for recognizing antigens bound to major histocompatibility complex (MHC) molecules. It is a heterodimer consisting of an alpha and beta chain in 95% of T-cells, while 5% of T-cells have TCRs consisting of gamma and delta chains. The TCR is a heteodimer composed of one α and one β chain of approximately equal length.

The α chain is of 40-60KD acidic glycoprotein and β chain is of 40-50KDda basic glycoprotein. Each chain has a short cytoplasmic tail but it is to small to be able to transduce an activation signal to the cell. Both α chain's and β chain have a transmembrane region comprised of hydrophobic amino acids by which the molecule is anchored in the cell membrane. Both α chains and β chains have a constant (C) region and a variable (V) region similar to the immunoglobulin chains. The variable region of both chains contains hypervariable regions that determine the specificity for antigen. Each T-cell bears a TCR of only one specificity (i.e. there is allelic exclusion).

The variable (V) region of α chains and β chain are 102 to 119 amino acid long and contains two cysteine residues responsible for forming intrachain disulphide loop that span about 60-70 amino acid residue. The variable domain of both the TCR α-chain and β-chain have three hypervariable or complementarity determining regions (CDRs), while the variable region of the β-chain has an additional area of hypervariability (HV4) that does not normally contact antigen and therefore is not considered a CDR (Figure 10.1). CDR3 is the main CDR responsible for recognizing processed antigen, although CDR1 of the alpha chain has also been shown to interact with the N-terminal part of the antigenic peptide while CDR1 of the beta chain interacts with the C-terminal part of the peptide. CDR2 is thought to recognize the MHC. CDR4 of the β-chain is not thought to participate in antigen recognition, but has been shown to interact with superantigens.

The constanct (C) region of α chains and β chain are 138 to 179 amino acids long. The constant domain of the TCR domain consists of short connecting sequences in which a cysteine residue forms disulfide bonds which forms a link between the two chains.

There are striking structural similarity between α and β chain of TCR and Ig polypeptide. Each chain of the TCR is a member of the immunoglobulin superfamily and possesses one N-terminal immunoglobulin (Ig)-variable (V) domain, one Ig-constant (C) domain, a transmembrane-spanning region and a short cytoplasmic tail at the C-terminal end.

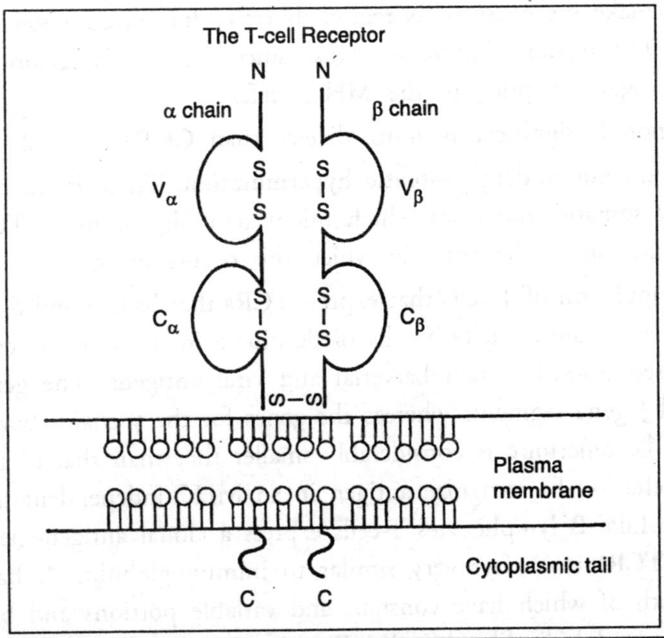

Fig. 10.1: Structure of TCR

Engagement of the TCR with antigen and MHC results in activation of its T lymphocyte through a series of biochemical events mediated by associated enzymes, co-receptors and specialized accessory molecules.

Organisation of TCR gene

The T-cell receptor genes were first identified by preparing complementary DNA (cDNA) clones of mRNA from clonal T-cell population. One approach used to isolate cDNA encoded TCR genes are substractive hybridization. The genetic basis for the generation of the vast array of antigen receptors on B-cells. The generation of a vast array of TCRs is accomplished by similar mechanism. The germline genes for the TCR β genes are composed of V, D and J gene segments that rearrange during T-cell development to produce many different TCR β chains. The germline genes for the TCR α genes are composed of V and J gene segments which rearrange to produce α chains. The specificity of the TCR β chains. The germline genes for the TCR α genes are composed of V and J gene segments which rearrange to produce α chains. The specificity of the TCR is determined by the combination of α and β chains. T-cell receptor genes are very similar to immunoglobulin genes in their organization. The TCR beta locus contains V, D and J segments like the IgH locus and the TCR alpha locus contains V and J segments like the Ig light chain loci. In fact the mechanism of gene rearrangement for T-cell receptors is essentially identical to Ig gene rearrangement. Mutations which lead to a failure of rearrangement affect both Ig and TCR; such mutations produce a lack of both T- and B-cells and therefore a complete lack of adaptive immunity and a severe immunodeficiency (SCID) which is naturally fatal.

There are two major differences between TCR and Ig receptor diversity

- T-cell receptors make more use of **N-region diversity** for introduction of new amino acids in the CDR3 region. This makes sense since the CDR3 region is believed to interact with antigenic peptide in the MHC cleft.
  The CDR3 region is significantly more diverse than CDR1, CDR2.
- T-cell receptors do not undergo **somatic hypermutation.** There is no mechanism for selection of any somatic mutations which might naturally occur in TCR genes and no affinity maturation in the sense in which this occurs in Ig.

There is a small population of T-cells that express TCRs that have γ and δ chains instead of α and β chains. These gamma/delta T-cells predominate in the mucosal epithelium and have a repertoire baised toward certain bacterial and viral antigens. The genes for the δ chains have V, D and J gene segments whereas the genes for the γ chains have only V and J gene segments but the repertoire is considerably smaller that than that of the alpha/beta T-cells. The gamms/delta T-cells recognize antigen in an MHC-independent manner unlike the alpha/beta T-cells. Like B lymphocytes T-cells express a clonal antigen-specific receptor. The T-cell receptor (**TCR**) is in fact very similar to immunoglobulin. It has two paired polypeptide chains both of which have constant and variable portions and both of which are composed of immunoglobulin-like domains. Two important differences are (a) TCR is monovalent (b) It exists only as a cell surface receptor and has no counterpart to secreted antibody.

The accessory molecules CD4 and CD8 define the effector function and MHC restriction of a T-cell. These molecules bind to MHC molecules at a different site to the TCR but form an important part of the TCR complex. In doing so they increase the affinity of interaction between the T-cell and the APC and probably also contribute to the signalling mechanism required for T-cell activation. Immature thymocytes begin life as emigrants from the bone marrow. When they **arrive in the** thymus these precursors do not express CD4,CD8 or T-cell receptor (**TCR**). **CD4 and CD8 are very useful markers** in thymocyte development and, since they express neither, these early thymocytes are said to be "double negative" cells.

T-cell development has many parallels with **B-cell development** but there is one very important difference which is that the T-cell repertoire is profoundly influenced by the MHC allotype of an individual through **selection.**

Developing T-cells (thymocytes) in fact undergo 3 different types of selection. The first critical stage in T-cell development is the rearrangement of the T-cell receptor beta chain.

This stage is very analogous to the rearrangement of the immunoglobulin heavy chain locus. As with the IgH locus the cell needs to make an in-frame functional rearranged TCRbeta chain to avoid death by apoptosis and to be given the signal to continue its development. Again, as with B-cells, the cell senses when a functional TCRbeta chain is made by expressing surrogate partners which can pair with the rearranged chain to form a **pre-T-cell receptor complex.** When a functional TCRbeta chain is made, these signals switch off beta chain rearrangement (thus providing **allelic exclusion** at the TCRbeta locus as well) and initiate a programme of differentiation which includes arrangement of the TCR alpha locus and expression of both the CD4 and CD8 molecules. Thus the first selective step in T-cell development selects **for** a functional TCRbeta rearrangement and cells which successfully pass this scrutiny become CD4+CD8+ "**double positive**" cells.

The other two selective steps occur at the double positive stage of thymocyte development. First the mature T-cell receptor must be expressed and this requires successful DNA rearrangement of the alpha chain locus. Once the mature TCR is expressed the TCR+ double positive thymocytes undergo two selective steps. These are called **negative selection** and **positive selection.** In negative selection clones bearing TCR which have a "high" affinity for self antigen are deleted, i.e. induced to undergo programmed cell death (apoptosis). In the context of TCR, self antigen means peptides from self proteins bound to MHC molecules. Negative selection has some parallel in B-cell development and ensures **self-tolerance.**

Positive selection is unique to T-cells. While this process is not yet fully understood it appears that cells are selected **for a "moderate" affinity** for self MHC-peptide complexes.

The TCR-CD3 Complex

The **T-cell receptor complex** with TCR-α and TCR-β chains, CD3 and ζ-chain accessory molecules and the co-receptor CD4. The transmembrane region of the TCR is composed of positively charged amino acids. It is thought that this allows the TCR to associate with other molecules like CD3, that possess three distinct chains (γ, δ and ε) in mammals and the ζ-chain. These **accessory** molecules have negatively charged transmembrane regions and are vital to propagating the signal from the TCR into the cell; the cytoplasmic tail of the TCR is extremely short, making it unlikely to participate in signalling. The CD3- and ζ-chains, together with the TCR, form what is known

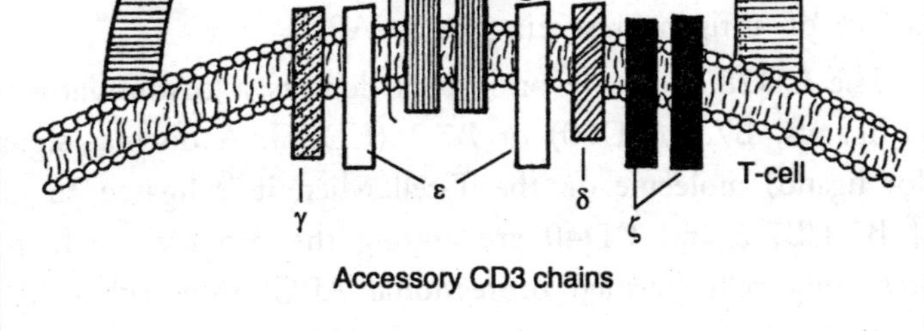

Fig. 10.2: Structure of TCR-CD3 complex

as the *T-cell receptor complex*. The TCR is part of a complex signalling machinery which includes the TCR-α β dimer, the accessory molecules CD4 or CD8 and a signal transduction module made up of various chains (CD3) (Figure 10.2).

Three member of complex are called CD3 protein sand include 25 to 28 kD glycosylated gamma chain, a 20 kD glycosylated delta chain, and 20 kD non glycosylated epsilon chain. These three most likely chains exist as monomer in TCR complex in addition 90% of TCR complex contains homodimer of 16 KD non glycosylated zeta chain.

TCR Co-Receptors

The signal from the T-cell complex is enhanced by simultaneous binding of the MHC molecules by a specific co-receptor.

- On helper T-cells, this co-receptor is CD4 that exclusively binds the class II MHC.
- On cytotoxic T-cells, this co-receptor is CD8 that is specific for class I MHC.

The co-receptor not only ensures the specificity of the TCR for the correctly presented antigen, it also allows prolonged engagement between the antigen-presenting cell and the T-cell and recruits essential molecules (e.g. LCK) inside the cell that are involved in the signalling of that activated T lymphocyte.

Engagement of the TCR by peptide antigen, in association with MHC gene products, leads to a series of intracellular biochemical events culminating in the transcription of new genes and cellular activation. The earliest identifiable intracellular change documented at present is the activation of one or more tyrosine kinases that phosphorylate first the CD3 chains themselves and subsequently other substrates. Subsequent to tyrosine kinase activation a series of secondary events have been observed to follow TCR engagement, including activation of serine/threonine kinases, activation of the GTP-binding protein p21ras and activation of transcription factors for receptors and growth factors such as the major T-cell growth factor **interleukin-2 (IL2)**. The CD4 and CD8 co-receptors bind a tyrosine kinase (p56lck) via their intracytoplasmic tail which plays a critical role in T-cell signalling.

COSTIMULATION

However TCR receptor binding is not sufficient to activate T-cells. The two signal hypothesis held that a second signal is required in addition to that through the antigen receptor. This model may even be too simple. There is evidence of a critical role for two distinct **costimulator molecules** on T-cells which interact with specific ligands on the surface of antigen-presenting cells (APC).

The **CD28** molecule on T-cells delivers a costimulatory signal upon engaging either of its ligands, **B7.1** (CD80) or **B7.2** (CD86). A distinct signal is transduced by the **CD40L** (for ligand) molecule on the T-cell when it is ligated to **CD40** (Fig.10.3). The expression of B7.1/B7.2 and CD40 are among the properties which distinguish specialised antigen presenting cells (termed **professional** APC) from other MHC positive cells.

A number of other molecules on the surface of APC may serve some role in costimulation, although their full role or mechanism of action is not clear.

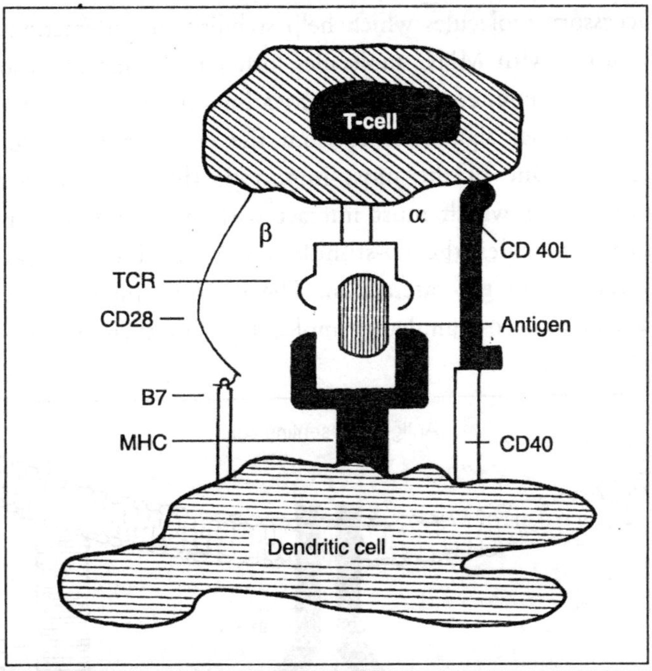

Fig. 10.3: Structure and interaction of Co receptor

Antigen presentation stimulates T-cells to become either "cytotoxic" CD8+ cells or "helper" CD4+ cells.

Generation of the TCR

Processes for TCR formation are similar to those described for B-cell antigen receptors, otherwise known as immunoglobulins. The TCR *alpha chain* is generated by VJ recombination while the *beta chain* is generated by V(D)J recombination (both involve a somewhat random joining of gene segments to generate the complete TCR chain). Similarly, generation of the TCR *gamma chain* involves VJ recombination while generation of the TCR *delta chain* occurs by V(D)J recombination. The intersection of these specific regions (V and J for the alpha or gamma chain, V D and J for the beta or delta chain p3n15) corresponds to the CDR3 region that is important for antigen-MHC recognition. Furthermore, it is the unique combination of the segments at this region, along with palindromic and random N- and P- nucleotide additions, which accounts for the great diversity in specificity of the T-cell receptor for processed antigen.

The interaction between the TCR and MHC molecules are not very strong. Accessory molecules are necessary to help stabilize the interaction. These include: 1) CD4 binding to Class II MHC, which ensures that Th cells only interact with APCs; 2) CD8 binding to class I MHC, which ensures that Tc cells can interact with target cells; 3) CD2 binding to LFA-3 and 4) LFA-1 binding to ICAM-1. The accessory molecules are invariant and do not contribute to the specificity of the interaction, which is solely determined by the TCR. The expression of accessory molecules can be increased in response to cytokine, which is one way that cytokines can modulate immune responses.

In addition to accessory molecules which help stabilize the interaction between the TCR and antigen in association with MHC molecules, other molecules are also needed for T-cell activation. Two signals are required for T-cell activation–one is the engagement of the TCR with Ag/MHC and the other signal comes from the engagement of co-stimulatory molecules with their ligands. One of the most important (but not the only) co-stimulatory molecule is CD28 on T-cells which must interact with B7-1 (CD80) or B7-2 (CD81) on APCs. Like accessory molecules the co-stimulatory molecules are invariant and do not contribute to the specificity of the interaction. The multiple interactions of TCR with Ag/MHC and the accessory and co-stimulatory molecules with their ligands have been termed the "immunological synapse".

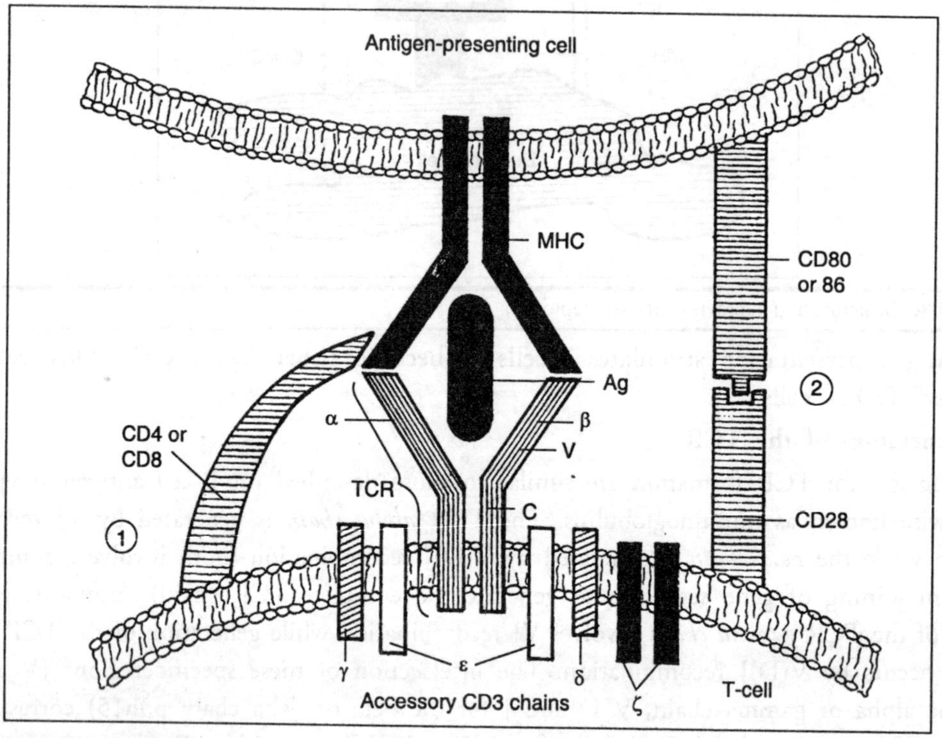

Fig. 10.4: Activation of T-cells

Not only is co-stimulation necessary for T-cell activation, a lack of co-stimulation may result in anergy (i.e. inability to respond to antigen) or downregulation of the response. Figure 10.4 shows the possible outcomes of a T-cell receiving one other signals necessary for activation. Engagement other TCR with Ag/MHC but no co-simulation results in anergy. Engagement of only the co-stimulatory molecule has no effect. Engagement of TCR with Ag/MHC and co-stimulatory molecules with their ligand results in activation. Engagement of the TCR with Ag/MHC and engagement of B7 ligand with CTLA-4, molecules similar to CD28, results in down-regulation of the response. CTLA-4/B7 interaction sends an inhibitory signal to the T-cell rather than an activating signal. This is one of the ways that immune responses are regulated.

CHAPTER 11

THE COMPLEMENT SYSTEM

INTRODUCTION

The complement system is composed of a group of serum proteins that play a major role in the **animal's** immune response. The complement system is the major factor of the humoral branch of the immune system. Complement is of group of serum protein that is activated by factor such as combination of antigen and antibody and result in variety of biological consequence including cell lysis and opsonization. The complement system is a complex system of serum proteins which interact in a cascade; many of the early components are serine proteases which activate each other sequentially.

The complement system acts in a cascade fashion, the activation of one components resulting in the activation of next. The complement cascade is made up of at least 17 complement proteins designated C1 to C9 in addition to factor B, factor D, factor H, factor I, C4b binding protein, C1 INH complex, S protein, and properdin. Collectively the complement proteins make up much of the globulin fraction of serum. Within plasma and other body fluids, complement proteins are in an inactive state. They usually are activated after the binding of antibodies to antigens and specifically directed against the target molecules identified by the antibodies to antibodies. Because complement activation involves binding of the component to antibody antigen complexes and to each other with their consequent removal from serum, this event is called complement fixation.

There are two pathways by which complement activation is initiated. The **classical pathway** is activated by antibody-antigen complexes. The **alternative pathway** is initiated when a previously activated complement component binds to the surface of a pathogen, where it is **protected**. Activation of complement has a number of important biological effects.

COMPONENTS OF COMPLEMENTS

The components of the classical pathway are called Cl and C2 etc. Where components are proteolytically cleaved, the products are referred to as C2a + C2b. Alternative Pathway specific components are given letters as names (factor B). The central component of the complement system is C3. C3 is an abundant serum protein (1.2mg/ml) which contains an unusual internal thiolester bond. In native C3 this bond is stable but this thiolester bond can become highly reactive as a result of conformational changes in the C3 protein structure. Generally the activation of C3 comes about as a result of proteolytic cleavage of the C3 molecule into two biologically active fragments.

The proteins and glycoproteins that compose the complement system are synthesised mainly by liver hepatocytes although significant amount of them are also produced by blood monocytes, tissue macrophages and epithelial cells of the gastrointestinal and genitourinary tracts. These components constitute 5% dry weight of the serum globulin fraction and most circulate in the serum in functionally inactive forms. Many components are proenzymes which are inactive until proteolytic cleavage, which are inactive site. The complement reaction sequence starts with an enzyme as cascade several of the activated components become inactivates shortly is they do not reach with the next component in the sequence.

Complement components are designated by numerical (C1-C9) by letter symbols [e.g. Factor D] or by trivial names e.g. homologous reaction factor. The peptide fragments formed by activation of a component are denoted by small letter. In most cases, the smaller fragment resulting from cleavage of a component is designated "a" and the larger fragment designated "b" e.g. C3a, C3b; note that C2 an extraction; C2a is the larger cleavage fragment.

The larger fragment binds to the target near the site of activation, and the small fragment defuse from the site of activation, and the smaller fragment defuse from the site and can initiate localized inflammatory responses by binding to specific receptors. The complement fragment increase with one another to form functional complexes. Those complexes that have enzymatic activity are designated by a bar over the number of symbol [e.g. $\overline{C4ba}$, $\overline{C3bBb}$].

CONVERGENCE OF THE CLASSICAL AND ALTERNATIVE PATHWAYS

The early phases of the classical and alternative pathways generate the key players in complement activation **C3 convertase.** In fact, a number of the components of the two pathways are related evolutionarily and structurally as well as funtionally. C4 and C3 are homologous as are C2 and factor B. The C3 convertases are **powerful amplifiers** of complement activation; up to 1000 C3 molecules are cleaved to C3b by each molecule of active C3 convertase. Each C3 convertase can be converted into a **C5 convertase** by the binding of a C3b molecule; this subsequently splits C5 into C5b and C5a.

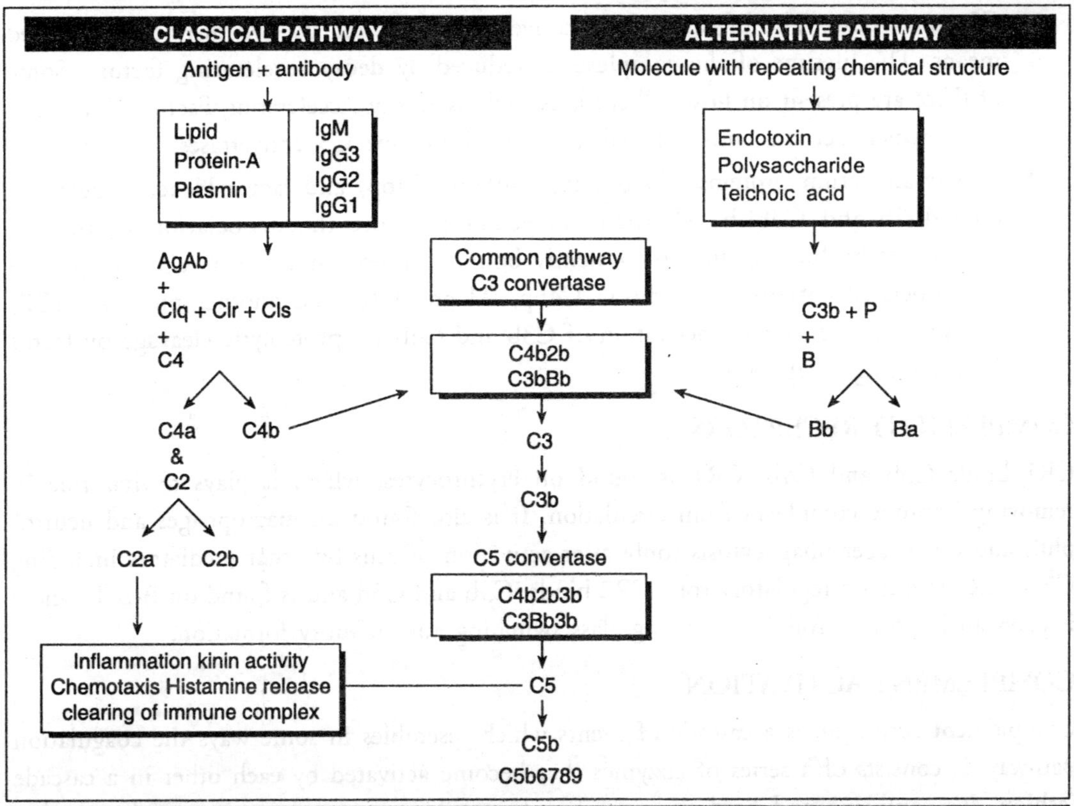

Fig. 11.1: Outline of classical and alternative pathway

TABLE 11.1
Homologous component in classical and alternative pathway

Function	AP	CP	Related?
initiating serine protease	factor D	C1s	not known
covalent attachment	3b	C4b	yes
C3/C5 cleaving serine protease	Bb	C2a	yes
inactivation	factor H	C4bp	yes

REGULATION OF COMPLEMENT ACTIVATION

Because of the potent damaging effects of complement activation the system is tightly regulated. Regulatory mechanisms act at three key points.

- C1: C1 inhibitor (C1 INH) acts in two ways. It binds free C1 in serum ([C1 INH] > [C1]) and inhibits spontaneous activation of C1; it is released on activation by immune complexes. It also limits the activation of C4 and C2 by binding and inhibiting the C1r and C1s proteases.

- **C3 convertases:** The lifetime of C3 convertases is limited by two distinct but related means. The lifetime of the complexes is reduced by **decay accelerating factors**. Some of these are present on host cell surfaces such as Decay Accelerating Factor (DAF) and CR1, which act on both the classical and alternative C3 convertases.
- Others are serum proteins, C4 binding protein (C4bp) and factor H act respectively on C4b2b and C3bBb. All these molecules promote the dissociation of the C3 convertases by binding to the covalently bound component (C4b or C3b) displacing the associated cofactor. In addition C4bp, fH and CR1 and another protein (MCP) catalyze the permanent inactivation of C3b and C4b via proteolytic cleavage by factor I (generating iC3b etc).

COMPLEMENT RECEPTORS

CR1 binds C3b and C4b. CR1 is found on erythrocytes, where it plays a vital role in removing immune complexes from circulation. It is also found on macrophages and neutrophils and can trigger phagocytosis (only after activation of cells by other mediators including C5a). CR1 also has a regulatory role. CR2 binds iC3b and C3d and is found on B-cells where it plays an important role in activating class switching and memory formation.

COMPLEMENT ACTIVATION

Complement activation is a cascade of events which resembles in some ways the coagulation pathway. It consists of a series of enzymes that become activated by each other in a cascade fashion. It's confusing and hard to remember, but still important because complement has an important role in clearing antigen.

This is done by the four major functions of complement activation:

1. Local vasodilation
2. Attraction of immune cells, especially phagocytes (chemotaxis)
3. Opsonization (or tagging) of foreign organisms for phagocytosis
4. Destruction of invading organisms by the membrane attack complex (MAC attack)

The central molecule is the C3 protein. It is an enzyme that is split into two fragments by components of either the classical pathway or the alternative pathway. The classical pathway is induced by antibodies, especially IgG and IgM, while the alternative pathway is nonspecifically stimulated by bacterial products like lipopolysaccharide (LPS). Briefly, the products of the C3 split include a small peptide C3a which is **chemotactic** for phagocytic immune cells and results in local **vasodilation** by causing the release of C5a fragment from C5. The other part of C3, C3b coats antigens on the surface of foreign organisms and acts to **opsonize** or 'tag' the organism for destruction. C3b also reacts with other components of the complement system to form a **membrane attack complex** (MAC) consisting of C5b, C6, C7, C8 and C9. A simplified version of the complement pathway is shown here.

The early steps in complement activation culminating in formation of C5b, can occur by the classical pathway, the alternative pathway or the lectin pathway. The final steps that lead to a membrane attack are the same in all pathways.

THE CLASSICAL PATHWAY

The classical activation pathway is triggered when a molecule of C1 binds the antibody-antigen complex. C1 is a complex of 1 C1q, 2C1r and 2C1s subunits; C1q has 6 binding sites for IgFc. The affinity of a single C1q binding site for 1 Fc is low so that at least two sites are required for stable binding of C1q and subsequent activation of C1r. This requires two juxtaposed IgGs or a distorted IgM, thus free IgG and IgM do not activate complement. Activated C1r cleaves C1s to generate the active C1s protease which in turn cleaves first C4 (to C4a +C4b) and, after C2 has bound C4b, C2 to generate **C4bC2a, the classical C3 convertase.** C4b contains a highly reactive (i.e. short-lived) **thiol-ester** bond which is capable of forming a covalent attachment to nearby proteins and carbohydrates on immune complexes, cell and bacterial surfaces.

Complement activation by the classical pathway commonly begins with the formation of soluble antigen-antibody complex or with binding of antibody to antigen on a suitable target such as a bacterial cell. IgM and certain subclasses of IgG human IgG2, and IgG3 can activate the classical complement pathway, as can certain non immunologic activators. The initial stage of activation involves C1, C2, C3 and C4, which are present in plasma in functionally inactive for because the components were named in order of their discovery and before their functional roles had been determined, the numbers in the named do not always reflects the order in which the react.

The complexing of antibody with antigen induces conformational changes in the Fc portion of the IgM molecules that expose a binding site for the C1 component of complement system. C1 in serum is macromolecular complex consisting C1q and two molecule of each C1r and C1s, together in a complex [$C1qr_2s_2$] stabilized by Ca^2+ ions. The C1q molecules is composed of 18 polypeptide chains that associate to form six collagen–like triple helical arms, the tips of which bind to exposed Clq binding sites in the CH_2 domain of the antibody molecule.

Each C1 molecule must bind, by its C1q globular heads, to at least two Fc sites for a stable C1-antibody interaction to occur. When pentameric IgM is bound to antigen on a target surface it assumes the so-called staple configuration in which at least three binding sites for C1q are exposed. Circulating IgM, however exist as a planar configuration in which the C1q-binding sites are not exposed. For this reason, circulating IgM cannot activate the complement cascade by itself. An IgM molecule, on the other hand, contains only a single, C1q-binding site in the CH_2 domain of the Fc, so that firm C1q binding is achieved only when two IgM molecules are within 30-40m of each other on a target surface or in a complex providing two attachment sites for C1q. This difference in the structure of IgM and IgG account for the observation that a single molecule of IgM bound to a red blood cell is enough to activate the classical complement pathway and lyse the red blood cell, whereas some 1000 molecules are of IgM are required it two molecules, randomly distributed are to end up close enough to each other to initiate C1q binding. The intermediates in the classical activation pathway are depicted (Fig 11.2). Binding of C1q to Fc binding sites induces a conformational change in C1r that converts C1r to an active serine protease

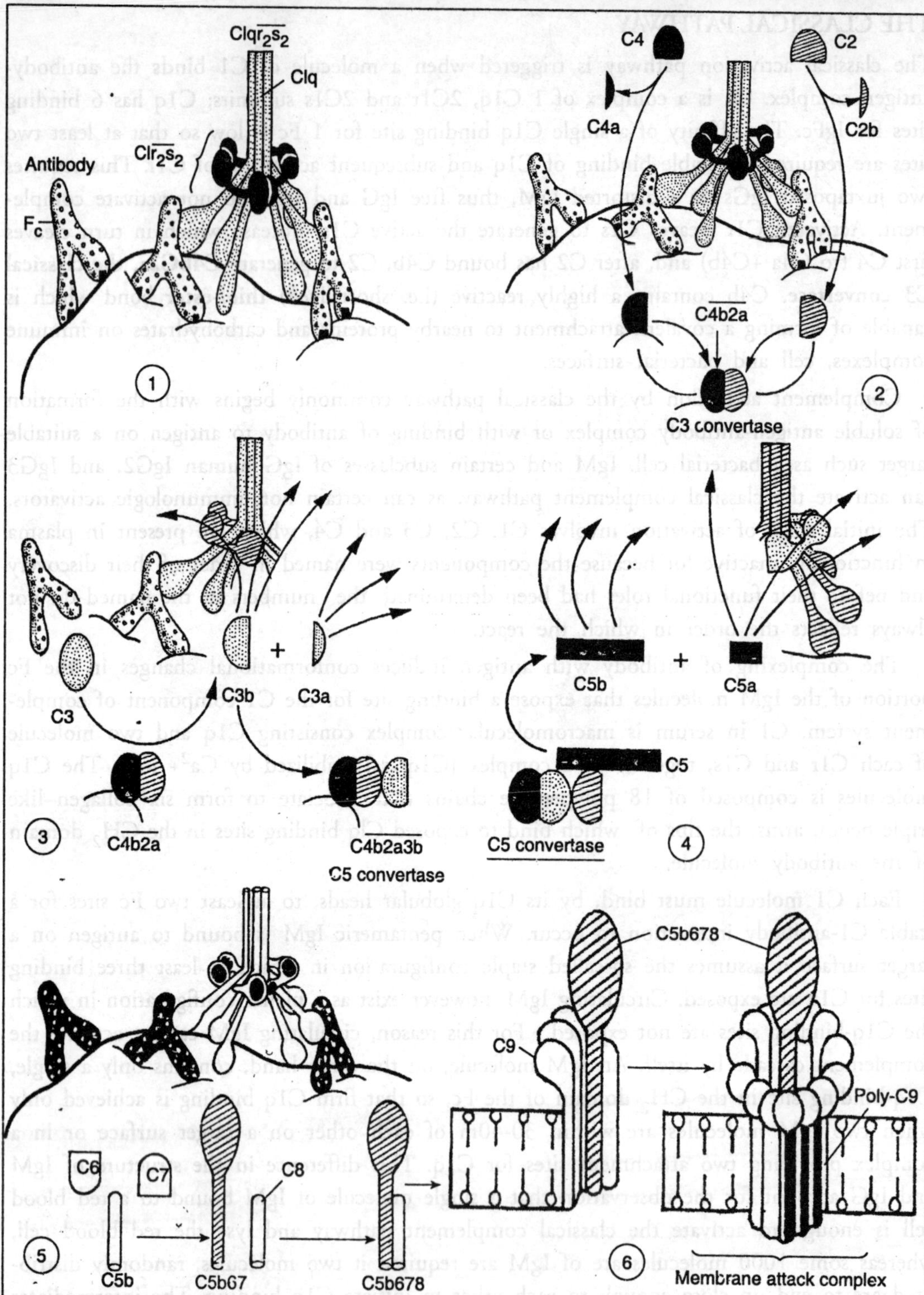

Fig. 11.2: Schematic diagram of intermediates in the classical pathway of complement activation

enzyme C1r, which then cleaves Cls to a similar active enzyme C1s, C1s has two substrates, C4 and C2. The C4 component is a glycoprotein containing three polypeptide chains alpha and beta. CC4 is activated when C1s hydrolyzes a small fragment (C4a) from the amino-terminus of the chain, exposing the binding site on the larger fragment (C4b). The C4b fragment attaches to the target surface in the vicinity of C1, and the C2 proenzymes then attacks to the proenzymes exposed binding site on C4b, where the C2 is then cleaved by the neighboring Cls; the smaller fragment C4b2a complex is called C3 convertase, referring to its role in converting the C3 enzyme into an enzymatically active form. The smaller fragments from C4 cleavage. C4a is an anaphylatoxins, which includes the smallest fragment of C4, C3 and C5. The native C3 component consists of two polypeptide chains, alpha and beta. Hydrolysis of a short fragment (C3a) from the amino terminus of the chain by the C3 convertase generates C3b.

A single C3 convertase molecule can generate over 200 molecules of C3b, resulting in tremendous amplification at this step of the sequence. Some of C3 binds to C4b2 to form a trimolecular complex C4b2a3b, called C5 convertase. The C3b component of the complex binds C5 and alters its conformation so that the C4b2a component can cleave C5 into C5a, which diffuses away and C5b, which attaches to C5 and initiates formation of the membrane attack complex in a sequence described later.

Some of the C3b generated by C3 convertase activity dose not associate with C4b2a; instead it diffuses away and then coats immune complexes and particulate antigens, functioning as an opsonin as discussed in a later section. C3b may also bind directly to cell membranes.

The Alternative Pathway

The alternative pathway (AP) is an innate immune mechanism which exploits the properties of C3. Initiation of the alternative pathway relies on the generation of an activated C3 molecule. There are various ways in this initiating activated C3 can be generated. These include proteolysis by enzymes derived from bacteria themselves and, via blood clotting enzymes, injury. However activation of C3 can occur spontaneously. This process proceeds constitutively at a slow rate. The activated thiolester will normally be rapidly hydrolyzed in solution. But if C3 binds via its ester bond to a cell surface or protein complex, it can recruit the AP serum protein factor B. Factor B, once bound to activated C3, is cleaved by another serum protein, factor D, and the resulting complex forms an enzyme which activates further C3 molecules by cleaving them to form the activated fragment C3b (plus C3a). At this point the alternative pathway becomes self-generating by a positive feedback loop, those activated C3b molecules which are not quenched by hydrolysis binding to nearby surfaces, recruiting factors B and D and generating the C3bBb complex which in turn cleaves more C3 etc. The complex C3bBb is known as the alternative C3 convertase. This complex is a short-lived one, but is stabilized on certain microbial surfaces by binding of the serum factor Properdin (P).

The alternative pathways is an antibody-independent means to generate bound C5b, the same product as the classical pathways generate in a sequence of events requiring antigen-antibody complexes for initiation. Because no antibody is required, the alternative pathway is a component of the innate immune system. This major pathway of complement activation involve four serum proteins; C3, factor B, factor D, and properdin. Unlike the classical, pathway, the alternative pathway is initiated without requiring antibody, in most cases by various cell surface constituent that are foreign to the host. For example, both Gram negative and Gram positive bacteria have cell-wall components that can activate the alternative pathway. The intermediates in the alternative pathway for generating C5b are depicted schematically in (Figure 11.3).

In the classical pathway, C3 is cleaved to C3a and C3b rapidly by the enzymatic activity of the C3 convertase. In the alternative pathway, serum CC3, which contains an unstable thiolester bond, is subjected to slow spontaneous hydrolysis to yield C3a and C3b. The C3b component can bind to foreign surface antigens (such as those on bacterial cells or viral particles). The membranes on the most mammalian cells have high levels of sailic acid, which contributes the rapid inactivation of bound C3b molecules on host cells, consequently this binding rarely leads to further reaction on the host cell membrane.

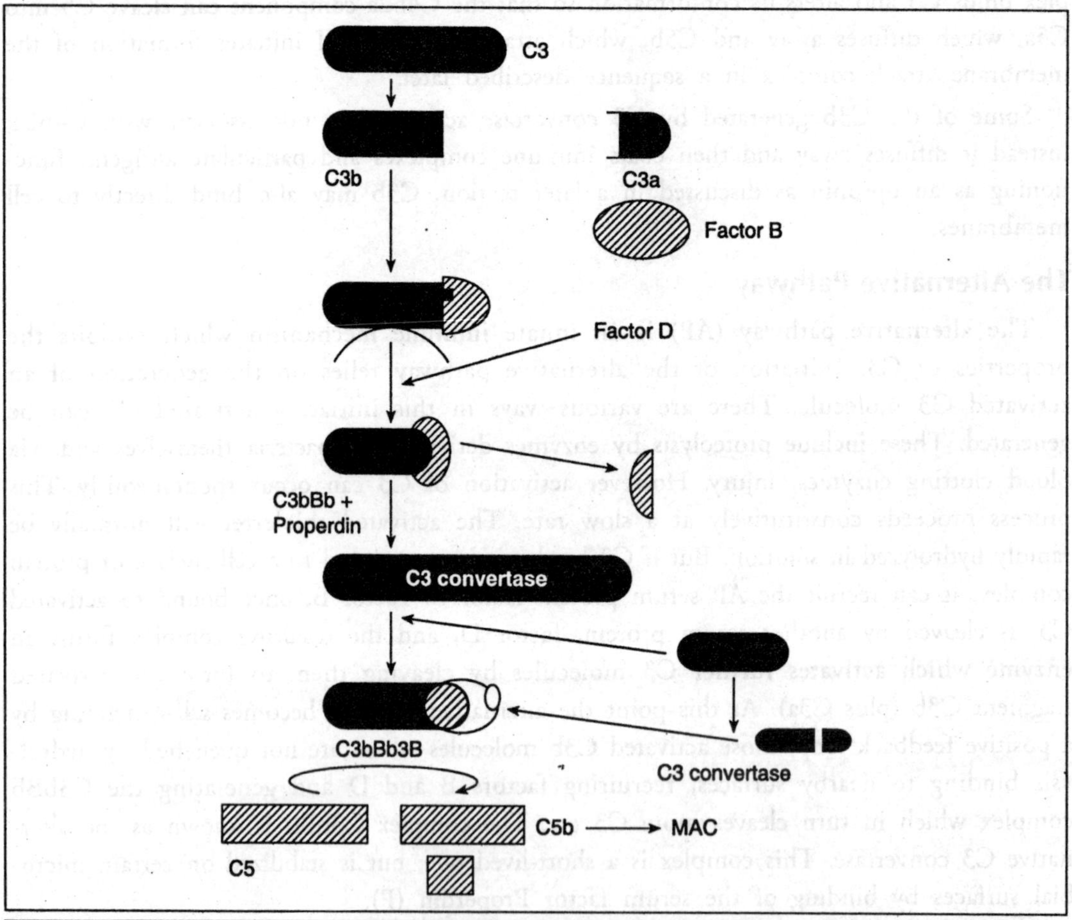

Fig. 11.3: Schematic diagrams of inter mediates in form of bound C5b by the alternative pathway of complement activation

The C3bBb complex is stabilized by binding of properdin. Membrane bound inter mediates are shown. Conversion of bound C5b to the membrane-attack complex occurs by the same sequence of reaction as in the classical pathway.

Because many foreign antigens surfaces e.g. bacterial cell walls, yeast cell wall, and certain viral envelopes have only low level of sailic acid, C3b bound to these surfaces remains active for a longer time. The c3b present on the surface on the foreign cells can bind another serum protein called factor B that serves as a substrate for an enzymatically active serum protein called 'Factor D' factor cleaves the C3b factor B, releasing a small fragment (Ba), which diffuse away, and generates C3bBb. The C3bBb complex has C3 convertase activity and thus is analogue to the C4b2a complex in classical pathway. The C3 convertase activity of C3bBb has a half-life of only 5 minutes unless the serum proteins properdin binds to it, stabilizing it and extending the half-life of this convertase activity to 30 min.

The C3bBb generated in the alternative pathway can activate unhydrolyzed C3 to generate more C3b auto catalytically. As a result the initial steps are repeated and amplified so that more than 2*10 molecules of C3b can be deposited on a antigenic surface in less than 5 minutes. The C3 convertase activity of C3bBb generates the C3bBb3b complex which exhibits C5 convertase activity, analogous to the C4b2a3b complex in the classical pathway.

The nonenzymetic C3b components can bind C5, and the Bb component subsequently hydrolyzes the bound C5 to generate C5a and C5b the latter binds to the antigenic surface.

Initiators of the alternative pathway of complement activation:

Pathogens and particles of microbial origin

- Many strains of Gram-negative bacteria.
- Lipopolysaccharides from Gram-negative bacteria.
- Many strains of Gram-positive bacteria.
- Teichoic acid from Gram-positive cell walls.
- Fungal and yeast cell walls (zymogene).
- Some virus and virus-infected cells.
- Some tumor cells.
- Parasites (trypanosomes).

Non-pathogens

- Human IgG, IgA, and IgE in complexes.
- Rabbit and guinea pig IgG in complexes.
- Cobra venom, heterologous erythrocytes, rabbit, mouse, chicken.
- Anionic polymers dextran sulfates.
- Pure carbohydrates (agarose, insulin).

TABLE 11.2

Complement components in the formation of C3 and C5 convertase

	Classical Pathway.	Lectin Pathway	Alternative pathway
Precursors Proteins	C4+c2	C4+c2	C3+Factor B
Activating Protease	– C15	MASP	Factor D
C3 convertase	C4b2a	C4b2a	C3bBb
C5 convertase	C4b2a3b	C4b2a3b	C3bBb3b
C5-binding Components	C3b	C3b	C3b

Terminal sequence (formation of membrane attack complex)

The terminal sequence of complement activation involves C5b, C6, C7, C8 and C9, which interact sequentially to form a macromolecular structure called the membrane attack complex (MAC). This complex displaces the membranes phospholipids, forming a large transmembrane channel that disrupts the membrane of the target cell and enables ions and small molecules to diffuse through it freely.

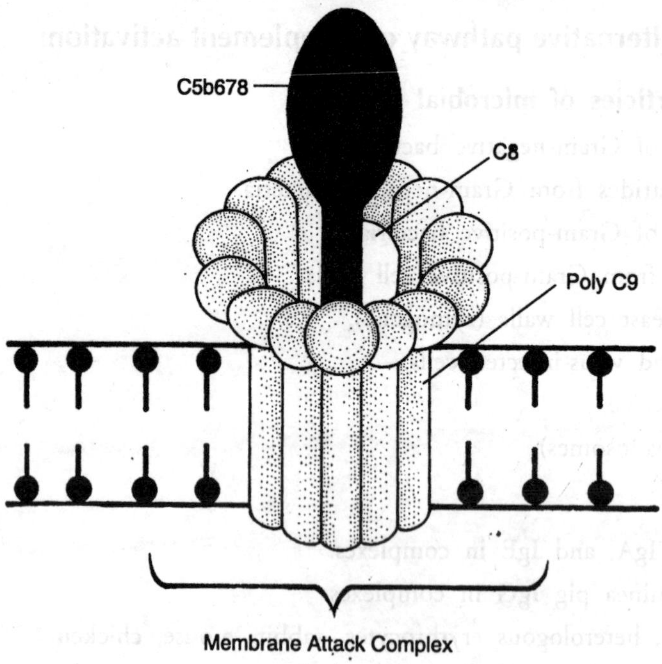

Fig. 11.4: The membrane attack complex

The membrane attack complex is a tabular structure that forms a transmembrane pore in the target cells plasma membrane. As, noted previously, in the classical, lectin alternative pathways, a C5 convertase cleavage C5, which contains two protein chains a and b. After binding of C5 to the nonenzymetic C3b component of the convertase, the amino terminus of the chain is cleaved. This generates the small C5a fragment, which diffuses away and the large C5b fragment, which bounds to the surface of the target cells and provides a binding site for the subsequent components of the MAC. The C5b component is extremely labile and is inactivated within 2 minutes unless C6 binds to it and stabilizes its activity. Binding of C8 to membrane bound C5b67 induces a conformational change in C8, so that to undergoes a hydrophilic, amphiphilic structural transition, exposing a hydrophobic region, which interacts with the plasma membrane. The C5b678 complex creates a small pore of 10A in diameter; formation of this pore can lead to lysis of red blood cells but not of nucleated cells. The final step in formation of the MAC is the binding and polymerization of C9, a perforin-like molecule, to the C5b678 complex.

As many as 10-17 molecules of C9 can be bound and polymerized by a single C5b678 complex. During polymerization the C9 molecule undergo a hydrophilic, amphiphilic transition, so that they also can insert into the membrane. The completed MAC which has the tubular form and functional pore size of 70-100A and consist of a C5b678 complex surrounded by a poly-C9 complex. Since one and small molecule can diffuse freely through the central channel of the MAC, the cell cannot maintain its osmotic stability and is killed by an influx water and loss of electrolytes. The components of the terminal sequence of the component pathways are summarized in this Table 11.4.

TABLE 11.3

Proteins involved in the formation of the membrane attack complex {MAC}

Component	Active protein split product	Immunologic function
C5	C5aC5B	Peptide mediator of chemotaxis and inflammation anaphylotoxin. Binds C6 to initiate form of MAC.
C6	C6	C5b6 binds C7.
C7	C7	After in amphiphilic transition of C5b67; the resulting complex inserts into the lipid bilayer.
C8	C8	C5b678 binds multiple C9 molecules, initiating their polymerization.
C9	C9	Polymerizes to complete formation of MAC pore.

Regulation of the complement system

Because the complement system is nonspecific and thus capable of attacking host cells, as well as microorganisms elaborate regulatory mechanisms have evolved to restrict its activity to designated targets. A general mechanism of regulatory components, which undergo spontaneous inactivation as they diffuse away from target cells. In addition the complement system includes a series of regulatory proteins that interactive various complement components.

TABLE 11.4

Proteins that regulates complement system

Protein	Type of Protein	Pathway affected	Immunologic function
C1 inhibitor (cllnh)	Soluble	Classical	Serine Protease inhibitor: causes clr2s2 to dissociates from clq.
C4b-binding Protein (c4bBp)	Soluble	Classical and Lectin	Block form of C3 convertase by binding C4b: cofactor cleavage of C3b by factor1.
Factor H	Soluble	Alternative	Block form of or convertase by binding C3b: cofactor for cleavage of C3b by factor 1.
Complement receptor type 1 (cri) Membrane Cofactor protein (MCP)	Membrane bound	Classical, alternative and lectin	Block form o C3 convertase by bind ing C4b or C3b; cofactor for factor 1-catalyzed cleavage of C4b or C3b.
Decay-accelerating. Factor (DAF)	Membrane bound	Classical, alternative and lectin	Accelerates dissociation of C4b2a & C3bBb.
Factor I	Soluble	Classical, alternative and lectin	Serine Protease; cleavage C4b or C3b using C4bP,cRI factor H, DAF, or MCP as factor.
S Protein	Soluble	Terminal	Binds Soluble C5b67 & prevents its insertion into cell membrane.
Homologous restriction factor (HRF) membrane inhibitor of reactive lysis (MIRL)	Membrane bound	Terminal	Binds to C5b678 on analogous cells, blocking binding of C9.
Anaphylatoxins in activator	Soluble	Effector	Inactivates, anaphylotoxin activity of C3a, C4a, and C5a by carboxy peptides N removal of c-terminal Arg.

Biological functions of complement system

Research on complement now includes more than 30 soluble and cell bound proteins. Biologic activities of this serum impact both innate and acquired immunity and reach for beyond the original observation of antibody-mediated lysis of bacteria and red blood cells. After initial activation the various complement components interact, in a high regulated cascade, to carry out a number of basic functions such as lysis of cells, bacteria and viruses. Complement serves as on important mediator of the humoral response by amplifying the response and converting it into an effective defense mechanism to destroy invading microorganisms and viruses. The MAC mediates cell lysis. While other complement component or split product participate in the inflammatory response, opsonizsation of antigen viral neutralization and clearance of immune complexes.

Many of the biological activities of the complement system depend on the binding of complement fragments to complement receptors which are exposed by various cells. In addition some complement receptors play and important role in regulating complement components and degrading them into inactive products. The complement receptors and their primary ligands, which include various, complement component and their proteolytic breakdown products.

Cell lysis

The complement system is generally quite effective in lysing Gram negative bacteria. However, some Gram negative bacteria and most Gram positive bacteria have mechanism for evading complement mediated damage. For example-A few Gram negative bacteria can develop resistance to complement mediated lysis that correlates with the virulence of the organisms. The membrane attack complex formed by complement activation is capable of lysing a broad spectrum of microorganism, viruses, erythrocyte, and nucleated cells. 5b binds and recruits C6 and C7 to the target surface. C7 and subsequently C8 change conformation to expose hydrophobic domains which insert in the lipid bilayer. The C5b678 complex catalyses the polymerisation of the final component C9 which forms a transmembrane pore of ~10nm diameter causing lysis of the cell. This macromolecular assembly is known as the membrane attack complex (MAC).

Inflammatory response

The inflammatory response can be triggered by non immune and immune factors. The first is mediated by IgE attachments to the surface of mast cells and basophil cells that lead to the activation of mast cells and basophil which results in release of histamine.

Histamine causes the most of the characteristics events of inflammatory responses. The second route involves the complement pathways. The complement cascade is often viewed in terms of the final outcome of cell lysis, but various peptides generated during formation of the MAC play a decisive role in the development of an effective inflammatory response. The C5a and, less potently, the C4a and C3a fragments are important inflammatory activators inducing vascular permeability, recruitment and activation of phagocytes. The smaller fragments resulting from complement C3a, C4a and C5a called anaphylatoxins, bind to receptor on mast cells and blood has basophils and induce degranulation with release of histamine and other pharmacologically active mediators. The anaphylatoxins induces smooth muscles contraction and increased vascular permeability.

Activation of the complement system thus results in influxes of fluid that carries antibody and phagocytic cells, to the site of antigen entry. The activities of there highly reactive an anaphylatoxins are regulated by a serum protease called carboxy peptidase N, which cleaves an Arg residue from the C terminus of the molecules. Resulting in what are termed the des-Arg resulting in what are termed the der-Arg forms. The des-Arg forms of C3a and

C4a are completely inactive while that are C5a retains about 10% of its ability to cause smooth muscle contraction.

Each of C3a, C5a and C5b67 can induce monocytes and neutrophils to adhere to vascular endothelial cells, extravasate through the endothelial lining cells, extravasate through the end endothelial lining of the capillary and migrate towords the site of the complement activating in tissues. C5a in most potent in mediating their processes, with Pico molar quantities beings effective. The role complement in leukocyte chemotaxis.

Opsonization

This phagocytic process can be greatly enhanced by opsonization. Opsonization (Greek *opson*, prepare victims for) is the process in which microorganisms or other particles are coated by antibody and or complement there by being prepared for "recognition, and ingestion by phagocytic cell. Opsonization which promotes the phagocytosis of particulate antigens, binding on specific complement receptors on cells of the immune system, triggering activation of immune responses such as inflammations and secretion of immunoregulatory mediators that amplify or later specific immune response. Opsonization or uptake of particulate antigens including bacteria by phagocytes activation of inflammatory response and clearance of circulating immune complexes by cells in the liver and spleen. Soluble complement proteins are schematically indicated by a triangle and receptors by a semicircle. No attempt is made to differentiate among individual components of the complement system. The C3b and, to a lesser degree, C4b molecules are **opsonins**. That is they coat foreign organisms either by the activation pathway or those already bound by antibody. Opsonisation of particles greatly enhances their phagocytosis by means of binding to specific complement receptors.

Opsonizing antibodies belong to the IgG1, IgG3 or IgM isotypes and have a site in the CH_3 domain that binds to an Fc receptor on the surface of macrophages. This binding forms a bridge between the phagocyte and the antigen. IgG or IgM antibodies also can activate the complement system most often through the classical pathway but sometime by the alternative pathway. The net result is to assembles the C3b convertase on the surface of the microorganism together with opsonic fragment C3b. Both neutrophils and microorganisms complement, and phagocytic cell bind together. If both antibody and C3b opsonize, binding is greatly enhanced.

Opsonizing antibody can be directed against surface component of bacteria. Components that can trigger deposition of C3b on the surface of microorganisms by the alternative complement pathway includes lipopolysaccharides of Gram negative bacteria and fungal polysaccharide. Finally some bacterial components, staphylococcal protein A or lipid A of endotoxin, can activate the classical complement pathway without the participation of antibody.

Viral neutralization

The complement system mediate viral neutralization by various mechanism. Some degree of neutralization is achieved through large aggregates. The fixation of classical pathway complement component C4b to the virus aids the neutralization process. Viral neutralization prevents a viral infection due to the inability to the virus to target cell.

IgG, IgM and IgA antibodies can bind to some viruses during their extracllular phase and inactivate them. This antibody mediated viral inactivation is called viral neutralization.

Solubilisation of immune complexes or (Immune complex clearance)

Complement has a very important role in solubilising and causing removal from the circulation of immune complexes. It does this by the binding of C4b and C3b, covalently bound to the immune complex, to **CR1 complement receptors** on red blood cells which transport the complexes to the liver and spleen where they give the complexes up to phagocytes for destruction.

The role of the complement system is clearing immune complexes can be seen in patients with the autoimmune disease systemic lupus erythematous (SLE). The individual produce large quantities of immune complexes and suffer tissue damage as a result of complement mediated lysis and the induction of type II and type III of hypersensitivity.

Although complement plays a significant role in development of tissue damage in SLE the paradoxical finding is that deficiencies in C1, C2, C4 and CRI predispose and individual into SLE; indeed 90% of individuals who completely lack C4 develop SLE. The complement deficiencies are through to interfere to effective solubilisation and clearance of immune complexes, the results is the persistence of these complex and subsequent tissue damage by the very system whose deficiency was to blame. The lower levels of CRI expressed on the erythrocytes of SLE patients also may interfere with the proper binding and clearance of immune complex.

COMPLEMENT DEFICIENCIES

Genetic deficiencies of almost all the components of the complement system have been found in humans. The phenotypes of these deficiencies tell us a good deal about the role of the complement system, though in the case of C1 (INH). It is complicated by the activity if C1 INH in inhibiting the serum proteases kallikrein and plasmin as well as C1s so that it's loss also dysregulates coagulation.

Genetic deficiencies have been described for each of the complement components. Homozygous deficiencies in any of the early components of the early pathway (Clq, Clr, Cls, C4 and to exhibit similar symptoms, notably a marked increase in immune complex disease such as systemic lupus erythematous, glomerulonephirities, and vasulitis. These deficiencies highlight the important role of the early complement reaction in generating C3b for solubilisation and clearance of immune complexes. In addition to immune complexes diseases, individuals with such complement deficiencies suffer from recurrent infections by pyogenic bacteria such as streptococci and staphylococci. These organisms are Gram positive and therefore resistant in any case to the lytic effects of the MAC.

Nonetheless, the early complement components ordinarily prevent recurrent infection by mediating a localized inflammatory response and opsonizing the bacteria. Deficiencies in factor D and properdin early components of the alternative pathway appear to be associated with Neisseria infections but not with immune—complex disease.

TABLE 11.5

Complement deficiency chart

Complement deficiencies

group	disease resulting	component	effect	genetics
I	immune complex deficiency	C1q C1r, C1s C2 C4	prevents activation of the classical pathway	AR AR AR AR
II	angioedema	C1INH	uninhibited acvitation C1 etc	AD
III	recurrent pyogenic infections	C3 factor H factor I	absence or consumption of C3	AR AR AR
IV	recurrent Neisserial infections	C5 C6 C7 C8	prevents formation of the MAC	AR AR AR AR
		properdin	inhibits AP activation	XL
		factor D	blocks AP activation	?
V	asymptomatic	C9	blocks full pore formation	AR
VI	PNH	GPI	deficiency in DAF, CD59, HRF	XL

AR = autosomal recessive; AD = autosomal dominant; XL = X linked

Individuals with homozygous deficiencies in the components involved in the MAC developed recurrent meningococcal and gonococci infections caused by Neisseria species. In normal individuals, these Grams negative bacteria are generally susceptible to complement mediated lysis or are cleared by the opsonizing activity of C3b. MAC deficient individuals with rarely have immune—complex disease, which suggest that they produce enough C3b to clear immune complexes. Interestingly, a deficiency in C9 results in no clinical symptoms, suggesting, a deficiency in some cases the entire MAC is not necessary for complement—mediated lysis.

Congenital deficiencies of complement regulatory proteins have also been reported. The C1 inhibitor (C1 Inh) regulars activation of the classical pathway by presenting excessive C4 and C2 activation by C1. Deficiency of C1 Inh is as autosomal dominant condition by C1. The deficiency gives rise to a disease called hereditary **angioedema** which monitors clinically as localized edema of the tissue. A number of the membrane bound regulatory components, including decay—accelerating factor (DAF) and homologous retraction factor (HRF) are anchored to the plasma membrane by glycosylphosphatidilinositol membrane anchors. As a consequence of these defects, red blood cells can be lysed by levels of complement that are much lower than normally require.

The lack of these inhibitors gives rise to increased lysis of red blood cells, causing chronic hemolytic anemia. The hemoglobin released by the broken red blood cells appears in the patient's urine.

CHAPTER

12

CYTOKINE

INTRODUCTION

Cytokines are group of proteins and peptides that are used in organisms as signalling compounds. The cytokine family consists mainly of smaller water-soluble proteins and glycoproteins (proteins with an added sugar chain) with a mass of between 8 and 30 kDa. Cytokines are produced by a wide variety of cell types (both haemopoietic and non-haemopoietic), and can have effects on both nearby cells or throughout the organism. Sometimes these effects are strongly dependent on the presence of other chemicals and cytokines. These chemical signals are similar to hormones and neurotransmitters, and are used to allow one cell to communicate with another. The hormones are released from specific organs into the blood and neurotransmitters are released by nerves, cytokines are released by many types of cells. They are particularly important in both innate and adaptive immune responses. Due to their central role in the immune system, cytokines are involved in a variety of immunological, inflammatory, and infectious diseases. However, not all their functions are limited to the immune system, as they are also involved in several developmental processes during embryogenesis.

When the immune system is fighting pathogens, cytokines signal immune cells such as T-cells and macrophages to travel to the site of infection. In addition, cytokines activate those cells, stimulating them to produce more cytokines.

Biologically active substances released by activated T lymphocytes are called as lymphokines. As most lymphokines exhibit multiple biological effects and the same effect may be caused by different lymphokines, their names lack precision. The term interleukin was therefore introduced for those products of leucocytes which exert a regulatory influence on other cells. Interferon's, growth factors and others were found to have similar effects. Cytokines are peptide mediators or intercellular messengers which regulate immunological, inflammatory and reparative host responses. They are highly potent hormone like substances, active even at (10 M) concentration. They differ from endocrine hormones in being produced not by specialized glands but by widely distributed cells and acting not systematically but locally near the producing cells (paracrine effect) or directly on the producing cells them selves (autocrine effect) they are, in general pleiotropic, having multiple effects on the growth and differentiation of various cell types. There is considerable overlap in the effects produced by different cytokines.

Cytokines are small secreted proteins which mediate and regulate immunity, inflammation, and hematopoiesis. They must be produced *de novo* in response to an immune stimulus. They generally (although not always) act over short distances and short time spans and at very low concentration. They act by binding to specific membrane receptors, which then signal the cell via second messengers, often tyrosine kinases, to alter its behavior (gene expression). Responses to cytokines include increasing or decreasing expression of membrane proteins (including cytokine receptors), proliferation, and secretion of effector molecules.

TYPES OF CYTOKINES

Cytokines have been variously named as lymphokines, interleukins, and chemokines, based on their presumed function, cell of secretion, or target of action. Because cytokines are characterized by considerable redundancy and pleiotropism, such distinctions, allowing for exceptions, are obsolete. Cytokines are a category of less-widely-known signaling proteins and glycoproteins that, like hormones and neurotransmitters, are used extensively in cellular communication. While hormones are secreted from specific organs to the blood, and neurotransmitters are related to neural activity, the cytokines are a more diverse class of compounds in terms of origin and purpose. They are produced by a wide variety of hematopoietic and non-hematopoietic cell types and can have effects on both nearby cells or throughout the organism, sometimes strongly dependent on the presence of other chemicals.

Lymphokines

Lymphokines are the cytokines made by lymphocytes, Monokine are the cytokines made by monocytes, Chemokine are the cytokines with chemotactic activities, and interleukin are the cytokines made by one leukocyte and acting on other leukocytes. Cytokines may act on the cells that secrete them (autocrine action), on nearby cells (paracrine action), or in some instances on distant cells (endocrine action).

Interleukin

The term *interleukin* was initially used by researchers for those cytokines whose presumed targets are principally leukocytes. It is now used largely for designation of newer cytokine molecules discovered every day and bears little relation to their presumed function. The vast majority of these are produced by T-helper cells.

Interleukin-1 (IL-1)

Interleukin-1 is another inflammatory cytokine produced by activated macrophages. Its effects are similar to that of TNF-α and it also helps to activate T-cells. Interleakin-1 in renamed as IL-1. It has two molecular forms alpha and beta. They are principally secreted by macrophages and other cell types. Their major functions are proliferation and differentiation of T, B and other cells, induce acute phase proteins and bone marrow cell proliferation.

Interleukin-2 (IL-2)

Interleukin-2 is principally secreted by T-cells. It's major function is to promote growth and differentiation of T- and B-cells, cytotoxicity of T and NK cells, secretion of lymphokines.

Interleukin 2 is produced by Th cells, although it can also be produced by Tc cells to a lesser extent. It is the major growth factor for T-cells. It also promotes the growth of B cells and can activate NKcells and monocytes. IL-2 acts on T-cells in an autocrine fashion. Activation of T-cells results in expression of IL-2R and the production of IL-2. The IL-2 binds to the IL-R and promotes cell division. When the T-cells are no longer being stimulated by antigen, the IL-2R will eventually decay and the proliferative phase ends.

Interleukin-3 (IL-3)

Interleukin-3 is principally secreted by T-cells. It stimulates multilineage hematopoiesis and is therefore known also as the multicolony stimulating factor (multi-(SF).

Interleukin-4 (IL-4)

Interleukin 4 is produced by macrophages and Th2 cells. It stimulates the development of Th2 cells from naïve Th cells and it promotes the growth of differentiated Th2 cells resulting in the production of an antibody response. It also stimulates Ig class switching to the IgE isotype.

Interleukin-4 is secreted by T-cells. It performs proliferation of B- and cytotoxic T-cells, increase IgGI and IgE production, enhance MHC class II molecule and IgE receptors.

Interleukin-5 (IL-5)

Interleukin-5 in secreted by T-cells. It causes proliferation of eosinophils, stimulate IgA and IgM production. Interleukin 5 is produced by Th2 cells and it functions to promote the growth and differentiation of B-cells and eosinophils. It also activates mature eosinophils.

Interleukin-6 (IL-6)

Interleukin-6 is secreted by Th, macrophages and fibroblasts. It promotes B-cell differentiation IgG production and acute phase proteins.

Interleukin-7 (IL-7)

Interleukin-7 is secreted by spleen and bone marrow stromal cells. It acts as B- and T-cell growth factor.

Interleukin-8 (IL-8)

Interleukin-8 is synthesized from macrophages and other cells. It acts as neutrophil chemotactic factor. IL-8 (interleukin-8) is the only chemokine originally named an interleukin.

Interleukin-9 (IL-9)

Interleukin-9 is secreted by T-cell. It act as T-cell growth factor and proliferates T cell.

Interleukin-10 (IL-10)

Interleukin 10 is produced by activated macrophages and Th2 cells. It is predominantly an inhibitory cytokine. It inhibits production of IFN-γ by Th1 cells, which shifts immune responses toward a Th2 type. It also inhibits cytokine production by activated macrophages and the expression of class II MHC and co-stimulatory molecules on macrophages, resulting in a dampening of immune responses.

Interleukin-10 is secreted by T-cell, B-cell and macrophoges. Interleukin-10 inhibit IFN production and mononuclear cell function.

Interleukin-11 (IL-11)

Interleukin-11 is secreted by bone marrow stromal cells. It induce acute phase proteins.

Interleukin-12 (IL-12)

Interleukin 12 is produced by activated macrophages and dendritic cells. It stimulates the production of IFN-γ and induces the differentiation of Th cells to become Th1 cells. In addition, it enhances the cytolytic functions of Tc and NK cells. Interleukin-12 secreted by T-cells. It activates NK cells.

Chemokines

The term *chemokine* refers to a specific class of cytokines that mediates chemoattraction (chemotaxis) between cells.

Pleiotropy

It is common for different cell types to secrete the same cytokine or for a single cytokine to act on several different cell types. Cytokines are redundant in their activity, meaning similar functions can be stimulated by different cytokines. Cytokines are often produced in a **cascade**, as one cytokine stimulates its target cells to make additional cytokines. Two or more cytokines acting together are called synergistic cytokines, the cytokines causing opposing activities called antagonistic cytokines. Their short half life, low plasma concentrations, pleiotropy, and redundancy all complicated the isolation and characterization of cytokines. Searches for new cytokines are now often conducted at the DNA level, identifying genes similar to known cytokine genes.

Cytokine function

Cytokine activities are characterized using recombinant cytokines and purified cell populations *in vitro*, or with **knock-out mice** for individual cytokine genes to characterize cytokine functions *in vivo*. Cytokines are made by many cell populations, but the predominant producers are helper T-cells (Th) and macrophages.

The largest group of cytokines stimulates immune cell proliferation and differentiation. This group includes interleukin 1 (IL-1), which activates T-cells; IL-2, which stimulates proliferation of antigen-activated T- and B-cells; IL-4, IL-5, and IL-6, which stimulate proliferation and differentiation of B cells; Interferon gamma (IFNã), which activates macrophages; and IL-3, IL-7 and granulocyte monocyte colony-stimulating factor (GM-CSF), which stimulate hematopoiesis.

Cytokines are critical to the functioning of both innate and adaptive immune responses. They are often secreted by immune cells which have encountered a pathogen as a way to activate and recruit more immune cells and increase the system's response to the pathogen. However, apart from their role in the development and functioning of the immune system, as well as their aberrant modes of secretion in a variety of immunological, inflammatory and infectious diseases, cytokines are also involved in several developmental processes during embryogenesis.

Other groups of cytokines include interferons and chemokines. **Interferons** IFN alpha and IFN-β inhibit virus replication in infected cells, while IFN-α also stimulates antigen-presenting cell MHC expression. **Chemokines** attract leukocytes to infection sites. Chemokines have conserved cysteine residues that allow them to be assigned to four groups. The groups, with representative chemokines, are C-C chemokines (RANTES, MCP-1, MIP-1alpha, and MIP-beta), C-X-C chemokines (IL-8), C chemokines (Lymphotactin), and CXXXC chemokines (Fractalkine). Some cytokines are predominantly inhibitory. For example, IL-10 and IL-13 inhibit inflammatory cytokine production by macrophages.

Helper T-cells have two important functions: to stimulate cellular immunity and inflammation and to stimulate B-cells to produce antibody. Two functionally distinct subsets of T-cells secrete cytokines which promote these different activities. Th1 cells produce IL-2, IFN-α, and IFN-beta which activate Tc and macrophages to stimulate cellular immunity and inflammation. Th1 cells also secrete IL-3 and GM-CSF to stimulate the bone marrow to produce more leukocytes. Th2 cells secrete IL-4, IL-5, IL-6, and IL-10, which stimulate antibody production by B-cells.

T-cells are initially activated as Th cells, which produce IL-2, IL-4 and IFN gamma. The nearby cytokine environment then influences differentiation into Th1 or Th2 cells. IL-4 stimulates Th2 activity and suppresses Th1 activity, while IL-12 promotes Th1 activities. Th1 and Th2 cytokines are antagonistic in activity. Th1 cytokine IFNα inhibits proliferation of Th2 cells, while IFNα and IL-2 stimulate B-cells to secrete IgG_{2a} and inhibit secretion of IgG_1 and IgE. Th2 cytokine IL-10 inhibits Th1 secretion of IFNα and IL-2; it also suppresses Class II MHC expression and production of bacterial killing molecules and inflammatory cytokines by macrophages. IL-4 stimulates B-cells to secrete IgE and IgG_1. The balance between Th1 and Th2 activity may steer the immune response in the direction of cell-mediated or humoral immunity.

The effect of a particular cytokine on a given cell depends on the cytokine, its extracellular abundance, the presence and abundance of the complementary receptor on the cell surface, and downstream signals activated by receptor binding; these last two factors can vary by cell type. Cytokines are characterized by considerable "redundancy", in that many cytokines appear to share similar functions. Generalization of functions is not possible with cytokines; nonetheless, their actions may be grouped as:

- autocrine, if the cytokine acts on the cell that secretes it
- paracrine, if the action is restricted to the immediate vicinity of a cytokine's secretion
- endocrine, if the cytokine diffuses to distant regions of the body (carried by blood or plasma) to affect different tissues.

It seems to be a paradox that cytokines binding to antibodies have a stronger immune effect than the cytokine alone. This may lead to lower therapeutic doses and perhaps fewer side-effects.

Overstimulation of cytokines can trigger a dangerous syndrome known as a cytokine storm; this may have been the cause of severe adverse events during a clinical trial of TGN1412.

Structural and functional classification

Cytokines are classified into four types:

- **The four α-helix bundle family,** member cytokines have three-dimensional structures with four bundles of α-helices. This family in turn is divided into three sub-families: the IL-2 subfamily, the interferon (IFN) subfamily, the IL-10 subfamily. The first of these three subfamilies is the largest. It contains several non-immunological cytokines including erythropoietin (EPO) and thrombopoietin (THPO). Also, four α-helix bundle cytokines can be grouped into *long-chain* and *short-chain* cytokines.
- **The IL-1 family,** which primarily includes IL-1 and IL-18
- **The IL-17 family,** which has yet to be completely characterized, though member cytokines have a specific effect in promoting proliferation of T-cells that cause cytotoxic effects
- Chemokines.

Cytokine receptors

In recent years, the cytokine receptors have come to demand the attention of more investigators than cytokines themselves, partly because of their remarkable characteristics, and partly because a deficiency of cytokine receptors has now been directly linked to certain debilitating immunodeficiency states. In this regard, and also because the redundancy and pleiomorphism of cytokines are, in fact, a consequence of their homologous receptors, many authorities are now of the opinion that a classification of cytokine receptors would be more clinically and experimentally useful.

Cytokines act on their target cells by binding specific membrane receptors. The receptors and their corresponding cytokines have been divided into several families based on their structure and activities. Hematopoietin family receptors are dimers or trimers with conserved cysteines in their extracellular domains and a conserved Trp-Ser-X-Trp-Ser sequence. Examples are receptors for IL-2 through IL-7 and GM-CSF. Interferon family receptors have the conserved cysteine residues but not the Trp-Ser-X-Trp-Ser sequence, and include the receptors for IFN alpha, IFN Beta and IFN gamma. Tumour necrosis factor family receptors have four extracellular domains; they include receptors for soluble TNF alpha and TNF beta as well as membrane-bound CD40 (important for B-cell and macrophage activation) and Fas (which signals the cell to undergo apoptosis). Chemokine family receptors have seven transmembrane helices and interact with G protein. This family includes receptors for IL-8, MIP-1 and RANTES. Chemokine receptors CCR5 and CXCR4 are used by HIV to preferentially enter either macrophages or T-cells.

Hematopoietin family

Hematopoietin cytokine receptors are the best characterized. They generally have two subunits, one cytokine-specific and one signal transducing. An example is the GM-CSF subfamily, where a unique alpha subunit specifically binds either GM-CSF, IL-3, or IL-5 with low affinity and a shared beta subunit signal transducer also increases cytokine-binding affinity. Cytokine binding promotes dimerization of the alpha and beta subunits, which then associate with cytoplasmic tyrosine kinases to phosphorylate proteins which activate mRNA transcription. GM-CSF and IL-3 act on hematopoietic stem cells and progenitor cells and activate monocytes. With IL-5, they also stimulate eosinophil proliferation and basophil degranulation. All three receptors phosphorylate the same cytoplasmic protein. Antagonistic GM-CSF and IL-3 activities can be explained by their competition for limited amounts of beta subunit.

The IL-2R subfamily of receptors for IL-2, IL-4, IL-7, IL-9, and IL-15 have a common signal-transducing gamma chain. Each has a unique cytokine-specific alpha chain. IL-2 and IL-15 are trimers, and share an IL-2R beta chain. Monomeric IL-2R alpha has low affinity for IL-2, dimeric IL-2R beta, gamma has intermediate affinity, and trimeric IL-2R(alpha, beta, gamma) binds IL-2 with high affinity. IL-2R alpha chain (Tac) is expressed by activated but not resting T-cells. Resting T cells and NK cells constitutively express low numbers of IL-2R beta, gamma. Antigen activation stimulates T-cell expression of high affinity IL-2R trimers as well as secretion of IL-2, allowing autocrine stimulation of T-cell proliferation in an antigen-specific manner. Antigen specificity of the immune response is also maintained by the close proximity of antigen-presenting B-cells and macrophages with their helper T-cells, so that cytokines are secreted in the direction of and close to the membrane of the target cell. **X-linked severe combined immunodeficiency** (X-*scid*) is caused by a defect in IL-2R family, gamma chain, which results in loss of activity from this family of cytokines.

Cytokine activity can be blocked by **antagonists**, molecules which bind cytokines or their receptors. IL-1 has a specific antagonist that blocks binding of IL-1 alpha and IL-1beta to their receptor. During immune responses, fragments of membrane receptors may be shed and then compete for cytokine binding. Microbes also influence cytokine activities.

For example, Vaccinia virus (Smallpox and Cowpox) encodes soluble molecules which bind IFNgamma, while Epstein-Barr virus (Infectious Mononucleosis) encodes a molecule homologous to IL-10 that suppresses immune function in the host.

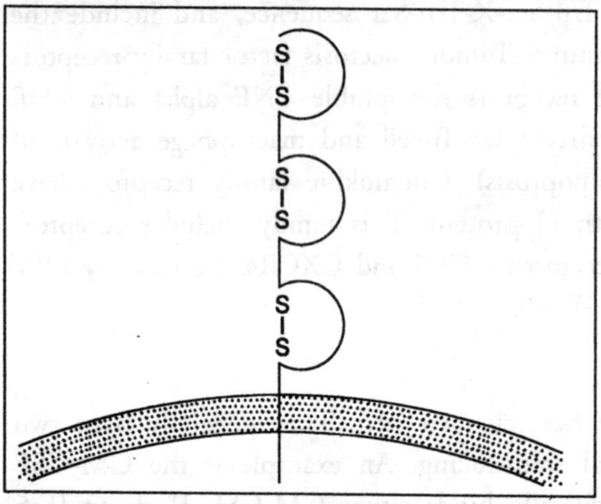

Fig. 12.1: Immunoglobulin (Ig) super family receptor

The TNF receptor family molecules CD40 and Fas bind cell surface ligands on effector T-cells: CD40L and FasL. CD40 is expressed on B-cell and macrophage plasma membranes. T-cell CD40L binding to B-cell CD40 stimulates B-cell proliferation and isotype switching. T-cell CD40L binding to macrophage CD40 stimulates macrophages to secrete TNF and become much more sensitive to IFN gamma. T-cell FasL binding to Fas leads to the activation of caspase proteases that initiate apoptosis of the cell expressing membrane Fas. Activated lymphocytes express Fas, so that FasL-positive Tc cells can regulate the immune response by eliminating activated cells. An immune deficiency disease linked to expression of a mutant Fas is characterized by over-proliferation of lymphocytes.

A classification of cytokine receptors based on their three-dimensional structure has, therefore, been attempted. Such a classification, though seemingly cumbersome, provides several unique perspectives for attractive pharmacotherapeutic targets.

- **Immunoglobulin (Ig) super family**, which are ubiquitously present throughout several cells and tissues of the vertebrate body, and share structural homology with immunoglobulins (antibodies), cell adhesion molecules, and even some cytokines. Examples: IL-1 receptor types.
- **Haemopoietic growth factor (type 1) family**, whose members have certain conserved motifs in their extracellular amino-acid domain. The IL-2 receptor belongs to this chain, whose α-chain (common to several other cytokines) deficiency is directly responsible for the x-linked form of severe combined immunodeficiency (X-SCID).
- **Interferon (type 2) family**, whose members are receptors for IFN α and β.
- **Tumour necrosis factors (TNF) (type 3) family**, whose members share a cysteine-rich common extracellular binding domain, and includes several other non-cytokine ligands like CD40, CD27 and CD30, besides the ligands on which the family is named (TNF).
- **Seven transmembrane helix family**, the ubiquitous receptor type of the animal kingdom. All G-protein coupled receptors (for hormones and neurotransmitters) belong to this family (Figure 12.3). Chemokine receptors, two of which act as binding proteins for HIV (CXCR4 and CCR5), also belong to this family.

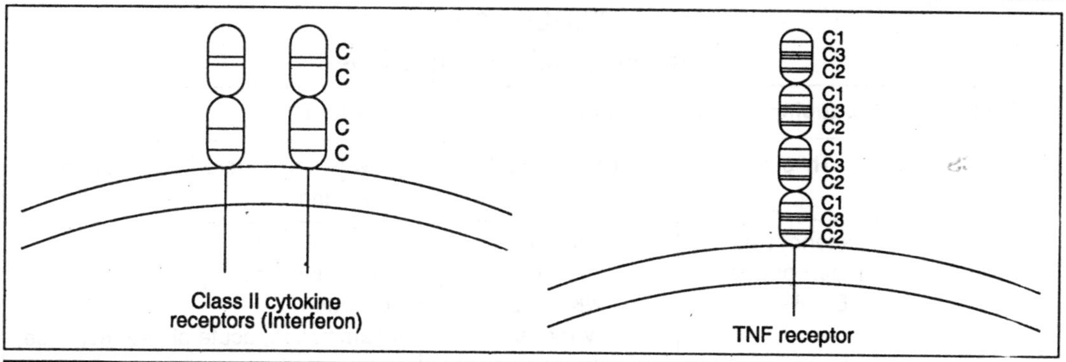

Fig. 12.2: Interferon and TNF (type 3) family receptor

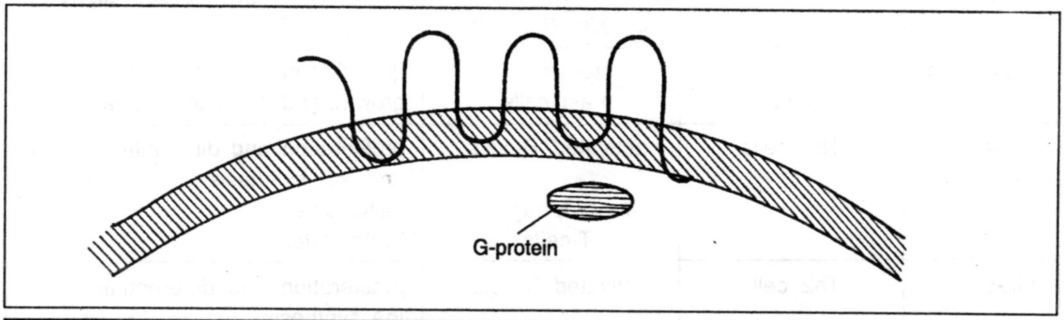

Fig. 12.3: G-protein coupled receptors

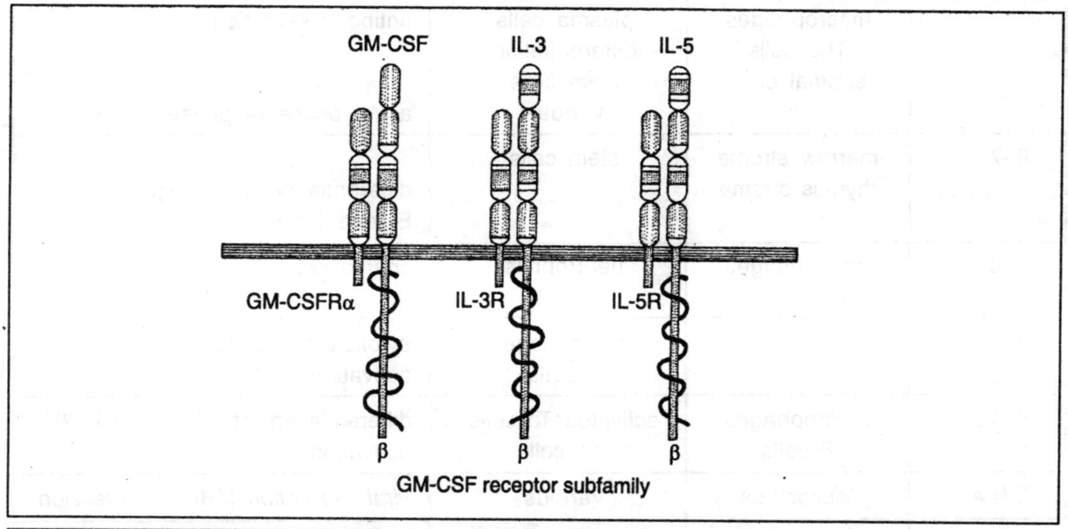

Fig. 12.4: GM-CSF receptors subfamily

Each cytokine binds to a specific cell-surface receptor. Subsequent cascades of intracellular signalling then alter cell functions. This may include the upregulation and/or downregulation of several genes and their transcription factors, in turn resulting in the production of other cytokines, an increase in the number of surface receptors for other molecules, or the suppression of their own effect by feedback inhibition.

TABLE 12.1
Cytokinese and their functions

Cytokine	Producing Cell	Target Cell	Function**
GM-CSF	Th cells	progenitor cells	growth and differentiation of mono cytes and DC
IL-1a IL-1b	monocytes macrophages B-cells DC	Th cells B-cells NK cells various	co-stimulation maturation and proliferation activation inflammation, acute phase response, fever
IL-2	Th1 cells	activated T- and B-cells, NK cells	growth, proliferation, activation
IL-3	Th cells NK cells	stem cells mast cells	growth and differentiation growth and histamine release
IL-4	Th2 cells	activated B-cells macrophages T-cells	proliferation and differentiation IgG_1 and IgE synthesis MHC Class II proliferation
IL-5	Th2 cells	activated B cells	proliferation and differentiation IgA synthesis
IL-6	monocytes macrophages Th2 cells stromal cells	activated B-cells plasma cells differentiation stem cells various	differentiation into plasma cells antibody secretion acute phase response
IL-7	marrow stroma thymus stroma	stem cells	differentiation into progenitor B- and T-cells
IL-8	macrophages endothelial cells	neutrophils	chemotaxis
IL-10	Th2 cells	macrophages B-cells	*cytokine production* activation
IL-12	macrophages B-cells	activated Tc cells NK cells	differentiation into CTL(with IL-2) activation
IFN-a	leukocytes	various	*viral replication* MHC I expression
IFN-b	fibroblasts	various	*viral replication* MHC I expression
IFN-g	Th1 cells, Tc cells, NK cells	various macrophages activated B-cells Th2 cells macrophages	*Viral replication* MHC expression Ig class switch to IgG_{2a} *proliferation* pathogen elimination
MIP-1a	macrophages	monocytes, T-cells	chemotaxis

MIP-1b	lymphocytes	monocytes, T-cells	chemotaxis
TGF-b	T-cells, monocytes	monocytes, macrophages activated macrophages activated B cells various	chemotaxis IL-1 synthesis IgA synthesis *proliferation*
TNFa	macrophages, mast cells, NK cells	macrophages tumour cells	CAM and cytokine expression cell death
TNF-b	Th1 and Tc cells	phagocytes	phagocytosis, NO production

Chemokines & Receptors

Chemokines are a large superfamily of mostly small, secreted chemotactic cytokines that function in leukocyte trafficking, recruitment and activation. They also play a critical role in many normal and pathophysiological processes such as allergic responses, infectious and autoimmune diseases, angiogenesis, inflammation, tumour growth and metastasis, and hematopoietic development. Chemokines are divided into subfamilies based on conserved amino acid (aa) sequence motifs. Most family members have at least four conserved cysteine residues that form two intramolecular disulfide bonds. The subfamilies are defined by the position of the first two cysteine residues.

C Chemokines & Receptors

The C (gamma) subfamily lacks the first and third cysteine residues. Lymphotactin (also known as SCM-1 alpha) and SCM-1 beta are currently the only two family members. Both have chemotactic activity for lymphocytes and NK cells.

CC Chemokines & Receptors

Cytokine is a term used to describe a broad range of structurally diverse molecular families and individual proteins best known for their many roles in immune system function. They are often pleiotropic molecules with diverse and cell type specific activities. Cytokine functions are mediated by binding specific receptors and their activities include regulating cell activation, hematopoiesis, apoptosis, cell migration, and cell proliferation. In this capacity, they are involved in virtually all aspects of both innate and adaptive immune responses. Leukocytes are a primary source of cytokines, although they may be produced by many other cell types as well.

The CC (beta) subfamily of chemokines comprise a subfamily of the chemokine superfamily and are defined by the arrangement of the first two of four invariant cysteine residues found in all chemokines. In CC chemokines, these two cysteines are adjacent, while in the CXC subfamily, they are separated by a single amino acid. Chemokines exhibit promiscuous binding to multiple seven-transmembrane, G-protein coupled CC chemokine receptors. Chemokines are typically small, secreted molecules that function in leukocyte trafficking, recruitment, and activation. They also play roles in normal and pathological processes including allergic responses, infectious and autoimmune disease, angiogenesis, inflammation, and tumour growth and metastasis.

CX3C Chemokines & Receptors

The lone CX3C (delta) subfamily member, fractalkine, has three intervening aa between the first two cysteine residues. It is a transmembrane protein with a chemokine domain attached to a long mucin-like stalk. Membrane-bound fractalkine has been shown to promote adhesion of leukocytes. The soluble chemokine domain of human fractalkine is chemotactic for T-cells and monocytes, while the soluble chemokine domain of mouse fractalkine chemoattracts neutrophils and T-lymphocytes, but not monocytes.

CXC Chemokines & Receptors

Chemerin, also known as tazarotene-induced gene 2, (TIG2), is a member of the cystatin superfamily. Members of this superfamily contain at least two intrachain disulfide bonds and an alpha-helical structure over a distance of about 100 amino acids. Chemerin is synthesized as a 163 aa precursor that contains a hydrophobic 20 aa N-terminal sequence, an intervening 137 aa cystatin-fold containing domain, and a six aa C-terminal prosegment. The 137 aa mature segment is known to bind to the G protein-coupled receptor, ChemR23.

Other chemokine-related ligands and receptors are as follows:

1. Chemokines and receptors
2. Common beta chain receptor family
3. Common gamma chain receptor family
4. EGF Family
5. FGF Family
6. IL-1 Ligands and IL-1R/TLR Family
7. IL-6 Family
8. IL-10/Interferon Family
9. IL-17 Family
10. Other Interleukins
11. TGF-beta Superfamily

CHAPTER 13

HYPERSENSITIVITY

INTRODUCTION

Hypersensitivity refers to undesirable (damaging, discomfort-producing and sometimes fatal) reactions produced by the normal immune system. Hypersensitivity reactions require a pre-sensitized (immune) state of the host. Frequently, a particular clinical condition (disease) may involve more than one type of reaction. An immune response evokes a battery of effectors molecules that act to remove antigen by various mechanisms. Generally, these effector molecules induce a subclinical, localized inflammatory response that eliminates antigen without extensively damaging the host's tissue. Under certain circumstance and however, this inflammatory response can have deleterious effects, resulting in significant tissue damage or even death. This inappropriate immune response is termed "Hypersensitivity or Allergy." Hypersensitivity reactions can be divided into four types: type I, type II, type III and type IV, based on the mechanisms involved and time taken for the reaction.

The hypersensitivity in most cases is commonly known as allergy. The antigen which causes the allergy is called as allergen. When an allergen first enters the body, it starts a process called as sensitization, instead of eliciting the normal process of immunization. The individual when firstly comes in contact with allergen, there is no clinical manifestation but second exposure of same allergen shows the allergic symptoms. The hypersensitivity refers to the acquired reactivity of body to subsequent exposure of an antigen which results pathological manifestation. The term hypersensitivity is used to describe immune responses which are damaging rather than helpful to the host. Nearly 40 years ago Gell and Coombs proposed a classification scheme which defined four types of hypersensitivity reactions. The first three are mediated by antibody, the fourth by T-cells. Although the word hypersensitivity implied an increased response, the response is not always and heightened but may, instead, be an inappropriate immune response to an antigen.

Hypersensitivity is of following four types

1. Type I—IgE mediated hypersensitivity.
2. Type II—Antibody mediated cytotoxic hypersensitivity.
3. Type III—Immune complex mediated hypersensitivity.
4. Type IV—T_{DTH} mediated hypersensitivity

Three types of such reaction are IgE mediated, antibody mediated and immune complex mediated. And fourth is the T_{DTH} delayed type of hypersensitivity. We currently refer to anaphylactic reactions within the humoral branch initiated by antibody as antigen-antibody complexes as immediate hypersensitivity, because the symptoms are manifest within minutes or hours after a sensitized recipient encounter and antigen.

1. IgE Mediated or Type I Hypersensitivity

Type-I hypersensitivity is also known as **immediate** or anaphylactic hypersensitivity. Anaphylaxis literally means opposites of protection and mediated by IgE antibodies. A type I hypersensitivity reaction induced by certain type of antigens are called as allergens and induces the humoral immune response. The allergen eliciting anaphylaxis include a great variety of substances like pollens, fibers, insects, venom, fungal spore, as well as various food material like egg, milk, fish etc. The reaction may involve skin (urticaria and eczema), eyes (conjunctivitis), nasopharynx (rhinorrhea, rhinitis), bronchopulmonary tissues (asthma) and gastrointestinal tract (gastroenteritis). The reaction may cause a range of symptoms from minor inconvenience to death. The reaction usually takes 15-30 minutes from the time of exposure to the antigen, although sometimes it may have a delayed onset (10-12 hours). Immediate hypersensitivity is mediated by IgE. The primary cellular component in this hypersensitivity is the mast cell or basophil. The reaction is amplified and/or modified by platelets, neutrophils and eosinophils. A biopsy of the reaction site demonstrates mainly mast cells and eosinophils.

The mechanism of reaction involves preferential production of IgE, in response to certain antigens (**allergens**). IgE has very high affinity for its receptor on mast cells and basophils. A subsequent exposure to the same allergen cross links the cell-bound IgE and triggers the release of various pharmacologically active substances (Fig.13.1). Cross-linking of IgE Fc-receptor is important in mast cell triggering. Mast cell degranulation is preceded by increased **Ca++ influx**, which is a crucial process; ionophores which increase cytoplasmic Ca^{++} also promote degranulation, whereas, agents which deplete cytoplasmic Ca^{++} suppress degranulation agents released from mast cells and their effects are listed in Table 13.1. Mast cells may be triggered by other stimuli such as exercise, emotional stress, chemicals (e.g. photographic developing medium, calcium ionophores, codeine, etc.), anaphylotoxin (e.g. C4a, C3a, C5a, etc.). These reactions, mediated by agents without IgE-allergen interaction, **are not hypersensitivity reactions** although they produce the same symptoms.

Hypersensitivity Type I will be familiar to most people and describes the rapid ('Immediate') allergic reaction. The symptoms produced by exposure of a sensitized person to antigen depend upon the site of contact. Hayfever (allergic rhinitis), eczema, asthma and urticaria all result from type I hypersensitivity. It is caused upon contact with antigen against which the host has pre-existing IgE antibody. IgE is present in very low levels in serum in most people (i.e. 5×10^{-8}gm) per ml. Its' half life in serum is only 2-3 days but much of the IgE in the body is bound to high affinity receptors (Fc epsilonRI), in the bound state the half-life is ~3 weeks. The high affinity Fc epsilonRI receptors are found on mast cells and basophils. Each cell has a high density of these receptors (40-250,000 per cell) so that a wide spectrum of antigen specificities is represented. The cells are activated by the **cross-linking** of the Fc epsilonRI receptors via antigen binding to the bound IgE molecules.

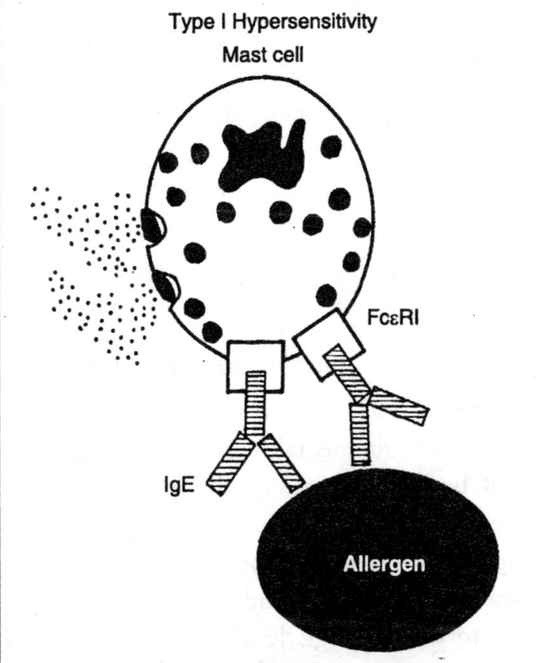

Fig. 13.1: Mechanism of anaphylactic reaction

Such cross-linking leads to rapid degranulation (60-300 secs) of the mast cells and the release of **primary inflammatory mediators** stored in the granules. These mediators cause all the normal consequences of an acute inflammatory reaction-increased vascular permeability, smooth muscle contraction, granulocyte chemotaxis and extravasation etc. Mast cell activation via Fc epsilonRI also leads to the production of two other type of mediators. These **secondary mediators**, unlike the stored granule contents, must be synthesised *de novo* and comprise arachidonic acid metabolites (prostaglandins and leukotrienes) and proteins (cytokines and enzymes).

TABLE 13.1

Shows primary inflammatory mediators and secondary inflammatory mediators

Molecule	**Effects** (Primary Inflammatory mediators)
Histamine	Vascular permeability, smooth muscle contraction
Serotonin	vascular permeability, smooth muscle contraction
ECF-A	eosinophil chemotaxis
NCF-A	neutrophil chemotaxis
proteases	mucus secretion, connective tissue degradation
Molecule	**Effects** (Secondary Inflammatory mediators)
Leukotrienes	vascular permeability, smooth muscle contraction
Prostaglandins	vasodilation, smooth muscle contraction
Bradykinin	vascular permeability, smooth muscle contraction
Cytokines	numerous effects inc. activation of vascular endothelium, eosinophil recruitment and activation

Mirroring the two types of mediators, one can see two components to the type I response. A very rapid early response occurs when you are challenged with an antigen to which you are sensitised which is apparent within a few minutes and maximal after about 20 minutes. If the challenge is cutaneous it produces the so-called 'wheal and flare'—raised patch surrounded by a pink effusion. The late response seen after some hours is characterised by cellular infiltrate which gives a hard but barely pigmented nodule in the case of skin.

That these responses are caused by distinct mediators can be shown with inhibitory drugs. Arachadonic acid metabolism inhibitors, such as indomethacin, block only the late response. Sodium cromoglycate which blocks mast cell activation and degranulation blocks both early and late responses.

Incidence and genetic susceptibility

Some 20-30% of the population exhibit type I hypersensitivity or atopic allergy to common environmental substances. There is a genetic component to atopic allergy such that if both your parents exhibit this susceptibility you are more than 2 × more likely to do so and if neither parent has manifest allergies you are less than half as likely to when compared to the population as a whole. Some individuals have multiple and severe allergies, typically both hayfever and eczema; these individuals are termed atopic and frequently have raised total serum IgE levels (10–100 × normal). There is a correlation between total [IgE] and atopy.

Allergens

Numerous ideas have been put forward as to what property might distinguish antigens which stimulate a sufficient IgE response to generate type I hypersensitivity (**allergens**) from those antigens which rarely or never do so. However no common property has yet been discerned. Below is a list of common allergens.

TABLE 13.2

Common antigen which causes Type I hypersensitivity

Pollens	Birch Tree, Ragweeds, oilseeds rape
Foods	Nuts, Eggs, Seafoods
Drugs	Penicillin, asprin
Insects Products	Bee venom, House Dust mite .
Animal hairs	Cat hair and dander

The clinical manifestations of type I reactions can range from serious life threatning conditions, such as systemic anaphylaxis and asthma to hay fever and eczema, which are merely annoying.

Systemic anaphylaxis

Systemic anaphylaxis is a shock like and often fatal state whose onset occurs within minutes of a type I hypersensitivity reaction. This was the response observed by Portier and

Richet in dogs after antigenic challenge. Systemic anaphylaxis can be induced in a variety of experimental animals and is seen occasionally in humans, each species exhibits characteristic symptoms, which reflect differences in the distribution of mast cells and in the biologically active contents of their granules. The animal model of choice for studying systemic anaphylaxis has been the guinea pig. Anaphylaxis can be induced in guinea pigs with ease, and its symptoms closely parallel those observed in humans.

While we are probably all familiar with the consequences of a localized type I reaction (anaphylaxis) which are unpleasant and annoying, the consequences of a generalized reaction are potentially fatal. Ingestion of nuts or seafood, insect bites (venom), and drug injection may all cause life-threatening reactions in highly sensitized individuals. Death in such cases is due to systemic release of vasoactive mediators leading to general vasodilation and smooth muscle contraction resulting in sudden loss of blood pressure, massive oedema and severe bronchiole constriction systemic anaphylaxis.

Active sensitization in guinea pigs is induced by a single injection of a foreign protein such as egg albumin, after an incubation period of about 2 weeks; the animal is usually challenged with an intravenous injection of the same protein. Within 1 minute, the animal becomes restless, its respiration becomes labored, and its blood pressure drops. As the smooth muscles of the gastrointestinal tract and bladder contact, the guinea pig defecates and urinates. Finally bronchiole constriction results in death by asphyxiation within 2-4 min of the injection. These events all stem from the systemic vasodilatation and smooth-muscle contraction brought on by mediators released in the course of the reaction. Post-mortem examination reveals that massive edema, shock, and bronchiole constriction are the major causes of death.

Systemic anaphylaxis in humans is characterized by a similar sequence of events, a wide range of antigen have been shown to trigger this reaction in susceptible humans, including the venom from bee, wasp, hornet and ant stings, drugs, such as penicillin, insulin and antitoxins and sea food and huts, it not treated quickly, these reactions can be total. Epinephrine is the drug of choice for systemic anaphylactic reactions. Epinephrine counteracts the effects of mediators such as histamine and the leukotrienes by releasing the smooth muscles and reducing vascular permeability. Epinephrine also improves cardiac output, which is necessary to prevent vascular collapse during an anaphylactic reaction. In addition, epinephrine increase cAMP levels in the mast cell, there by blocking further deregulation.

Localized anaphylaxis (Atopy)

In localized anaphylaxis, the reaction is limited to a specific target tissue or organ, often involving epithelial surfaces at the site of allergen entry. The tendency to manifest localized anaphylactic reactions is inherited and is called atopy. Atopic allergies, which afflict at least 20% of the population in developed, countries, include a wide range of IgE mediated disorders, including allergic rhinitis (hay fever), asthma, atopic dermatitis (ecezema), and food allergies.

Allergic rhinitis

The most common atopic disorder affecting 10% of the V.S. population I allergic rhinitis, commonly known as hay fever. This results from the reaction of airborne allergens with sensitized mast cells in the conjunctivae and nasal mucosa to induce the release of pharmacologically active mediators from mast calls. These mediators than cause localized vasodilation increased capillary permeability. The symptoms include watery exudation of the conjunctivae, nasal mucosa, and upper respiratory tract, as well as sneezing and coughing.

Asthma

Another common manifestation of localized anaphylaxis is asthma. In some cases, air borne or blood-borne allergens, such as pollens, dust, fumes, insect products, or viral antigens, trigger an asthmatic attack (allergic asthma) in other cases an asthmatic attack can be induced by cold, apparently independently or allergen stimulation (intrinsic asthma). Like hay fever, asthma is triggered by degranulation of mast cells with release of mediators, but instead of occurring in the nasal mucosa, the reaction developed in the lower respiratory tract. The resulting contraction of the bronchial smooth muscle leads to bronchoconstriction. Airway exterva, mucus secretion, and inflammation contribute to the bronchial constriction and to air way obstruction asthmatic patients may have abnormal levels of receptors for neuropeptids.

Most clinicians view asthma as primarily an inflammatory disease. The asthmatic response can be divided into early and late responses. The early response occurs within minutes of allergen exposure and primarily involves histamine, leukotrienes (LTC_4), and vasodilation. The late response occurs hour later and involves additional mediators, including PL-4, PL-5, IL-16, TNF-r; eosinophil chemotactic factors (ECF), and platelet activating factor (PAF). The overall effects of these mediators is to increase endothelial cell adhesion as well as to recruit inflammatory cells including eosinophils and neutrophils into the bronchial tissue.

2. Type II hypersensitivity

Type II hypersensitivity is also known as cytotoxic hypersensitivity and may affect a variety of organs and tissues. The antigens are normally endogenous, although exogenous chemicals (hapten) which can attach to cell membranes can also lead to type II hypersensitivity. Drug-induced hemolytic anaemia, granulocytopenia and thrombocytopenia are such examples. The reaction time is minutes to hours. Type II hypersensitivity is primarily mediated by antibodies of the IgM or IgG classes and complement. Phagocytes and K cells may also play a role (ADCC).

This type of hypersensitivity involves IgG antibodies and complement system and results in cell destruction. IgM may also take part in cell damaging reaction. This type of reaction is best exemplified by blood transfusion reaction, in which host antibodies are react with foreign antigen on incompatible transfused blood cells and mediate destruction of these cells. The second class of damaging reactions is caused by specific antibody binding to cells or tissue antigens. The antibodies are of the IgM or IgG classes and cause cell destruction by Fc dependent mechanisms either directly or by recruiting complement via the classical pathway. Except where the reaction is autoimmune, the target cells are foreign to the host.

Type II hypersensitive reactions involve antibody-mediated destruction of cells. These types of reactions, in which host antibodies react with foreign antigens on the incompatible transfused blood cell and mediate destruction of these cells. Antibody can mediate cell destruction by activating the complement system to create pores in the membrane of the foreign cell destruction by antibody-dependant cell-mediate cytotoxicity (ADCC).

In this process, cytotoxic cells with Fc receptor bind to the Fc region of antibodies on target cells and promote killing of the cells. Antibody bound to a foreign cell also can serve as an opsonin, enabling phagocytic cell with Fc or C3b receptors to bind and phagocytose antibody coated cell.

A large number of proteins and glycoproteins on the membrane of red blood cells are encoded by different genes each of which has a number of alternative alleles. An individual possessing one allelic forms on transfused blood as foreign and mount an antibody been induced by natural exposure to similar antigenic determinants on a variety of microorganisms through to be normal flora of the gut. This is the case the ABO blood group antigens.

Transfusion of blood into a recipient possessing antibodies to one of the blood group antigens can result in a transfusion reaction.

The clinical manifestations of transfusions reactions result from massive intra vascular hemolysis of the transfused red blood cells by antibody plus complement. These manifestations may be either immediate or delayed. Reactions that begin immediately are most commonly associated with ABO blood group incompatibilities, which lead to complement-mediated lysis triggered by the IgM isohemaglutinins, within hours, free hemoglobin can be detected in the plasma it is filtered through the kidneys, resulting in hemoglobin urine. Some of the hemoglobin gets converted to bilirubin, which at high revels is toxic. Typical systems include fever, chills, nausea, clotting within blood vessels, pain in the lower back, hemoglobin in the urine. Treatment involver prompt termination of the transfusion and maintenance of urine flow with a diuretic, because the accumulation of hemoglobin the kidney and cause a cute tubular necrosis.

Delayed hemolytic transfusion reactions generally occurs in the individuals. Who have received repeated transfusions of ABO- compatible blood that incompatible for other blood group antigens. The reactions develop between 2 and 6 days after transfusion and reflect the secondary nature of the reactions. The tramesfased blood induces clonal selection and production of IgG against a variety of blood group membrane antigens. The most common blood group antigens that induce delayed transfusion reactions are Rh, Kidd, Kell, and Duffy.

The predominant isotope involved in the reactions is IgG, which is less effective than IgM in activating complement for this reason, complement–mediated lysis of the transfused red blood cells is incomplete, and many of the transfused cells are destroyed at extravascular sites by agglutination, opsonization, and subsequent phagocytosis by macrophages. Symptoms include fever, low hemoglobin, increased bilirubin, mild jaundice, and anaemia, tree hemoglobin is usually hot detected in the plasma or urine in these reactions because RBC destruction occurs in extravascular sites.

Transfusion reaction

In practice this means type II hypersensitivity reactions are usually only seen in blood transfusion recipients and patients with certain autoimmune diseases. The classic ABO incompatibility reaction is a type II, with IgM antibodies causing complement lysis of erythrocytes. Rhesus disease (or haemolytic disease of the newborn) is a special example since the IgG antibodies which cause destruction of foetal red blood cells by antibody dependent cellular cytotoxicity (ADCC) are passively acquired by the host via the placenta. Because of cross-matching transfusion reactions are now rare, nevertheless sometimes IgG can exist for minor blood group antigens at a level sufficient to cause destruction of the transfused cells but too low to detect *in vitro.*

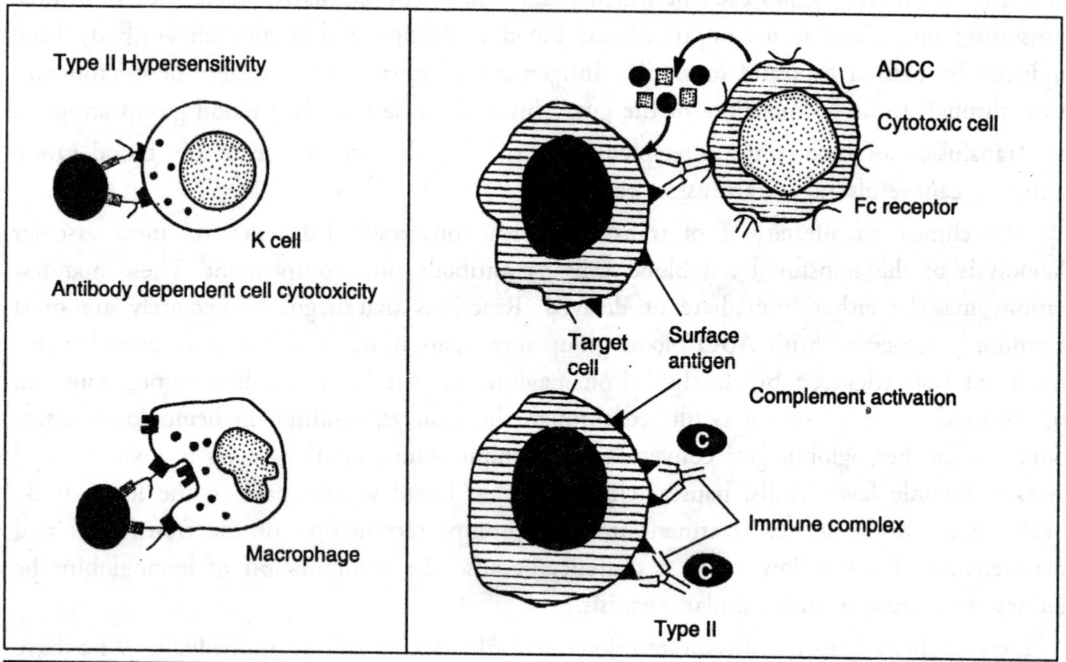

Fig. 13.2: Type II hypersensitivity reactions

3. Type III Hypersensitivity

Type III hypersensitivity involves the formation of immune complexes. Normally these complexes are removed by monocytes of reticuloendothelial system. In presence of excess amount of some antigen, antigen-antibody complex may not be successfully removed. Their accumulation can lead to a hypersensitivity reaction from complement that triggers a various inflammatory processes. This inflammation causes damaged, especially of blood vessels, basement membrane of kidney, joints and skin.

Type III hypersensitivity is also known as immune complex hypersensitivity. The reaction may be general (e.g. serum sickness) or may involve individual organs including skin (e.g. systemic lupus erythematosus, Arthus reaction), kidneys (e.g. lupus nephritis), lungs (e.g. aspergellosis), blood vessels (e.g. polyarteitis), joints (e.g. rheumatoid arthritis) or other organs. This reaction may be the pathogenic mechanism of diseases caused by many microorganisms.

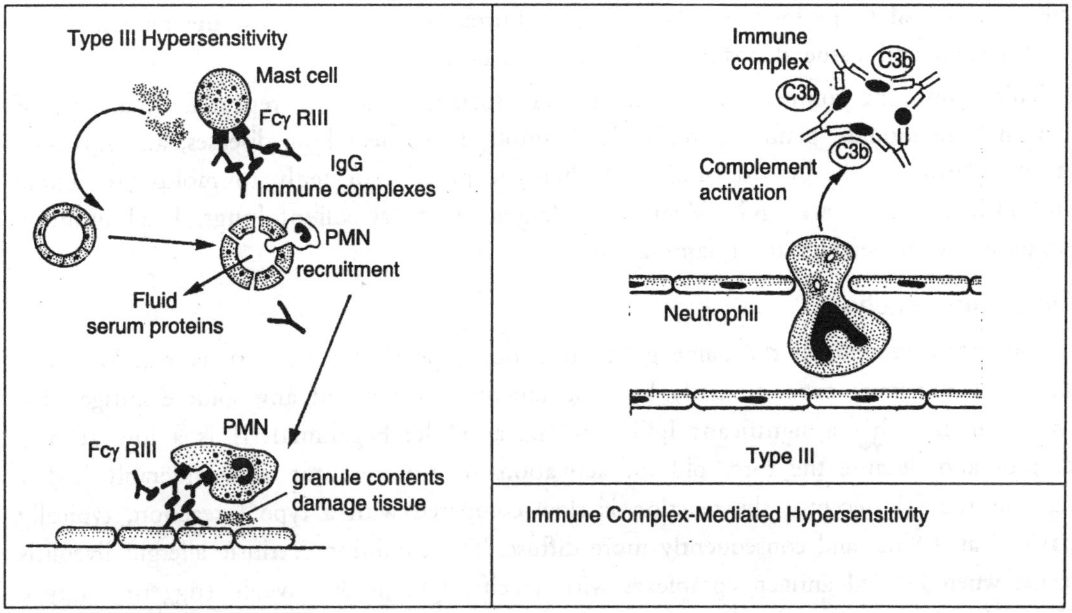

Fig. 13.3: Immune complex formations (type III reaction)

The reaction may take 3-10 hours after exposure to the antigen (as in Arthus reaction). It is mediated by soluble immune complexes. They are mostly of the IgG class, although IgM may also be involved. The antigen may be exogenous (chronic bacterial, viral or parasitic infections), or endogenous (non-organ specific autoimmunity: e.g. systemic lupus erythematosus, SLE). The antigen is soluble and not attached to the organ involved. Primary components are soluble immune complexes and complement (C3a, 4a and 5a). The damage is caused by platelets and neutrophils. The lesion contains primarily neutrophils and deposits of immune complexes and complement. Macrophages infiltrating in later stages may be involved in the healing process (Fig. 13.3).

The affinity of antibody and size of immune complexes are important in production of disease and determining the tissue involved. **Diagnosis** involves examination of tissue biopsies for deposits of Ig and complement by immunofluorescence. The immunofluorescent staining in type III hypersensitivity is granular (as opposed to linear in type II such as seen in Goodpasture's syndrome). The presence of immune complexes in serum and depletion in the level of complement are also diagnostic. Polyethylene glycol-mediated turbidity, binding of C1q and Raji cell test are utilized to detect immune complexes. Treatment includes anti-inflammatory agents.

The diseased resulting from type III hypersensitivity reaction can be placed into three groups.

First, a persistent viral, bacterial, protozoan infection, together with weak antibody response, leads to chronic immune complex formation and eventually deposition of these complexes in host tissue.

Second, a continued production of autoantibodies to self antigen during autoimmune diseases can lead to prolong immune complex formation. This overload the reticuloendothelial system and tissue deposition of complexes occurs.

Third, immune complex can form at body surface, following repeated inhalation of allergen from moulds, plants, animals. For example, in farmer's lung diseases, an individual has circulating antibodies to fungi after being exposed repeatedly to mold hay. These antibodies primarily are IgG. When the allergen enters alveoli of lungs, local immune complexes form, leading to inflammation.

The arthus reaction

The arthus reaction is the name given to a local type III hypersensitivity reaction. It is easy to demonstrate experimentally by subcutaneous injection of any soluble antigen for which the host has a significant IgG titre. Because the FcgammaRIII is a low affinity receptor and because the threshold for activation via this receptor is considerably higher than for the IgE receptor the reaction is slow compared with a type I reaction, typically maximal at 4-8hrs, and consequently more diffuse. The condition extrinsic allergic alveolitis occurs when inhaled antigen complexes with specific IgG in the alveoli, triggering a type III reaction in the lung, for example in 'pigeon fanciers lung' where the antigen is pigeon proteins inhaled via dried faeces. Complement is not required for the arthus reaction, but may modify the symptoms.

Injection of an antigen intradermally or subcutaneously into an animal that has high levels of circulating antibody, Specific for that antigen leads to formation of localized immune complexes, which mediates can acute arthus reaction within 4-8h. Microscopic examination of the tissue reveals neutrophils adhering to the vascular endothelium and than migrating into the tissue at the site of immune complex deposition. Is the reaction develops, localized tissue and vascular damage results in an accumulation of fluid [edema] and red blood cells [erythema] at the site. The severity of the reaction can vary from mild swelling and redress to tissue necrosis. After an insect bite, a sensitive individual may have rapid, localized type I reaction at the site. Often, some 4-5h later, a typical. Arthus reaction also develops at the site with pronounced erythema and edema. Intrapulmonary arthus-type reactions induced by bacterial spores, fungi, of dried fecal proteins can also cause pnemonitis.

Generalized or systemic reactions

The presence of sufficient quantities of soluble antigen in circulation to produce a condition of antigen excess leads to the formation of small antigen-antibody complexes which are soluble and poorly cleared. In the normal animal these complexes fix complement but experiments in animals genetically deficient in C3 or C4 have shown that complement is not required for pathology to be observed following antibody-antigen complex challenge. The major pathology is due to complex deposition which seems to be exacerbated by increased vascular permeability caused by mast cell activation via FcgammaRIII as above. The deposited immune complexes trigger neutrophils to discharge their granule contents with consequent damage to the surrounding endothelium and basement membranes. The complexes may be deposited in a variety of sites such as skin, kidney and joints. Common examples of generalised type III reactions are post-infection complications such as arthritis and glomerulonephritis.

4. Type IV Hypersensitivity

Delayed type of immunity is the form of cell mediated immunity in which ultimate effector cell is mononuclear phagocyte. This type of immunity is part of primary defense mechanism against intracellular bacteria such as *Listeria monocytogenes* and mycobacteria. In delayed type of hypersensitivity, antigen activating T-cells secrets cytokine, which have several effects. Some cytokine activates venular endothelial cells to recruit monocyte from blood at the site of antigen challenge. Other monocyte converts monocyte into activated macrophages that serves to eliminate the antigen. The T-cells mediate the DTH are usually CD4+Th1 cells, but cytokines produced by CD8+T-cells can initiate the same reaction. This is the only class of hypersensitive reactions to be triggered by antigen-specific T-cells.

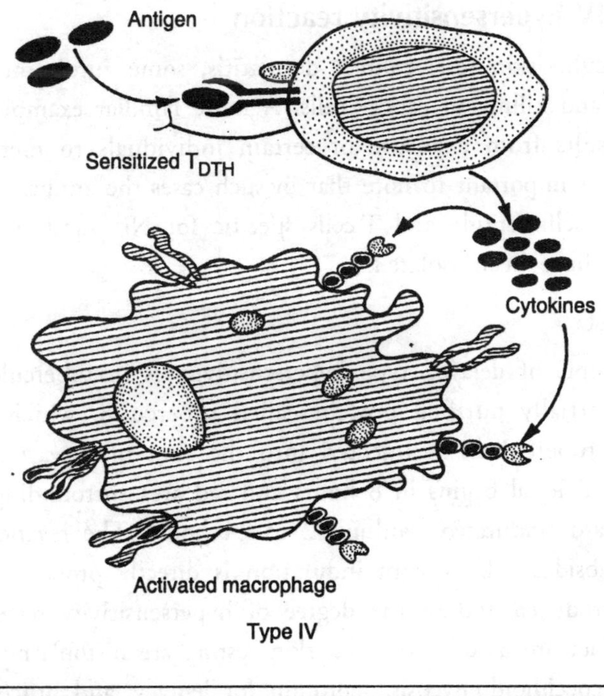

Fig. 13.4: Type IV Hypersensitivity

We would now call these T_H1 cells but they were originally termed T_{DTH} after the alternative name for this reaction-delayed type hypersensitivity.

Delayed type hypersensitivity results when an antigen-presenting cell, typically a tissue dendritic cell which has picked up antigen, processed it and displayed appropriate peptide fragments bound to class II MHC is contacted by an antigen specific T_H1 cell patrolling the tissue. The resulting activation of the T-cell produces cytokines such as chemokines for macrophages, other T-cells and, to a lesser extent, neutrophils as well as TNF beta and IFN gamma. The consequences are a cellular infiltrate in which mononuclear cells (T-cells and macrophages) tend to predominate. It is usually maximal in 48-72 hours.

The problem which this explanation faces is the rarity of antigen-specific T-cells. Despite the fact that "memory T-cells", unlike naive T-cells, do circulate through tissues, there is some doubt that a single T-cell could initiate the event. The answer to this conundrum may lie in the recent observations that at least some Type IV reactions absolutely require the presence of 'natural' IgM antibody for initiation.

Due to the nature and kinetics of the reaction it is still believed that activation of memory T_H1 cells is primarily responsible for propagating the reponse, but initiation may require IgM and probably also complement. One theory is that limited IgM-antigen complexes in local capillaries may lead to a limiting, localised complement activation within the vessel activating the vascular endothelium and thus recruiting inflammatory cells including memory T-cells.

Example of Type IV hypersensitivity reaction

It includes tuberculin skin test, contact dermatitis, some autoimmune diseases, transplantation rejection, and killing of cancer cells. A more familiar example is contact hypersensitivity which results from exposure of certain individuals to metal salts and small reactive chemicals. It is important to note that in such cases the antigen must be a complex of the hapten and a self peptide and T-cells specific for Ni^{++} and reactive chemicals eg dinitrofluorobenzene have been isolated.

Tuberculin skin test

The classical example of delayed type hypersensitivity is in tuberculosis. In tuberculin hypersensitivity, a partially purified protein called tuberculin, which is obtained from bacillus that causes tuberculosis is injected into the skin of forearm. The response in tuberculin positive individual begins in 8 hours and red area surrounding the injection site becomes firm and hard (indurated) within 12 to 24 hours. The reaction reaches its pick in 48 hours then subsides. The size of induration is directly proportional to amount of antigen that was introduced and to the degree of hypersensitivity tested individual. The other microbial product are used in type IV skin testing are histoplasmin for histoplasmosis, coccidioidin for coccidioidomycosis, lepromin for lepracy, and brucellergen for brucellosis.

Allergic contact dermatitis.

Allergic contact dermatitis is caused by hapten that combines with protein in skin to form allergen that elicits immune response. The haptens are the antigenic determinant and skin proteins are the carrier's molecule for hapten. The example of theses hapten include cosmetics, plant material catechol molecule from poison ivy and poison oak, metal and jewelary (specially contains nickel).

TABLE 13.3
Summary of hypersensitivity classification

TYPE	DESCRIPTIVE NAME	INITIATION TIME	MECHANISM	EXAMPLES
I	IgE-mediated hypersensitivity	2-30 mins	Ag induces cross-linking of IgE bound to mast cells with release of vasoactive mediators	Systemic anaphylaxis, Local anaphylaxis, Hay fever, Asthma, Eczema
II	Antibody-mediated cytotoxic hypersensitivity	5-8hrs	Ab directed against cell-surface antigens mediates cell destruction via ADCC or complement	Blood transfusion reactions, Haemolytic disease of the newborn, Autoimmune Haemolytic anaemia
III	Immune-complex mediated hypersensitivity	2-8hrs	Ag-Ab complexes deposited at various sites induces mast cell degranulation via FcgammaRIII, PMN degranulation damages tissue	Arthus reaction (Localised); Systemic reactions disseminated rash, arthritis, glomerulonephritis
IV	Cell-mediated hypersensitivity	24-72hrs	Memory TH1 cells release cytokines that recruit and activate macrophages	Contact dermatitis, Tubercular lesions

CHAPTER 14

IMMUNOTOXIN

INTRODUCTION

An immunotoxin is a chemical that can cause immune system malfunction with exposure. When the immune system function is suppressed (immunosuppression) there is an increased susceptibility to infectious diseases and cancers. Polychlorinated-biphenyls (PCB), asbestos and herbicide DDT are known immunosuppressants. Immunotoxins can also cause autoimmune disease, where an overactive immune system attacks the bodies cells.

Immunotoxins offer the potential to selectively target cells of the immune system in order to promote the goal of tolerance to organ transplants. Conceptually, immunotoxins take advantage of the exquisite specificity of antibodies to selectively target drug delivery and the potency of toxins to kill the target cells. A variety of such agents have been developed for experimental use. Some immunotoxins and chimeric toxins have been designed to target cancer cells. Unfortunately, the toxicity of some of these agents has prevented their further clinical development. An additional problem has been the immunogenicity of the toxin. The ideal toxin molecule for use in an immunotoxin construct would have minimal toxicity and poor immunogenicity.

The targeting component of an immunotoxin is most frequently a monoclonal antibody reactive with the desired specificity but it may also be a receptor or ligand for a receptor on the surface of the targeted cell. Monoclonal antibodies can be used as whole molecules or as enzymatically cleaved fragments, which are less immunogenic but have shorter half-lives in vivo. Genetically-engineered fusion proteins containing immunoglobulin heavy chain constant regions confer a longer half-life, and humanized constructs can be engineered to eliminate almost all immunogenicity.

Toxins used in immunotoxin constructs are derived from bacteria, fungi, and plants, and most function by inhibiting protein synthesis. Toxins such as diphtheria toxin (DT) and *Pseudomonas* exotoxin (PE) prevent protein synthesis by an effect on elongation factor 2 (EF-2). They are synthesized as single chain proteins which are modified post-translationally into two-chain molecules joined by disulfide bonds. The interchain disulfide bond is necessary for the cytotoxic effect of the toxin. Toxins have separate functional domains committed to target cell binding, translocation across membranes, and inhibition of protein synthesis. In order to be effective, however, immunotoxin must be internalized and route to the appropriate intracellular compartment for translocation of their attached toxin into the cytosol. The targeting moiety and toxin are joined by a crosslinker which is stable extracellularly but labile intracellularly so that the toxin can function in the cytosol.

Immunotoxins are protein toxins connected to a cell binding ligand of immunologic interest. Classically, beginning 35 years ago, immunotoxins were created by chemically conjugating an antibody to a whole protein toxin, or, for more selective activity, by using a protein toxin devoid of its natural binding domain. Immunologic proteins that are smaller than monoclonal antibodies (MAbs), like growth factors and cytokines, have also been chemically conjugated and genetically fused to protein toxins. While some do not consider growth factor toxin fusions or conjugates to be immunotoxins, these newer immunotoxins, like classical immunotoxins, bind to target cells and contain a toxin that kills cells.

Immunotoxins represent a class of magic bullet in which the toxin of plant or bacterial origin is attached to an antibody or growth factor. After binding to the cell surface, the toxin is internalized and kills the cell by catalytic inhibition of protein synthesis. One such immunotoxin consists of anti idiotype antibody conjugated with ricin, a toxin of such devastating potency that one molecule entering the cell is lethal. Such immunotoxins need to have a reasonable half-life in the circulation, to be able to penetrate into tumours and not bind significantly to nontumor cells. Immunotoxins composed of tumour-specific monoclonal antibodies coupled to lethal toxins are potentially valuable therapeutic reagents. The toxins used in preparing immunotoxins include ricin, *Shigella* toxin, and diphtheria toxin, all of which inhibit protein synthesis. These toxins are so potent that a single molecule has been shown to kill a cell. Each of these toxins consists of two types of functionally distinct polypeptide components, an inhibitory (toxin) chain and one or more binding chains, which interact with receptors on cell surfaces; without the binding polypeptide(s) the toxin cannot get into cells and therefore is harmless. An immunotoxins is prepared by replacing the binding polypeptide(s) with a monoclonal antibody that is specific for a particular tumour cell. In theory, the attached monoclonal antibody will deliver the toxin chain specifically to tumour cells, where it will cause death by inhibiting protein synthesis. The initial clinical responses to such immunotoxins in patients with leukemia, lymphoma, and some other types of cancer have shown promise, and research to develop and demonstrate their safety and effectiveness is underway.

Role of immunotoxins

Immunotoxins are a new class of antitumour agents consisting of tumour-selective ligands (generally monoclonal antibodies [MoAbs]) linked to highly toxic protein molecules that have been modified to remove their normal tissue-binding domains (Fig. 14.1). These immuno-conjugates combine the potency of the parent toxin with the specificity of the attached ligand. Toxins used in the construction of immunotoxins belong to a group of peptides that catalytically inhibit the elongation step of protein synthesis, and include ricin, abrin, pokeweed antiviral protein, gelonin, *Pseudomonas* exotoxin A, diphtheria toxin, and alpha-sarcin. To synthesize immunotoxins, the normal cell-binding function must be removed by chemical cleavage or modification, or in the case of toxins that have been cloned, genetic engineering used to delete amino acids critical to cell binding. Covalent linkage of toxin to ligand generally involves a disulfide or thioether bond, though recently, recombinant toxin molecules with ligands that are genetically engineered into the protein have been made.

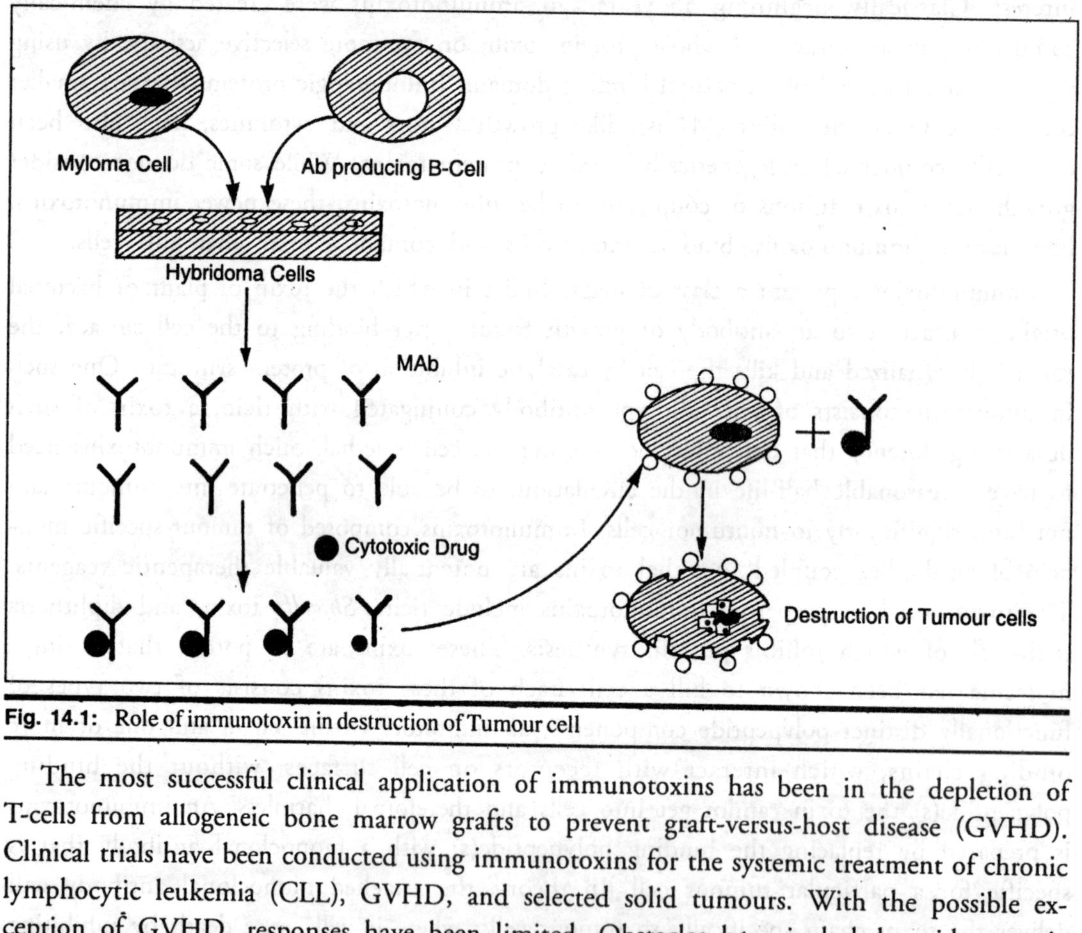

Fig. 14.1: Role of immunotoxin in destruction of Tumour cell

The most successful clinical application of immunotoxins has been in the depletion of T-cells from allogeneic bone marrow grafts to prevent graft-versus-host disease (GVHD). Clinical trials have been conducted using immunotoxins for the systemic treatment of chronic lymphocytic leukemia (CLL), GVHD, and selected solid tumours. With the possible exception of GVHD, responses have been limited. Obstacles have included rapid systemic clearance, poor delivery to extravascular tumour deposits, and humoral immune responses to the immunotoxin. Research to overcome these problems is in progress and should lead to a better definition of the role of immunotoxins in the therapy of malignancies.

Therapeutic immunotoxins are compounds that are being developed for cancer and virus treatment. These immunotoxins are chimeric protein that are linked to a toxin. The protein is a modified antibody that targets a specific cell type. When the protein binds to that cell, it is taken in and the toxin kills the cell.

Plant toxins

Plant toxins are of two types, hemitoxins or class I ribosome-inactivating proteins and holotoxins also referred to as class II ribosome-inactivating proteins (Fig. 14.2). Plant holotoxins include ricin, abrin, mistletoe lectin, and modeccin. Hemitoxins, or class I ribosome-inactivating proteins, include PAP, saporin, bryodin 1, bouganin, and gelonin. Holotoxins contain both binding and catalytic domains, whereas hemitoxins contain only catalytic domains. Plant toxins have been shown to prevent the association of elongation factor-1 and -2 (EF-1 and EF-2) with the 60s ribosomal subunit by removing the base of A in 28s rRNA. Ricin also removes the neighboring base G.

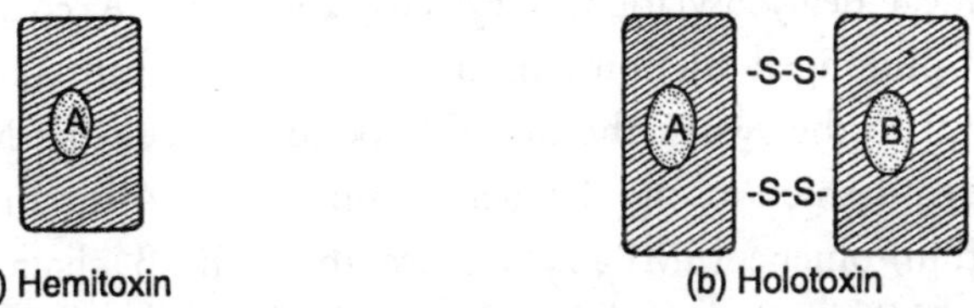

Fig. 14.2: (a) Structure of Hemitoxin containing Catelytic or activity domain (A); (b) Structure of Holotoxin containing Catalytic or activity domain (A) and binding domain (B) interlinked with disulphide linkage

Mechanism of action of plant toxins

Exactly how plant toxins move from the cell surface to the cytosol is unknown; the process probably differs for each plant toxin. Apoptosis has been shown to be involved in cell death induced by plant toxins. Only the enzymatic domain of both holo- and hemitoxins translocates to the cytosol, so the binding domains of holotoxins must be removed by reduction of the disulfide bond prior to translocation. The cytotoxicity of both plant and bacterial toxins is optimal when the catalytic domain alone translocates to the cytosol. A binding domain can be translocated to the cytosol if placed within the catalytic domain, but cytotoxic activity is significantly reduced. In an attempt to construct a fusion toxin containing RTA from which free A-chain could be generated, interleukin-2 (IL-2) was fused to recombinant RTA through a linker that contained a proteolytic cleavage site for DT or clotting factor Xa. Although the recombinant toxin could be cleaved extracellularly, it could not selectively target cells since the ligand and toxin were no longer connected. Later, IL-2 was fused to a mutant of PAP, but the fusion toxin was not purified and was not very cytotoxic. Ligands fused to plant toxins have produced recombinant toxins with significant cytotoxic activity, including 1 containing a CD40 single-chain antibody and bryodin 1, 1 containing urokinase binding domain and saporin, and 1 containing human fibroblast growth factor and saporin. For these molecules, it is not known whether (1) the recombinant toxin entered the cytosol of target cells intact, or (2) the ligand was unstable after internalization, permitting the catalytic domain alone to translocate to the cytosol. The ability of even stable ligands to predictably separate from the catalytic domain is an important feature of recombinant toxins and a unique feature among all toxins provided by the bacterial toxins PE and DT. The intracellular transport of ricin is dependent on sorting receptors that cycle between the endoplasmic reticulum (ER) and the terminal compartments of the Golgi. It has been shown that glycolipids that bind ricin may be transported from endosomes to the Golgi and that the lysine-aspartic acid-glutamic acid-leucine (KDEL) ER retention sequence, if added to ricin, enhances the delivery of this plant toxin to the cytosol. Originally, antibodies were chemically conjugated through a disulfide bond to the catalytic subunits of holotoxins such as ricin or abrin, each of which had been removed from its binding domain by reduction. Even without its binding domain, however, ricin A chain (RTA) was taken up nonspecifically by macrophages and hepatic nonparenchymal Kupffer cells. This uptake was due to glycosylated side residues of RTA binding to mannose receptors on the liver. The most successful technique for reducing nonspecific uptake of RTA was through chemical deglycosylation. Deglycosylated ricin A chain (dgA) immunotoxins had significantly prolonged lifetimes in mice, leading to an improved therapeutic index. Half-lives improved further when the disulfide bond between the MAb and the toxin was formed in a hindered fashion using the derivatizing agent 4-succinimidyloxycarbonyl-a-methyl-a (2-pyridyldithio)toluene (SMPT). Because the ricin B chain facilitates the cytotoxicity of RTA-containing immunotoxins, whole ricin has been targeted after blocking its oligosaccharide binding sites to prevent normal cell binding. These sites on ricin were blocked with ligands prepared by chemical modification of glycopeptides containing triantennary N-linked oligosaccharides. The resulting blocked ricin (bR) was then chemically conjugated to antibodies to make immunotoxins.

Bacterial Toxins

There are two types of bacterial toxins, *Pseudomonas* endotoxin (PE) and Diphtheria endotoxin (DT).

Pseudomonas endotoxin (PE)

Pseudomonas endotoxin (PE) is a single-chain 613-amino-acid protein containing 3 functional domains. Domain Ia (amino acids 1-252) is the binding domain, domain II (amino acids 253-364) is the translocating domain, and domain III (amino acids 400-613) contains the adenosine diphosphate ribosylating enzyme that inactivates elongation factor 2 in the cytosol, resulting in cell death. Domain Ib separates domains II and III and contains amino acids 365 to 399. *Pseudomonas* endotoxin (PE 38) is a 38 kDa truncated form of *Pseudomonas* endotoxin (PE) containing amino acids 253 to 364 and 381 to 613.

Fig. 14.3: Structure of *Pseudomonas* endotoxin (PE)

Full-length 613-amino-acid PE, as shown in above figure, is a single-chain protein containing 3 functional domains. Domain Ia (amino acids 1-252) is the binding domain, domain II (amino acids 253-364) is responsible for translocating the toxin to the cytosol, and domain III (amino acids 400-613) contains the ADP-ribosylating enzyme that inactivates EF-2 in the cytosol. The catalytic process of ADP ribosylation has been shown to involve residues His440 and Glu553. His440 binds nicotinamide adenine dinucleotide (NAD) via Adenosine monophosphate (AMP) ribose. The carboxyl group of the Glu553 side chain, through a water-mediated hydrogen bond with Tyr481 and Glu546, allows Tyr481 to bind NAD through a ring-stacking mechanism. The function of domain Ib (amino acids 365-399) is unknown.

Diphtheria endotoxin (DT)

Diphtheria endotoxin (DT) is 535 amino acids in length and is composed of the enzymatic A domain (amino acids 1-193) and the binding B domain (amino acids 482-535). The translocation or transmembrane (T) domain is located in between (Fig. 14.4). The truncated form of DT used in recombinant toxins is DT388 in DT388-GM-CSF (DTGM) or DAB_{389} in denileukin diftitox, each of which contain methionine followed by the first 388 amino acids of DT.

Fig. 14.4: Structure of Diphtheria endotoxin (DT)

DT is a single-chain protein 535 amino acids in length. It is composed of an enzymatic A domain (amino acids 1-193) and a binding B domain (amino acids 482-535). A third domain, which is the translocation or transmembrane (T) domain, is located in the center of the molecule. Based on DT's 3-dimensional structure in the presence and absence of NAD, DT is thought to undergo these steps to kill cells: (1) DT is proteolytically cleaved outside the cell between Arg193 and Ser194, which is within a disulfide loop formed by Cys186 and Cys201. (2) DT binds on the cell surface via residues 482 to 535 to a complex of heparin-binding epidermal growth factor (EGF)-like growth factor precursor and CD9. (3) DT internalizes into an endosome and unfolds at low pH, and the disulfide bond linking amino acids 186 and 201 is reduced. (4) The TH8 (amino acids 326-347) and TH9 (amino acids 358-376) domains form a hairpin, which inserts into the membrane of the endosome and forms a channel through which the enzymatic fragment translocates to the cytosol, probably from early endosomes. (5) In the cytosol, NAD binds to the active-site cleft of DT (amino acids 34-52), and the ADP ribose of NAD is transferred to EF-2. (6) As with PE, cell death is facilitated by apoptosis.

Mechanism of action of bacterial toxins

Both PE and DT enzymatically ADP-ribosylate EF-2 in the cytosol. They each catalyze the ADP-ribosylation of histidine-699 of EF-2, which is post-translationally modified to a diphthimide residue. Despite their similar action, PE and DT differ greatly in their amino acid sequence, and in fact PE's enzymatic domain is near the carboxyl terminus, while DT's is near the amino terminus. Conversely, PE's binding domain is near its amino terminus, and DT's is near its carboxyl terminus.

The recombinant immunotoxin

The recombinant immunotoxin RFB4(dsFv)-PE38 (BL22) is composed of the VL from the MAb RFB4 disulfide bonded to a fusion of VH with PE38. The disulfide bond connecting VH and VL is formed between 2 cysteine residues replacing Arg44 of VH and Gly100 of VL. For single-chain recombinant toxins containing DT, the ligand is at the carboxyl rather than at the amino terminus of the toxin. The single-chain recombinant immunotoxin anti-Tac(Fv)-PE38 (LMB-2) contains the variable heavy domain (VH) of the anti-Tac monoclonal antibody (MAb) fused via the peptide linker (G4S)3 to the variable light domain (VL), which in turn is fused to PE38.

CHAPTER

15

INFLAMMATION

INTRODUCTION

Inflammation (Latin, *inflammatio*, to set on fire) is the complex biological response of vascular tissues to harmful stimuli, such as pathogens, damaged cells, or irritants. It is a protective attempt by the organism to remove the injurious stimuli as well as initiate the healing process for the tissue. Inflammation is not a synonym for infection. Even in cases where inflammation is caused by infection it is incorrect to use the terms as synonyms: infection is caused by an exogenous pathogen, while inflammation is the response of the organism to the pathogen.

Inflammation is a complex response to local injury or other trauma; it is characterized by redness, swelling heat and pain. Inflammation involves various immune system cells and numerous mediators. Inflammation is a complex stereotypical reaction of the body expressing the response to damage of its cells and vascularized tissues. In avascular tissues, e.g. in normal cornea, the true inflammation does not occur. The discovery of the detailed processes of inflammation has revealed a close relationship between inflammation and the immune response. The five basic symptoms of inflammation—redness (rubor), swelling (tumour), heat (calor), pain (dolor) and deranged function (functio laesa) have been known since the ancient Greek and Roman era. These signs are due to extravasation of plasma and infiltration of leukocytes into the site of inflammation. Early investigators considered inflammation, a primary host defense system. From this point of view inflammation is the key reaction of the innate immune response but in fact, inflammation is more than this, since it can lead to death, as in anaphylactic shock, or debilitating diseases, as in arthritis and gout.

Inflammation is a response of a tissue to injury, often injury caused by invading parasites. It is characterized by increased blood flow to the tissue causing, increased temperature, redness, swelling, and pain. A bacterial infection initiates inflammation through several interconnecting mechanisms. The "nonself" surface of bacteria allows the **complement system** to be activated through the "alternative pathways". Specific surface molecules of the bacteria, called Pathogen-Associated Molecular Patterns (PAMPs), bind to Toll-like receptors (TLRs) on or in a variety of leukocytes.

A human or animal must defend itself against multitude of different pathogens including viruses, bacteria, fungi, and protozoan and metazoan parasites as well as tumours and a number of various harmful agents which are capable to derange its homeostasis. For this, a plenty of effect or mechanisms capable of defending the body against such antigens and agents have developed and these can be mediated by soluble molecules or by cells. If infection occurs as a consequence of the tissue damage, the innate and, later, the adaptive immune systems are triggered to destroy the infectious agent. In the absence of inflammation, wounds and infections would never heal and progressive destruction of the tissue would compromise the survival of the organism. However, inflammation which runs unchecked can also lead to a host of diseases, such as hay fever, atherosclerosis, and rheumatoid arthritis. It is for this reason that inflammation is normally tightly regulated by the body.

Inflammatory components

The cellular component involves leukocytes, which normally reside in blood and must move into the inflamed tissue via extravasation to aid in inflammation. Some act as phagocytes, ingesting bacteria, viruses, and cellular debris. Others release enzymatic granules which damage pathogenic invaders. Leukocytes also release inflammatory mediators which develop and maintain the inflammatory response. Generally speaking, acute inflammation is mediated by granulocytes, while chronic inflammation is mediated by mononuclear cells such as monocytes and lymphocytes.

Various modes of inflammation

According to different criteria, inflammatory responses can be divided into several categories. The criteria include: Time-hyperacute (peracute), acute, subacute, and chronic inflammation; The main inflammatory manifestation are alteration, exudation, proliferation; The degree of tissue damage is superficial and profound (bordered, not bordered); Characteristic picture are nonspecific, and specific. Immunopathological mechanisms includes allergic (reaginic) inflammation, inflammation mediated by cytotoxic antibodies, inflammation mediated by immune complexes and delayed-type hypersensitivity reactions. Different modes of inflammations are described below.

- **Granulomatous inflammation:** It is characterized by the formation of granulomas, they are the result of a limited but diverse number of diseases, which include among others tuberculosis, leprosy, and syphilis.
- **Fibrinous inflammation:** Inflammation resulting in a large increase in vascular permeability allows the blood vessels to pass through fibrin. If an appropriate *procoagulative* stimulus is present, such as cancer cells, a fibrinous exudate is deposited. This

is commonly seen in serous cavities, where the conversion of fibrinous exudate into a scar can occur between serous membranes, limiting their function.

- **Purulent inflammation:** Inflammation resulting in large amount of pus, which consists of neutrophils, dead cells, and fluid. Infection by pyogenic bacteria such as staphylococci is characteristic of this kind of inflammation. Large, localised collections of pus enclosed by surrounding tissues are called abscesses.
- **Serous inflammation:** Characterised by the copious effusion of non-viscous serous fluid, commonly produced by mesothelial cells of serous membranes, but may which also be derived from blood plasma. Skin blisters exemplify this pattern of inflammation.
- **Ulcerative inflammation:** Inflammation occurring near an epithelium can result in the necrotic loss of tissue from the surface, exposing lower layers. The subsequent excavation in the epithelium is known as an ulcer.

Mechanism of inflammation

Inflammation is the body's reaction to invasion by an infectious agent, antigen challenge or even just physical, chemical or traumatic damage. The mechanism for triggering the response the body to injury is extremely sensitive. Responses are to tissue damage that might not normally be thought of as injury, for example when the skin is stroked quite firmly or if some pressure is applied to a tissue. In addition, the body has the capacity to respond to both minor injuries such as bruising, scratching, cuts, and abrasions, as well as to major injuries such as severe burns and amputation of limbs.

Depending on the severity of the tissue damage resulting from an injury, the integrity of the skin or internal surfaces may be breached and damage to the underlying connective tissue and muscle, as well as blood vessels can occur. In this situation infection can, and frequently does result because the normal barrier to the entry of harmful organisms has been broken. It is obviously most important that the body can respond to injury by healing and repairing the damaged tissue, as well as by eliminating the infectious agents that may have entered the wound and their toxins. It is also important that the appropriate response to the tissue damage and infection can be made: it is no use bringing all of the body's defenses into action to repair a minor scratch, just as one would not expect a single mechanism to be able to deal with the sudden loss of a limb or a major infection.

The inflammatory reaction is phylogenetically and ontogenetically the oldest defense mechanism. The cells of the immune system are widely distributed throughout the body, but if an infection or tissue damage occurs it is necessary to concentrate them and their products at the site of damage. Three major events occur during this response:

1. An increased blood supply to the tissue "in danger". It is performed by vasodilation. The inflamed tissue looks like containing greater number of vessels.
2. Increased capillary permeability caused by retraction of the endothelial cells. This permit larger molecules than usual to escape from the capillaries, and thus allows the soluble mediators of immunity to reach the site of inflammation.

3. Leukocytes migrate out of the capillaries into the surrounding tissues. In the earliest stages of inflammation, neutrophils are particularly prevalent, but later monocytes and lymphocytes also migrate towards the site of infection.

For the possibility of surrounding tissue damage, inflammatory responses must be well ordered and controlled. The body must be able to act quickly in some situations, for example to reduce or stop the lost of blood, whereas tissue repair and reconstruction can begin a little later. Therefore, a wide variety of interconnected cellular and humoral (soluble) mechanisms are activated when tissue damage and infection occur. On the other hand if the injury is negligible, the body must have mechanisms which are able to stop the tissue damage when the injury agent was removed.

The development of inflammatory reactions is controlled by cytokines, by products of the plasma enzyme systems (complement, the coagulation clothing, kinin and fibrinolytic pathways), by lipid mediators (prostaglandins and leukotrienes) released from different cells, and by vasoactive mediators released from mast cells, basophils and platelets. These inflammatory mediators controlling different types of inflammatory reaction differ. Fast-acting mediators, such as vasoactive amines and the products of the kinin system, modulate the immediate response. Later, newly synthesized mediators such as leukotrienes are involved in the accumulation and activation of other cells. Once leukocytes have arrived at a site of inflammation, they release mediators which control the later accumulation and activation of other cells. However, inflammatory reactions initiated by the immune system, the ultimate control is exerted by the antigen itself, in the same way as it controls the immune response itself. For this reason, the cellular accumulation at the site of chronic infection, or in autoimmune reactions (where the antigen cannot ultimately be eradicated), is quite different from that at sites where the antigenic stimulus is rapidly cleared. The nervous system can also participate in the control of inflammation, especially axon reflexes, but inflammation may be realized in degenerated tissues as well.

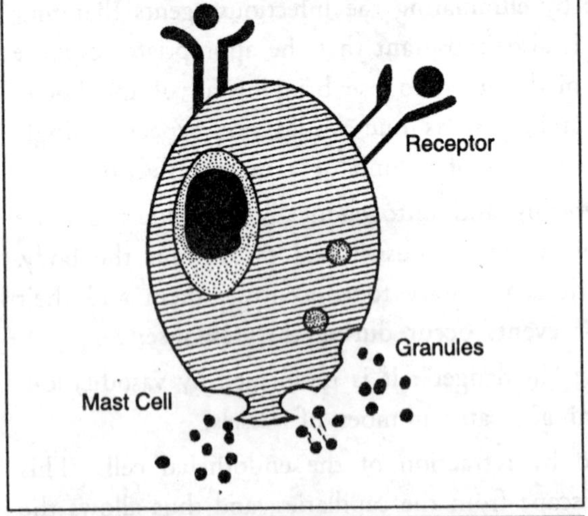

Fig. 15.1: Structure of Mast Cell

Mast cells

Mast cells are found in the tissues. Their cytoplasm is loaded with granules containing mediators of inflammation. Their surface is coated with a variety of receptors which, when engaged by the appropriate ligand, trigger exocytosis of the granules (Figure 15.1). Mast cells appear to be key players in the initiation of inflammation. Their Toll-like receptors trigger exocytosis when they interact with PAMPs like the **lipopolysaccharide** (LPS or "endotoxin") of Gram-negative bacteria (TLR-4) the peptidoglycan of Gram-positive bacteria (bind TLR-2). Their receptors for complement fragments trigger exocytosis when they bind C3a and C5a bacteria coated with C3b activated mast cells release literally dozens of potent mediators; some immediately as they discharge their granules some later as they synthesize them by new gene transcription. These mediators are active in either (or, in some cases, both) **recruiting** all the types of white blood cell to the site monocytes that become macrophages when they leave the blood and enter the tissue neutrophils antigen-presenting dendritic cells all kinds of lymphocytes: B-cells and T-cells, leading to an adaptive immune response; NK cells (an effector cell in innate immunity) eosinophils **activating** many of these recruited cells to produce their own mediators of inflammation.

Reactive oxygen species (ROS) are produced by activated phagocytes: macrophages and neutrophils. They are toxic for microorganisms but can also lead to tissue injury. Chemokines are **chemotactic cytokines**; that is, secreted proteins that attract other leukocytes into the area. Several have been identified. Large amounts of TNF-α are quickly released by stimulated mast cells. All the cells involved in inflammation have receptors for TNF-α, and are activated by it to synthesize more on their own. This positive feedback quickly amplifies the response. The granules of mast cells are loaded with histamine and their exocytosis releases this potent mediator. Histamine increases the blood flow to the area and the leakage of fluid and proteins from the blood into the tissue space. Thus the quick release of histamine is largely responsible for the redness and swelling associated with inflammation. Macrophages and monocytes are the main source of this cytokine. IL-1 has both paracrine effects on cells in the vicinity, e.g. causing them to produce tissue factor and thus triggering the blood clotting cascade. Stimulating the synthesis and secretion of a variety of other interleukins helping to activate T-cells and thus initiate an adaptive immune response. Endocrine effects is carried in the blood throughout the body decreasing blood pressure, inducing fever. IL-1 causes fever by stimulating the release of **prostaglandins**, which act on the temperature control center of the hypothalamus.

Inflammasomes

IL-1 is synthesized from a larger precursor that is cleaved by a. Caspase-1 is part of two (or more) multi-protein complexes in the cytosol of macrophages and neutrophils that are called inflammasomes. Inflammasomes are activated by several different products produced by invading bacteria. Some of these are first "seen" by toll-like receptors (TLRs) thus providing a link between the innate immune system and inflammation.

Types of inflammation

Inflammation can be classified as either *acute* or *chronic*. When it is acute, it occurs as an immediate response to trauma (an injury or surgery) usually within two hours. When it is chronic, the inflammation reflects an ongoing response to a longer-term medical condition, such as arthritis.

Acute inflammation

Acute inflammation is the initial response of the body to harmful stimuli and is achieved by the increased movement of plasma and leukocytes from the blood into the injured tissues. Acute inflammation is a short-term process which is characterized by the classic signs of inflammation—swelling, redness, pain, heat, and loss of function—due to the infiltration of the tissues by plasma and leukocytes. It occurs as long as the injurious stimulus is present and ceases once the stimulus has been removed, broken down, or walled off by scarring (fibrosis). The Inflammation is not the same as infection. Infections are caused by bacteria, fungus, and viruses, and infections sometimes produce inflammation. However, infection and inflammation are treated very differently. Your foot and ankle surgeon can best determine the cause of your inflamed tissue.

Inflammatory mechanism

The process of acute inflammation is initiated by the blood vessels local to the injured tissue, which alter to allow the exudation of plasma proteins and leukocytes into the surrounding tissue. A cascade of biochemical events propagates and matures the inflammatory response, involving the local vascular system and various cells within the injured tissue. The increased flow of fluid into the tissue causes the characteristic swelling associated with inflammation since the lymphatic system doesn't have the capacity to compensate for it, and the increased blood flow to the area causes the reddened color and increased heat. The blood vessels also alter to permit the extravasation of leukocytes through the endothelium and basement membrane constituting the blood vessel. Once in the tissue, the cells migrate along a chemotactic gradient to reach the site of injury, where they can attempt to remove the stimulus and repair the tissue.

Meanwhile, several biochemical cascade systems, consisting of chemicals known as plasma-derived inflammatory mediators, act in parallel to propagate and mature the inflammatory response. These include the complement system, coagulation system and fibrinolysis system. Finally, down-regulation of the inflammatory response concludes acute inflammation. Removal of the injurious stimuli halts the response of the inflammatory mechanisms, which require constant stimulation to propagate the process. Additionally, many inflammatory mediators have short half lives and are quickly degraded in the tissue, helping to quickly cease the inflammatory response once the stimulus has been removed.

Types of acute inflammation

The acute inflammatory response involves both localized and systemic response.

Localized inflammatory response

The localized response identified by the symptoms like swelling (tumour), redness (rubor), heat (calor) and pain (dolor) and loss of function. There is instant increase in vascular diameter (vasodilatation) just after the tissue injury occurred. It results in increase in volume of blood in the area and reduction in flow of blood. The increased blood volume heats tissues causes it to redden. Vascular permeability also increases leading to leakage of fluid from blood vessels. This leads to accumulation of fluid in the tissue is called as edema. The exudation of fluid from blood stream activates the kinin, clotting and fibrinolytic system. The many vascular changes occurs due to these mediators which induce the vasodilatation and increased vascular permeability. Within the few hour of onset of theses vascular changes neutrophill adhere to endothelial cells, and migrate out of the blood into the tissue spaces. These neutrophil phagocytose invading pathogen and release mediator that contribute the inflammatory response.

Systemic acute phase response

The acute phase response consists of rapid adjustment of plasma protein composition in response to injurious stimuli, including infection, burns trauma and neoplasia. Several different protein are rise in concentration such as C reactive protein, alpha2 microglobulin, clotting protein fibrinogen, serum amyloid A protein, whereas other falls such as albumin and transferrin. The C-reactive protein is a prototype acute phase protein whose serum level increases 1000 fold during acute phase response. It is composed of five identical polypeptide held together by noncovalent reaction. The C-reactive protein binds to a wide variety of microorganism and activates complement, resulting deposition of opsonin C3b on the surface of microorganisms. Phagocytic cells which express the C3b receptor can readily phagocytose C3b coated microorganisms. The precise function of acute phase response is largely unknown. Although the acute phase response is characterized by a rapid onset, it can persist in setting of chronic inflammation. In some patients with chronic inflammatory diseases persistence elevation of serum amyloid protein may lead to deposition of this protein in interstitium tissues. Such deposited protein in the form of fibril rich beta plated sheets structure, can interface with normal organ function e.g. myocardial contraction and glomurular filtration. Such patients are said to develop amyloidosis.

Chronic inflammation

Inflammation can become chronic. In certain settings the acute process, characterized by neutrophil infiltration and edema, gives way to a predominance of mononuclear phagocytes and lymphocytes. This probably occurs to some degree with the normal healing process but becomes exaggerated and chronic when there is ineffective elimination of foreign materials as in certain infections (e.g. tuberculosis) or following introduction of foreign bodies (e.g. asbestos) or deposition of crystals (e.g. urate crystals).

Chronic inflammation is seen under conditions of delayed hypersensitivity. Prolonged inflammation, known as *chronic inflammation*, leads to a progressive shift in the type of cells which are present at the site of inflammation and is characterised by simultaneous destruction and healing of the tissue from the inflammatory process. Chronic inflammation is a pathological condition characterised by concurrent active inflammation, tissue destruction, and attempts at repair. Chronic inflammation is not characterised by the classic signs of acute inflammation listed above. Instead, chronically inflamed tissue is characterised by the infiltration of mononuclear immune cells (monocytes, macrophages, lymphocytes, and plasma cells), tissue destruction, and attempts at healing, which include angiogenesis and fibrosis.

Chronic inflammation results from persistence of antigen due to infection with certain resistant microorganism or various pathologic conditions. Two cytokines, INF-gamma and

TNF-alpha are the central to the chronic inflammatory response, which often involves granuloma formation and tissue wasting. Chronic inflammation is often associated with fusion of mononuclear cells to form multinucleated gigant cells, which eventually become *granuloma*.

Chronic inflammatory response may accompany allergies, autoimmune diseases, microbial infection, transplants, and burns. The HEV like regions have been observed in the number of chronic inflammatory diseases in human, including rheumatoid arthritis, crohn's diseases, graves diseases and diabetes mellitus.

Endogenous causes include persistent acute inflammation. Exogenous causes are varied and include bacterial infection, especially by *Mycobacterium tuberculosis*, prolonged exposure to chemical agents such as silica, tobacco smoke, or autoimmune reactions such as rheumatoid arthritis. In acute inflammation, removal of the stimulus halts the recruitment of monocytes (which become macrophages under appropriate activation) into the inflamed tissue, and existing macrophages exit the tissue via lymphatics. However in chronically inflamed tissue the stimulus is persistent, and therefore recruitment of monocytes is maintained, existing macrophages are tethered in place, and proliferation of macrophages is stimulated.

Chronic inflammation is also a frequent cause of cancer. Liver cancer is often the sequel to years of inflammation caused by infection by hepatitis B and C viruses. Lung cancer often is the end stage of years of chronic inflammation caused by inhaled irritants, of which tobacco smoke is the most reliable. Cervical cancer can follow chronic infection and inflammation caused by papilloma viruses, chlamydiae bladder, colon, pancreas, stomach, and other cancers may similarly be the final stage of years of inflammation. The strong link between chronic inflammation and cancer should not be surprising when you consider that the reactive oxygen species (ROS) liberated during inflammation are powerful. DNA-damaging agents; increased mitosis in response to inflammation puts more cells at risk of mutations as they replicate their DNA during S phase; apoptosis, the programmed death of damaged cells, is suppressed in inflamed tissue. So cells with precancerous genetic mutations, which should have committed suicide, live on grow into a full-blown cancer.

Inflammatory mediators

A complex variety of mediators is involved in acute phase responses. Some act directly on the smooth muscle wall surrounding the arterioles to alter blood flow, other act on the venules to cause contraction on the endothelial cells with transient opening of interendothelial junction and consequent transduction of plasma. The migration of leukocyte from the blood stream is facilitate by mediators which upregulate the expression of adherence molecule on the both endothelial and white cells and other which lead the leukocyte to inflamed site through chemotaxis.

Lipid mediators

The phospholipids in the membrane of several types e.g. macrophages, monocytes, neutrophils and mast cells are degraded into arachidonic acid and lyso-platelet activating factor. It is later converted into platelet activating factor (PAF). Leukotrienes and prostaglandins are the potent mediators of inflammation are derivatives of arachidonic acid (AA) a 20-carbon unsaturated fatty acid produced from membrane phospholipids. The principal pathways of arachidonic acid metabolism are the 5-lipoxygenase pathway, which produces a collection of leukotrienes (LT). There are four leukotrienes e.g. LTB4, LTC4, LTD4, and LTE4. Three of these together makeup what was formerly called slow reacting substance of anaphylaxis (SRS-A), these mediators induce smooth muscle contraction. LTB4 is potent chemoatractant of neutrophil. The cyclooxygenase (COX) pathway, which produces prostaglandin H_2 (PGH_2) by different cells e.g. monocyte and macrophages. PGH_2 serves as the substrate for two enzymatic pathways: one leading to the production of several prostaglandins (PG); the other leading to the production of thromboxanes (Tx). The prostaglandin have divers physiologic effects, including vascular permeability, increased vasodialation, induction of neutrophil chemotaxis. The thromboxane causes platelet aggregation.

Leukotrienes and prostaglandins

These potent mediators of inflammation are derivatives of **arachidonic acid** (AA) a 20-carbon unsaturated fatty acid produced from membrane phospholipids. The principal pathways of arachidonic acid metabolism are the **5-lipoxygenase** pathway, which produces a collection of **leukotrienes** (LT) and the **cyclooxygenase** (COX) pathway, which produces prostaglandin H_2 (**PGH_2**). PGH_2 serves as the substrate for two enzymatic pathways: one leading to the production of several **prostaglandins** (PG); the other leading to the production of **thromboxanes** (Tx) (Fig. 15.2).

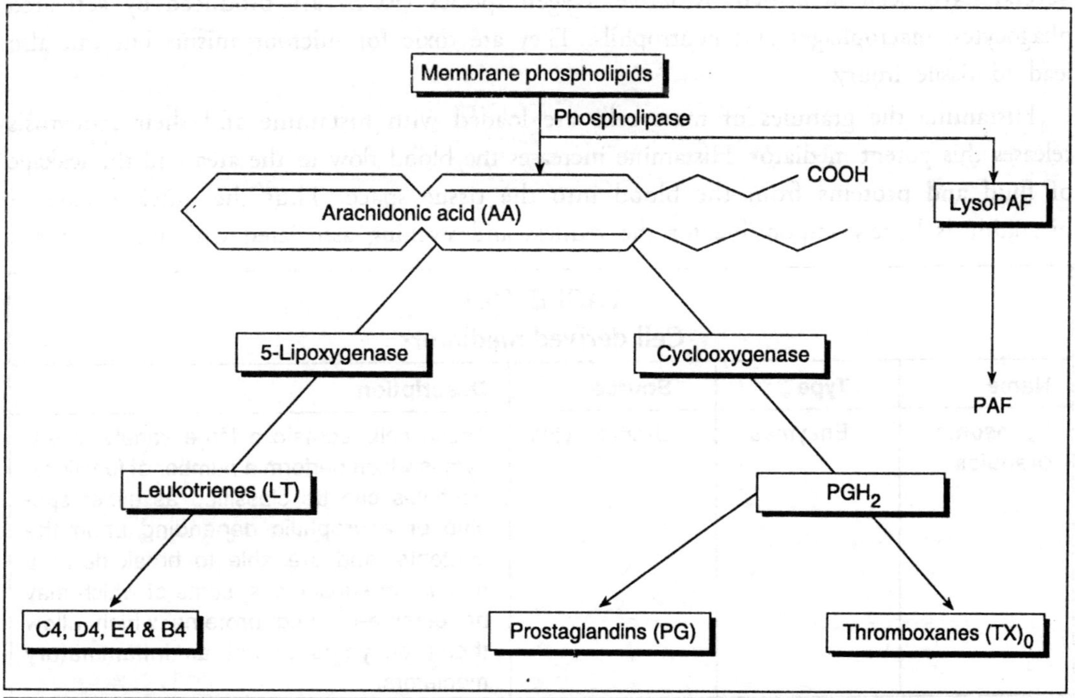

Fig. 15.2: The breakdown of membrane phospholipids generates important mediator of inflammation

Cell derived mediators

There are several cell derived mediator involved in inflammatory response are shown in table. Mast cells are found in the tissues. Their cytoplasm is loaded with granules containing mediators of inflammation. Their surface is coated with a variety of receptors which, when engaged by the appropriate ligand, trigger exocytosis of the granules. Mast cells appear to be key players in the initiation of inflammation. Their toll-like receptors trigger exocytosis when they interact with PAMPs like the **lipopolysaccharide** (LPS or "endotoxin") of Gram-negative bacteria (TLR-4) the peptidoglycan of Gram-positive bacteria (bind TLR-2). Their receptors for complement fragments trigger exocytosis when they bind C3a and C5a bacteria coated with C3b activated mast cells release literally dozens of potent mediators; some immediately as they discharge their granules some later as they synthesize them by new gene transcription. These mediators are active in either (or, in some cases, both) **recruiting** all the types of white blood cell to the site monocytes that become macrophages when they leave the blood and enter the tissue neutrophils antigen-presenting dendritic cells all kinds of lymphocytes: B-cells and T-cells, leading to an adaptive immune response; NK cells (an effector cell in innate immunity) eosinophils **activating** many of these recruited cells to produce their own mediators of inflammation.

Large amounts of TNF-α are quickly released by stimulated mast cells. All the cells involved in inflammation have receptors for TNF-α, and are activated by it to synthesize more on their own. This positive feedback quickly amplifies the response. Chemokines are **chemotactic cytokines**; that is, secreted proteins that attract other leukocytes into the area. Several have been identified. Reactive oxygen species (ROS) are produced by activated phagocytes: macrophages and neutrophils. They are toxic for microorganisms but can also lead to tissue injury.

Histamine the granules of mast cells are loaded with histamine and their exocytosis releases this potent mediator. Histamine increases the blood flow to the area and the leakage of fluid and proteins from the blood into the tissue space. Thus the quick release of histamine is largely responsible for the redness and swelling associated with inflammation.

TABLE 15.1
Cell derived mediators

Name	Type	Source	Description
Lysosome granules	Enzymes	Granulocytes	These cells contain a large variety of enzymes which perform a number of functions. Granules can be classified as either specific or azurophilic depending upon the contents, and are able to break down a number of substances, some of which may be plasma-derived proteins which allow these enzymes to act as inflammatory mediators.

Contd. to next page

Histamine	Vasoactive amine	Mast cells, basophils, platelets	Stored in preformed granules, histamine is released in response to a number of stimuli. It causes arteriole dilation and increased venous permeability.
IFN-γ	Cytokine	T-cells, NK cells	Antiviral, immunoregulatory, and anti-tumour properties. This interferon was originally called macrophage-activating factor, and is especially important in the maintenance of chronic inflammation.
IL-8	Chemokine	Primarily macrophages	Activation and chemoattraction of neutrophils, with a weak effect on monocytes and eosinophils.
Leukotriene B4	Eicosanoid	Leukocytes	Able to mediate leukocyte adhesion and activation, allowing them to bind to the endothelium and migrate across it. In neutrophils, it is also a potent chemoattractant, and is able to induce the formation of reactive oxygen species and the release of lysosome enzymes by these cells.
Nitric oxide	Soluble gas	Macrophages, endothelial cells, some neurons	Potent vasodilator, relaxes smooth muscle, reduces platelet aggregation, aids in leukocyte recruitment, direct antimicrobial activity in high concentrations.
Prostaglandins	Eicosanoid	Mast cells	A group of lipids which can cause vasodilation, fever, and pain.
TNF-α and IL-1	Cytokines	Primarily macrophages	Both affect a wide variety of cells to induce many similar inflammatory reactions: fever, production of cytokines, endothelial gene regulation, chemotaxis, leukocyte adherence, activation of fibroblasts. Responsible for the systemic effects of inflammation, such as loss of appetite and increased heart rate.

Plasma derived mediators

The plasma contains four interconnected mediator producing system. When tissue damage occurs the following system is activated to form web interacting system that generates number of mediators of inflammation.

The kinin system

The Coagulation system

The Fibrinolytic system

The Complement system

The kinin system

It is an enzymatic cascade, this system works when plasma clotting factor activated following tissue injury. Once plasma clotting factor activates then it activate the prekallikrein which cleaves kininogens to produce bradykinin. The bradykinin is the inflammatory mediator is a vasoactive protein which is able to induce vasodilation, increase vascular permeability, cause smooth muscle contraction, and induce pain. The kallikrein also act directly on the complement system which cleaves C5 into C5a and C5b. The C5a is the complement component which induces the mast cell degranulation resulting in the release of several inflammatory mediators from the mast cells.

The coagulation system

The coagulation system is also the sequential event of reaction, that triggered by damage of blood vessels yield the large quantity of thrombin. Thrombin cleaves the soluble plasma protein fibrinogen in the tissue fluids to produce insoluble fibrin, and fibrinopeptides. The insoluble fibrin strands crisscross one another or which aggregates to form a blood clot, which serves as barrier to spread infection. Thrombin can also bind to cells via the PAR1 receptor to trigger several other inflammatory responses, such as production of chemokines and nitric oxide. The fibrinopeptides also acts as inflammatory mediators which induces the neutrophil chemotaxis and increased vascular permeability.

TABLE 15.2
Plasma derived mediators

Name	Produced by	Description
Bradykinin	Kinin system	A vasoactive protein which is able to induce vasodilation, increase vascular permeability, cause smooth muscle contraction, and induce pain.
C3	Complement system	Cleaves to produce *C3a* and *C3b*. C3a stimulates histamine release by mast cells, thereby producing vasodilation. C3b is able to bind to bacterial cell walls and act as an opsonin, which marks the invader as a target for phagocytosis.
C5a	Complement	Stimulates histamine release by mast cells, thereby producing vasodilation. It is also able to act as a chemoattractant to direct cells via chemotaxis to the site of inflammation.
Factor XII (*Hageman Factor*)	Liver	A protein which circulates inactively, until activated by collagen, platelets, or exposed basement membranes via conformational change. When activated, it in turn is able to activate three plasma systems involved in inflammation: the kinin system, fibrinolysis system, and co agulation system.

Contd. to next page

Membrane attack complex	Complement system	A complex of the complement proteins C5b, C6, C7, C8, and multiple units of C9. The combination and activation of this range of complement proteins forms the *membrane attack complex*, which is able to insert into bacterial cell walls and causes cell lysis with ensuing death.
Plasmin	Fibrinolysis system	Able to break down fibrin clots, cleave complement protein C3, and activate Factor XII.
Thrombin	Coagulation system	Cleaves the soluble plasma protein fibrinogen to produce insoluble fibrin, which aggregates to form a blood clot. Thrombin can also bind to cells via the PAR1 receptor to trigger several other inflammatory responses, such as production of chemokines and nitric oxide.

The fibrinolytic system

Plasmin is the key enzyme used in this system, which is formed by conversion of plasminogen. Plasmin is a proteolytic enzyme able to break down fibrin clots, cleave complement protein C3, and activate Factor XII. Plasmin also contribute in activating the classical complement pathway. Removal of fibrin clot from injured tissue is achieved by the fibrinolytic system.

The complement system

Complement proteins are the important mediator of inflammation. The anaphylatoxins are binds to the receptor on the mast cells leads to degranulation of mast cell which releases the histamine and other pharmacologically active mediator. These mediator induce the smooth muscle contraction and increases the vascular permeability. It is also able to act as a chemoattractant to direct cells via chemotaxis to the site of inflammation.

A complex of the complement proteins C5b, C6, C7, C8, and multiple units of C9. The combination and activation of this range of complement proteins forms the *membrane attack complex*, which is able to insert into bacterial cell walls and causes cell lysis with ensuing death.

Treatment of inflammation

There are two main types of treatment of inflammation first is steroidal and second is nonsteroidal. Inappropriate inflammation can be treated with steroids like the glucocorticoid cortisol nonsteroidal antiinflammatory drugs (NSAIDs) like aspirin and ibuprofen (e.g. Motrin, Advil). A number of proteins produced by recombinant DNA technology also used for the treatment of inflammation.

Nonsteroidal anti-inflammatory drugs

The nonsteroidal anti-inflammatory drugs achieve their effects by blocking the activity of cyclooxygenase. In addition to reducing the fever and pain of inflammation, NSAIDs also inhibit clotting. They do this by interfering with the synthesis of thromboxane A_2 in platelets. This is the reason that aspirin is given to patients undergoing angioplasty; many men take a baby aspirin a day in the hope of avoiding heart attacks.

But regular use of NSAIDs has a downside: a tendency to develop ulcers in the stomach and duodenum. The body produces several different forms of cyclooxygenase (COX), including COX-1, which is involved in pain, clotting, and protecting the stomach; COX-2, which is involved in the pain produced by inflammation. Most of the NSAIDs inhibit them both. However, some newer drugs, the so-called **COX-2 inhibitors,** such as rofecoxib (Vioxx) celecoxib (Celebrex) are much more active against COX-2 than COX-1.COX-2 inhibitors are effective against inflammation and seem to avoid damage to the gastro intestinal tract. But, unfortunately, they increase the risk of blood clots—which can cause heart attacks and strokes—because they do not block the synthesis of thromboxane A_2 by platelets (which contain only COX-1). So people depending on NSAIDs for their heart protective effects must monitor any use of COX-2 inhibitors carefully.

Therapeutic proteins

Recombinant DNA and monoclonal antibody technology have produced some new therapies that are being enlisted in the battle against damaging inflammation. **Recombinant protein C is** help the body to dissolve the tiny clots that are triggered during inflammation. An IL-1 antagonist that binds and inactivates the IL-1 **receptor.** A soluble version of the TNF-α **receptor** binds TNF-α preventing it from carrying out its many inflammatory actions. **Infliximab** binds to tumour necrosis factor-alpha (TNF-α). Shows promise against some inflammatory diseases such as rheumatoid arthritis.

Leukocyte extravasation

Neutrophils migrate from blood vessels to the inflamed tissue via chemotaxis, where they remove pathogens through phagocytosis and degranulation. Various leukocytes are critically involved in the initiation and maintenance of inflammation. These cells must be able to get to the site of injury from their usual location in the blood, therefore mechanisms exist to recruit and direct leukocytes to the appropriate place. The process of leukocyte movement from the blood to the tissues through the blood vessels is known as *extravasation.* Neutrophils extravasate from blood vessels to the site of tissue injury or infection during the innate immune response.

Leukocyte extravasation is the movement of leukocytes out of the circulatory system, towards the site of tissue damage or infection. This process forms part of the innate immune response, involving the recruitment of non-specific leukocytes. Monocytes also use this process in the absence of infection or tissue damage during their development into macrophages.

Leukocyte extravasation occurs mainly in post-capillary venules, where haemodynamic shear forces are minimised. This process can be understood in several steps, outlined below as "chemoattraction", "rolling adhesion", "tight adhesion" and "(endothelial) transmigration" (Fig. 15.3). It has been demonstrated that leukocyte recruitment is halted whenever any of these steps is suppressed.

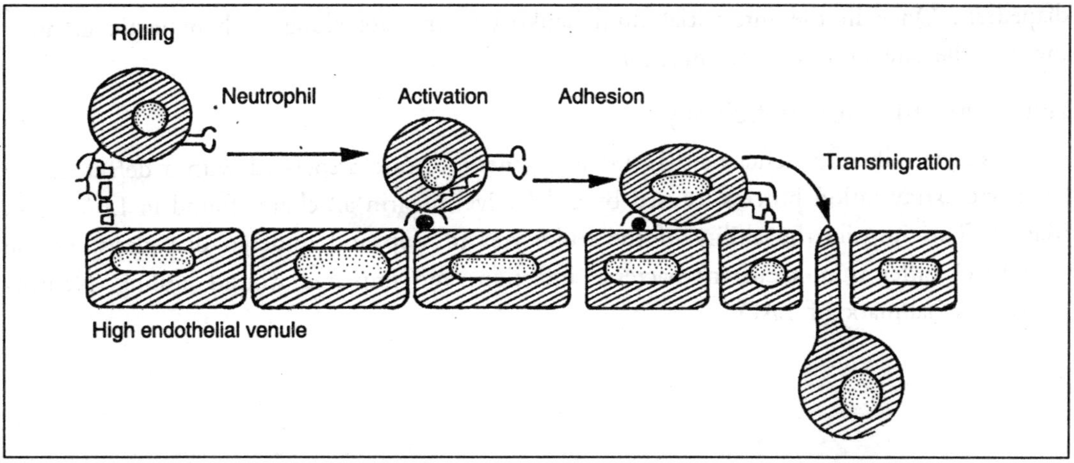

Fig. 15.3: Leukocyte extravasation

Chemoattraction

Upon recognition of and activation by pathogens, resident macrophages in the affected tissue release cytokines such as IL-1, TNF-α and chemokines. IL-1 and TNF-α cause the endothelial cells of blood vessels near the site of infection to express cellular adhesion molecules, including selectins. Circulating leukocytes are localised towards the site of injury or infection due to the presence of chemokines.

Rolling adhesion

Like velcro, selectin ligands on circulating leukocytes bind selectins on the inner wall of the vessel, with marginal affinity. This causes the leukocytes to slow down and begin rolling along the inner surface of the vessel wall. During this rolling motion, transitory bonds are formed and broken between selectins and their ligands.

Tight adhesion

At the same time, chemokines released by macrophages activate the rolling leukocytes and cause surface integrin molecules to switch from the default low-affinity state to a high-affinity state. This is assisted through juxtacrine activation of integrins by chemokines and soluble factors released by endothelial cells. In the activated state, integrins bind tightly to complementary receptors expressed on endothelial cells, with high affinity. This effects the immobilisation of the leukocytes, despite the shear forces of the ongoing blood flow.

Transmigration

The cytoskeletons of the leukocytes are reorganised in such a way that the leukocytes are spread out over the endothelial cells. In this form, leukocytes extend pseudopodia and pass through gaps between endothelial cells. Transmigration of the leukocyte occurs as PECAM proteins, found on the leukocyte and endothelial cell surfaces, interact and effectively pull the cell through the endothelium. The leukocytes secrete proteases that degrade the basement membrane, allowing them to escape the blood vessel–a process known as diapedesis. Once in the interstitial fluid, leukocytes migrate along a chemotactic gradient towards the site of injury or infection.

Leukocyte adhesion deficiency

Leukocyte adhesion deficiency (LAD) is a genetic disease associated with a defect in the leukocyte extravasation process, caused by a defective integrin â2 chain (found in LFA-1 and Mac-1). This impairs the ability of the leukocytes to stop and undergo diapedesis. People with LAD suffer from recurrent bacterial infections and impaired wound healing. Neutrophilia is a hallmark of LAD.

CHAPTER

16

INTERFERON

INTRODUCTION

Interferons (IFNs) are natural proteins produced by the cells of the immune system of most vertebrates in response to challenges by foreign agents such as viruses, bacteria, parasites and tumour cells. Interferons belong to the large class of glycoproteins known as cytokine. Interferons assist the immune response by inhibiting viral replication within other cells of the body.

While aiming to develop an improved vaccine for smallpox, two Japanese virologists, Yasu-ichi Nagano and Yasuhiko Kojima working at the Institute for Infectious Diseases at the University of Tokyo, noticed that rabbit-skin or testis previously inoculated with UV-inactivated virus exhibited inhibition of viral growth when re-infected at the same site with live virus. They hypothesised that this was due to some inhibitory factor, and began to characterize it by fractionation. The UV-irradiated viral homogenates using an ultracentrifuge. They published these findings in 1954 in the French journal now known as "Journal de la Societe". While this paper demonstrated that the activity could be separated from the virus particles, it could not reconcile the antiviral activity demonstrated in the rabbit-skin experiments, with the observation that the same supernatant led to the production of antiviral antibodies in mice. A further paper in 1958, involving triple-ultracentrifugation of the homogenate demonstrated that the inhibitory factor was distinct from the virus particles, leading to trace contamination being ascribed to the 1954 observations.

Meanwhile, the British virologist Alick Isaacs and the Swiss researcher Jean Lindenmann, at the National Institute for Medical Research in London, noticed an interference effect caused by heat-inactivated influenza virus on the growth of live influenza virus in chicken egg membranes in a nutritive solution chorioallantoic membrane. They published their results in 1957; in this paper they coined the term 'interferon', and today that specific interfering agent is known as a 'Type I interferon'.

Nagano's work was never fully appreciated in the scientific community; possibly because it was printed in French, but also because his *in vivo* system was perhaps too complex to provide clear resûlts in the characterisation and purification of interferon. As time passed, Nagano became aware that his work had not been widely recognised, yet did not actively seek revaluation of his status in field of interferon research. As such, the majority of the credit for discovery of the interferon goes to Isaacs and Lindenmann, with whom there is no record of Nagano ever having made personal contact.

Types of interferon

Several different types of interferon are now approved for use in humans. More recently, the FDA approved pegylated interferon-alpha, in which polyethylene glycol is added to make the interferon last longer in the body. (Pegylated interferon-alpha-2b was approved in January 2001; pegylated interferon-alpha-2a was approved in October 2002). The pegylated form is injected once weekly, rather than three times per week for conventional interferon-alpha. Used in combination with the antiviral drug ribavirin, pegylated interferon produces sustained cure rates of 75% or better in people with genotype 2 or 3 hepatitis C (which is easier to treat) but still less than 50% in people with genotype 1 (which is most common in the U.S. and Western Europe). Interferon-beta (Interferon beta-1a and Interferon beta-1b) is used in the treatment and control of multiple sclerosis. By an as-yet-unknown mechanism, interferon-beta inhibits the production of Th1 cytokines and the activation of monocytes.

There are three major classes of interferons that have been described for humans according to the type of receptor through which they signal:

- **Interferon type I:** All type I IFNs bind to a specific cell surface receptor complex known as the IFN-á receptor (IFNAR) that consists of IFNAR1 and IFNAR2 chains.
- **Interferon type II:** Binds to IFNGR.
- **Interferon type III:** Signal through a receptor complex consisting of IL10R2 (also called CRF2-4) and IFNLR1 (also called CRF2-12)

Functions of interferon

1. Interferons in general have several effects in common.
2. The JAK-STAT signalling pathway is the best-characterised and commonly accepted IFN signalling pathway.
3. They are antiviral and possess antioncogenic properties, macrophage and natural killer lymphocyte activation, and enhancement of major histocompatibility complex glycoprotein classes I and II, and thus presentation of foreign (microbial) peptides to T-cells.
4. In a majority of cases, the production of interferons is induced in response to microbes such as viruses and bacteria and their products (viral glycoproteins, viral RNA,

bacterial endotoxin, bacterial flagella, CpG sites), as well as mitogens and other cytokines, for example interleukin 1, interleukin 2, interleukin-12, tumour necrosis factor and colony-stimulating factor, that are synthesised in the response to the appearance of various antigens in the body.

5. Their metabolism and excretion take place mainly in the liver and kidneys.
6. They rarely pass the placenta but they can cross the blood-brain barrier.

Viral induction of interferons

All classes of interferon are very important in fighting RNA virus infections. However, their presence also accounts for some of the host symptoms, such as sore muscles and fever. They are secreted when abnormally large amounts of dsRNA are found in a cell. dsRNA is normally present in very low quantities. The dsRNA acts like a trigger for the production of interferon (via Toll Like Receptor 3 (TLR 3) a pattern recognition receptor of the innate immune system which leads to activation of the transcription factor IRF3 and late phase NF kappa Beta). The gene that codes for this cytokine is switched on in an infected cell, and the interferon synthesized and secreted to surrounding cells.

As the original cell dies from the cytolytic RNA virus, these thousands of viruses will infect nearby cells. However, these cells have received interferon, which essentially warns these other cells that there's a wolf in the pack of sheep. They then start producing large amounts of a protein known as protein kinase R (or PKR). If a virus infects a cell that has been "pre-warned" by interferon, it is like charging into a hail of bullets for the virus. The PKR is indirectly activated by the dsRNA (actually by 2'-5' oligoadenylate produced by the 2'-5' oligoadenylate-synthetase which is produced due to TLR3 activation), and begins transferring phosphate groups (phosphorylating) to a protein known as eIF-2, a eukaryotic translation initiation factor. After phosphorylation, eIF2 has a reduced ability to initiate translation, the production of proteins coded by cellular mRNA. This prevents viral replication and inhibits normal cell ribosome function, killing both the virus and the host cell if the response is active for a sufficient amount of time. All RNA within the cell is also degraded, preventing the mRNA from being translated by eIF2 if some of the eIF2 failed to be phosphorylated.

Furthermore, interferon leads to upregulation of MHC I and therefore to increased presentation of viral peptides to cytotoxic CD8 T-cells, as well as to a change in the proteasome (exchange of some beta subunits by b1i, b2i, b5i—then known as the immunoproteasome) which leads to increased production of MHC I compatible peptides.

Interferon can cause increased p53 activity in virus infected cells. It acts as an inducer and causes increased production of the p53 gene product. This promotes apoptosis, limiting the ability of the virus to spread. Increased levels of transcription are observed even in cells which are not infected, but only infected cells show increased apoptosis. This increased transcription may serve to prepare susceptible cells so they can respond quickly in the case of infection. When p53 is induced by viral presence, it behaves differently than it usually does. Some p53 target genes are expressed under viral load, but others, especially those that respond to DNA damage, aren't. One of the genes that is not activated is p21, which can promote cell survival. Leaving this gene inactive would help promote the apoptotic effect. Interferon enhances the apoptotic effects of p53, but it is not strictly required. Normal cells exhibit a stronger apoptotic response than cells without p53.

Additionally, interferon has been shown to have therapeutic effect against certain cancers. It is probable that one mechanism of this effect is p53 induction. This could be useful clinically: Interferons could supplement or replace chemotherapy drugs that activate p53 but also cause unwanted side effects[1]. Some of these side effects can be serious, severe and permanent.

Virus resistance to interferons

In a study of the blocking of interferon (IFN) by the Japanese Encephalitis Virus (JEV), a group of researchers infected human recombinant IFN-alpha with JEV, DEN-2, and PL406, which are all viruses, and found that some viruses have manifested methods that give them a way around the IFN-alpha/beta response. The viruses need to master these methods so they can have the ability to carry on viral replication and production of new viruses. The ways that viruses find a way around the IFN response is through the inhibition of interferon signalling, production, and the blocking of the functions of IFN-induced proteins.

It is not unusual to find viruses encoding for a multiple number of mechanisms to allow them to elude the IFN response at many different levels. While doing the study with JEV, Lin and his coworkers found that with IFN-alpha's inability to block JEV means that JEV may be able to block IFN-alpha signalling which in turn would prevent IFN from having STAT1, STAT2, ISGF3, and IRF-9 signalling. DEN-2 also significantly reduces interferon ability to active JAK-STAT. Some other viral gene products that have been found to have an effect on IFN signalling include EBNA-2, Polyomavirus large T antigen, EBV EBNA1, HPV E7, HCMV, and HHV8. "Several poxviruses encode a soluble IFN receptor homologue that acts as a decoy to inhibit the biological activity of IFN," and that activity is for IFN to "bind to their cognate receptors on the cell surface to initiate a signalling cascade, known as the janus kinase (JAK)-signal transducer and activation of transcription(Stat) pathways. For example, in a study done by a group of researcher, they found that the B18R protein, which acts as a type 1 IFN receptor and is produced by the vaccinia virus, it was found that the B18R protein inhibited IFN's ability to begin the phosphorylation of JAK1 which reduced the antiviral effect of IFN.

Some viruses can encode proteins that bind to dsRNA. In a study where the researchers infected human U cells with reovirus-sigma3 protein and then, using the western blot test, they found that reaovirus-sigma3 protein does bind to dsRNA. Along with that, another study in which the researchers infected mouse L cells with vaccinia virus E3L found that E3L encodes the p25 protein that binds to dsRNA. Without double stranded RNA (dsRNA), because it is bound to by the proteins, it is not able to create IFN-induced PKR and 2'-5' oligoadenylate-synthetase making IFN ineffective. It was also found that JEV was able to inhibit IFN-alpha's ability to activate or create ISGs such as PKR. PKR was not able to be found in the JEV infected cells and PKR RNA levels were found to be lower in those same infected cells, and this disruption of PKR can occur, for example, in cells infected with flavaviruses.

The H5N1 influenza virus, also known as bird flu, has been shown to have resistance to interferon and other anti-viral cytokines. This is part of the reason for its high mortality rates in humans. It is resistant due to a single amino acid mutation in non-structural protein 1 (NS1), the precise mechanism of how this confers immunity is unclear (reference is lethal H5N1 influenza viruses escape host anti-viral cytokine responses, Sang Heui Seo, Nature Med, 2002).

Route of administration

When used in the systemic therapy, IFN-α and IFN-α are mostly administered by an intramuscular injection. The injection of interferons in the muscle, in the vein, or under skin is generally well tolerated. Interferon alpha can also be induced with small imidazo-quinoline molecules by activation of TLR7 receptor. Aldara (Imiquimod) cream works with this mechanism to induce IFN alpha and IL12 and approved by FDA to treat Actinic Keratosis, Superficial Basal Cell Carcinoma, and External Genital Warts.

Adverse effects

All known adverse effects are usually reversible and disappear a few days after the therapy has been finished. The most frequent adverse effects are flu-like symptoms: increased body temperature, feeling ill, fatigue, headache, muscle pain, convulsion, dizziness, hair thinning, and depression. Erythema, pain and hardness on the spot of injection are also frequently observed. Interferon therapy causes immunosuppression and can result in some infections manifesting in unusual ways.

Application

Interferon is used in the treatment and control of the neurological disorder multiple sclerosis, an autoimmune disorder. As their natural function, interferons have antiviral, antiseptic and antioncogenic properties when administered as drugs. Interferon therapy is used (in combination with chemotherapy and radiation) as a treatment for many cancers. More than half of hepatitis C patients treated with interferon respond with better blood tests and better liver biopsies. There is some evidence that giving interferon immediately following infection can prevent hepatitis C; however, people infected by hepatitis C often do not display symptoms of HCV until months or years later. Administered intranasally in very low doses, interferon is extensively used in Eastern Europe and Russia as a method to prevent and treat viral respiratory diseases such as cold and flu. However, mechanisms of such action of interferon are not well understood; it is thought that doses must be larger by several orders of magnitude to have any effect on the virus. Consequently, most Western scientists are skeptical of any claims of good efficacy.

Interferons available in the market

There are several interferon commercial available in market for medicine purpose. Rebif, liquid form of interferon beta 1a, Avonex, lyophilized form of interferon beta 1a, cinnovex, generic/biosimilar form of interferon beta 1a (Avonex). Betaseron, interferon beta 1b, Roferon A. regular interferon-alpha2a, intron, regular interferon-alpha2b, PEGASYS, pegylated interferon alpha 2a, berlex, interferon beta 1b, pegintron, pegylated interferon alpha 2b. reiferon etard, pegylated interferon alpha 2a, immunotherapy, immunosuppression, immunosuppressive drugs, PEGASYS, Cinnovex, PEGLYLation.

CHAPTER

17

MAJOR HISTOCOMPATIBILITY COMPLEX (MHC)

INTRODUCTION

The Major Histocompatibility Complex (MHC) comprises a stretch of tightly linked cluster of genes that encodes protein associated with intracellular recognition and antigen presentation to T lymphocyte. Major histocompatibility complex (MHC), whose products play a role in intercellular recognition and indiscrimination between self and non-self. The MHC is a region of multiple loci that play major roles in determining whether transplanted tissue is accepted as self (histocompatible) or rejected as to reign (histoincompatible).

The MHC is a collection of genes arrayed within a long continuous stretch of DNA on chromosome 6 in humans and on chromosome 17 in mice, in human MHC referred as HLA complex and in mice as H-2 complex.

The MHC plays a central role in the development of both humoral and cell—mediated immune responses. There are 50,000-100,000 MHC molecules on a typical cell. Most 'normal' MHC molecules are occupied by self peptides. The requirements for binding to a particular allele are met by ~1/1000-1/10000 random peptides. The typical population of ~100,000 MHC class I molecules of a single allotype on a normal cell displays >1000 different peptides. Individual peptide-MHC complexes are represented in widely different amounts from 1-5000 molecules/cell (mean~100).

Gene products encoded in the major histocompatibility complex (MHC) were first identified as being important in rejection of transplanted tissues. Furthermore, genes in the MHC were found to be highly polymorphic (i.e. in the population there were many different allelic forms of the genes). Studies with inbred strains of mice showed that genes in the MHC were also involved in controlling both humoral and cell-mediated immune responses. For example, some strains of mice could respond to a particular antigen but other strains could not and these strains differed only in one or more of the genes in the MHC. Subsequent studies showed that there were two kinds of molecules encoded by the MHC–Class I molecules and class II molecules. Class I molecules were found on all nucleated cells whereas class II molecules were found only on antigen-presenting cells, (APCs) which included dendritic cells, macrophages, B-cells and a few other types. It was not until the discovery of how the T-cell receptor (TCR) recognizes antigen that the role of MHC genes in immune responses was understood. The TCR was shown to recognize antigenic peptides in association with MHC molecules. T-cells recognize portions of protein antigens that are bound non-covalently to MHC gene products. Cytotoxic T-cells (Tc) recognize peptides bound to class I MHC molecules and helper T-cells (Th) recognize peptides bound to class II MHC molecules. The three dimensional structures of MHC molecules and the TCR have been determined by X-ray crystallography so that a clear picture of how the TCR, MHC gene products and antigen interact has emerged.

Histocompatibility (transplantation) antigens: Antigens, on tissues and cells, which determine their rejection when grafted between two genetically different individuals; **Major histocompatibility (MHC) antigens:** Histocompatibility antigens which cause a very strong immune response and are most important in rejection; **MHC complex:** Group of genes on a single chromosome encoding the MHC antigens; **HLA (human leukocyte antigens):** MHC antigens of man (first detected on leukocytes, H-2 **antigens:** MHC antigens of mouse; **Xenograft:** Grafts between members of different species (also known as heterologous or xenogeneic grafts or hetero-grafts); **Allograft:** Grafts between two members of the same species (also known as allogeneic grafts or homo-grafts); **Isograft:** Grafts between members of the same species with identical genetic makeup (identical twins or inbred animals); **Haplotype:** a group of genes on a single chromosome.

MHC ANTIGENS

The **human leukocyte antigen** system (HLA) is the name of the human major histocompatibility complex. This group of genes resides on chromosome 6, and encodes cell-surface antigen-presenting proteins and many other genes. The major HLA antigens are essential elements in immune function:

- Different classes have different functions
  - **class I** antigens (A, B & C)—Present peptides from inside the cell (including viral peptides if present)
  - **class II** antigens (DR, DP, & DQ)—Present phagocytosed antigens from outside of the cell to T-lymphocytes

HLAs also have a role in disease defense, reproduction (may be involved in mate selection), cancer (may be protective or fail to protect), human disease, in autoimmunity-known to mediate many autoimmune diseases. HLAs as antigens responsible for organ transplant rejection. Aside from the genes encoding the 6 major antigens, there are a large number of other genes, many involved in immune function located on the HLA complex. Diversity of HLA in human population is one aspect of disease defense, and, as a result, the chance of two unrelated individuals having identical HLA molecules on all loci is very low.

HLA functions

The proteins encoded by HLAs are the proteins on the outer part of body cells that are (effectively) unique to that person. The immune system uses the HLAs to differentiate self cells and non-self cells. Any cell displaying that person's HLA type belongs to that person (and therefore is not an invader).

In infectious disease

When a foreign pathogen enters the body, specific cells called antigen-presenting cells (APCs) engulf the pathogen through a process called phagocytosis. Proteins from the pathogen are digested into small pieces (peptides) and loaded onto HLA antigens (specifically MHC class II). They are then displayed by the antigen-presenting cells for certain cells of the immune system called T-cells, which then produce a variety of effects to eliminate the pathogen.

Through a similar process, proteins (both native and foreign, such as the proteins of viruses) produced inside most cells are displayed on HLA antigens (specifically MHC class I) on the cell surface. Infected cells can be recognized and destroyed by components of the immune system (specifically CD8+ T-cells).

In graft rejection

Any cell displaying some other HLA type is "non-self" and is an invader, resulting in the rejection of the tissue bearing those cells. Because of the importance of HLA in transplantation, the HLA loci are among of the most frequently typed by serology or PCR relative to any other autosomal alleles.

In autoimmunity

HLA types are inherited, and some of them are connected with autoimmune disorders and other diseases. People with certain HLA antigens are more likely to develop certain autoimmune diseases, such as Type I diabetes, ankylosing spondylitis, celiac disease, SLE (Systemic Lupus Erythematosus), myasthenia gravis, inclusion body myositis and Sjogren's syndrome. HLA typing has led to some improvement and acceleration in the diagnosis of celiac disease and Type 1 diabetes; however for DQ2 typing to be useful it requires either high resolution B1*typing (resolving *0201 from *0202), DQA1*typing, or DR serotyping. Current serotyping can resolve, in one step, DQ8. HLA typing in autoimmunity is being increasingly used as a tool in diagnosis. In celiac disease it is the only effective means of discriminating between 1st degree relatives who are at risk from those who are not at risk, prior to the appearance of sometimes irreversible symptoms such an allergies and secondary autoimmune disease.

In cancer

Some HLA mediated diseases are directly involved in the promotion of cancer. Gluten sensitive enteropathy is associated with increased prevalence of enteropathy-associated T-cell lymphoma, and DR3-DQ2 homozygotes are within the highest risk group with close to 80% of gluten sensitive EATL cases. More often; however, HLA molecules play a protective role, recognizing the increase in antigens that were not tolerated because of low levels in the normal state. Abnormal cells may be targeted for apoptosis mediating many cancers before clinical diagnosis. Prevention of cancer may be a portion of heterozygous selection acting on HLA.

HLA specificities are identified by a letter for locus and a number (A1, B5, *etc.*), and the haplotypes are identified by individual specificities (e.g., A1, B7, Cw4, DP5, DQ10 DR8). Specificities which are defined by genomic analysis (PCR), are named with a letter for the locus and a four digit number (e.g. A0101, B0701, C0401, etc.). Specificities of Mouse MHC (H-2) are identified by a number. Since laboratory mice are inbred, each strain is homozygous and has a unique haplotype. The MHC haplotype in these strains is designated by a >small= letter (a, b, d, k, q, s, etc.). For example, the MHC haplotype of Balb/c, an inbred strain of mouse, is H2d.

MHC HAPLOTYPES

MHC genes are inherited as a group (**haplotype**), one from each parent. Thus, a heterozygous human inherits one paternal and one maternal haplotype, each containing three class-I (B, C and A) and three class II (DP, DQ and DR) loci. A heterozygous individual will inherit a maximum of 6 class I specificities similarly, the individual will also inherit DP and DQ genes and express both parental antigens. Since the class II MHC molecule consists of two chains (α and β), with some antigenic determinants (specificities) on each chain, and DR α- and β-chains can associate in *cis* (both from the same parent) or *trans* (one from each parent) combination, an individual can have additional DR specificities. Also, there are more than one functional DR α-chain genes. Hence, many DR specificities can be found in any one individual.

Haplotypes, normally, are inherited intact and hence antigens encoded by different loci are inherited together (e.g. A2; B27; Cw2; DPw6; DQw9; DRw2). However, on occasions, there is crossing over between two parental chromosomes, thereby resulting in new **recombinant** haplotypes. Thus any specificity encoded by one locus, may combine with specificities from other loci. This results in a vast heterogeneity in the MHC make-up in a given population.

MHC antigens are expressed on the cell surface in a **co-dominant** manner: products of both parental genes are found on the same cells. However, not all cells express both class I and class II antigens. While class I antigens are expressed on all nucleated cells and platelets (and red blood cells in the mouse), the expression of class II antigens is more selective. It is expressed on B lymphocytes, a proportion of macrophages and monocytes, skin associated (Langerhans) cells, dendritic cells and occasionally on other cells.

The MHC class I antigens are detected by serological assays (Ab and C). Tissue typing sera for the HLA were obtained, in the past, from **multiparous** women who were exposed to the child=s paternal antigens during the parturition and subsequently developed antibodies to these antigens. More recently they are being produced by the monoclonal antibody technology. With most laboratories switching to PCR for tissue typing, the use of serology is rapidly diminishing.

MIXED LEUKOCYTE REACTION (MLR)

Detection of MHC can be takes place by mixed leukocyte reaction. It has been observed that lymphocytes from one donor, when cultured with lymphocytes from an unrelated donor, are stimulated to proliferate. It has been established that this proliferation is due to a disparity in the MHC class II (DR) antigens and T-cells of one individual interact with allogeneic MHC class-II antigen bearing B-cells, dendritic cells, Langerhans cells, etc. This reactivity was termed as **mixed leukocyte reaction** (MLR) and has been used for typing some of MHC class II antigens. The test lymphocytes were mixed with irradiated or mitomycin-C treated homozygous leukocytes, containing B-lymphocytes and monocytes (stimulator cells). In culture (over 4-6 days), T-cells responder cells) recognize the foreign class II antigen and undergo transformation (DNA synthesis and enlargement: blastogenesis) and proliferation (mitogenesis). These changes were recorded by the addition of radioactive (tritiated, 3H) thymidine into the culture and monitoring its incorporation into DNA. Most modern laboratories, however, are switching to the PCR technology for tissue typing using specific probes for MHC specificities.

INDUCTION CYTOTOXIC T LYMPHOCYTES

Another consequence of the MHC antigen and T-cell interaction is the induction of cytotoxic T-lymphocytes. When T-lymphocytes are cultured in the presence of allogeneic lymphocytes, in addition to undergoing mitosis (MLR), they also become cytotoxic to cells of the type that stimulated MLR. Thus, T-lymphocytes of 'x' haplotype cultured over 5-7 days with B lymphocytes of 'y' haplotype will undergo mitosis and the surviving T-lymphocytes become cytotoxic to cells of the 'y' haplotype. The induction of mitosis in MLR requires disparity of only class II antigens whereas the induction of cytotoxic T-lymphocytes (CTL) requires disparity of both class I and class II antigens. However, once cytotoxic cells have been induced, the effector cytotoxic cells recognize only class I antigens to cause cytotoxicity.

GVH REACTION

Common manifestations of GVH reaction are diarrhoea, erythema, weight loss, malaise, fever, joint pains, etc. and ultimately death. Histoincompatible lymphoid cells when injected into an immunosuppressed host are readily accepted. However, the immunocompetent T-lymphocytes among the grafted cells recognize the alloantigens and in response, they proliferate and progressively cause damage to the host tissues and cells. This condition is known as **graft-versus-host** (GVH) disease and is often fatal.

An immunocompetent host recognizes the foreign antigens on grafted tissues (cells), and mounts an immune response, which results in rejection (host vs. graft reaction). On the other hand if an immunocompromised host is grafted with foreign immunocompetent lymphoid cells, the immunoreactive T-cells in the graft recognize the foreign antigens on the host tissue and cause their damage (graft vs. host reaction).

The duration of graft survival follows the order, xeno- < allo- < iso- = autograft. The time of rejection also depends on the antigenic disparity between the donors and recipient. While the MHC antigens are the major contributors in rejection, the minor histocompatibility antigens also play a significant role. Rejection due to disparity in several minor histocompatibility antigens may be as quick or quicker than rejection mediated by an MHC antigen. Like in other immune responses, there is immunological memory and secondary response in graft rejection. Thus, once a graft is rejected by a recipient, a second graft from the same donor, or a donor with the same histocompatibility antigens, will be rejected in a much shorter time.

THE MHC GENE COMPLEX

The MHC complex contains a number of genes, which control several antigens, most of which influence allograft rejection. These antigens and their genes can be divided into three major classes: class I, class II and class III. The class I and class II antigens are expressed on cells and tissues whereas as class III antigens are associated with proteins in serum and other body-fluids (e.g. C4, C2, factor B, TNF). Antigens of class III gene products have no role in immune responses that determine graft survival. The human MHC is located on chromosome 6. It consists of two major loci, **K** and **D**. There are three classes of MHC molecules.

MHC- Class I, MHC- Class II, MHC- Class III molecules.

MHC- Class I molecule

Class I MHC molecules are composed of two polypeptide chains, along á chain and a short â chain called α2-microglobulin. The α chain has four regions. First, a cytoplasmic region, containing sites for phosphorylation and binding to cytoskeletal elements. Second, a transmembrane region containing hydrophic amino acids by which the molecule is anchored in the cell membrane. Third, a highly conserved α3 immunoglubilin-like domain to which CD8 binds. Fourth, a highly polymorphic peptide binding region formed from the α1 and α2 domains.

The α2-microglobulin associates with the α chain and helps maintain the proper conformation of the molecule.

The class I gene complex contains three major loci of highest significance, **B**, **C** and **A** and some undefined loci of less significance. Each these loci codes for a polypeptide, α-chain that contains antigenic determinants that are **polymorphic**. MHC-Class I molecules contain a large chain associated noncovalently with the much small β-2-microglobulin molecule (Figure 17.1). Each α-chain associates with a β-2 microglobulin molecule (β-chain), encoded by a gene outside the MHC complex. The alpha chain is a polymorph transmembrane glycoprotein of about 45 Kda. α-2-microglobulin is of about 12Kda. Association of the chain with β2 microglobulin is required for expression of class I molecules on cell membrane, the chain is anchored in plasma membrane by its hydrophobic transmembrane segment and hydrophilic cytoplasmic tail.

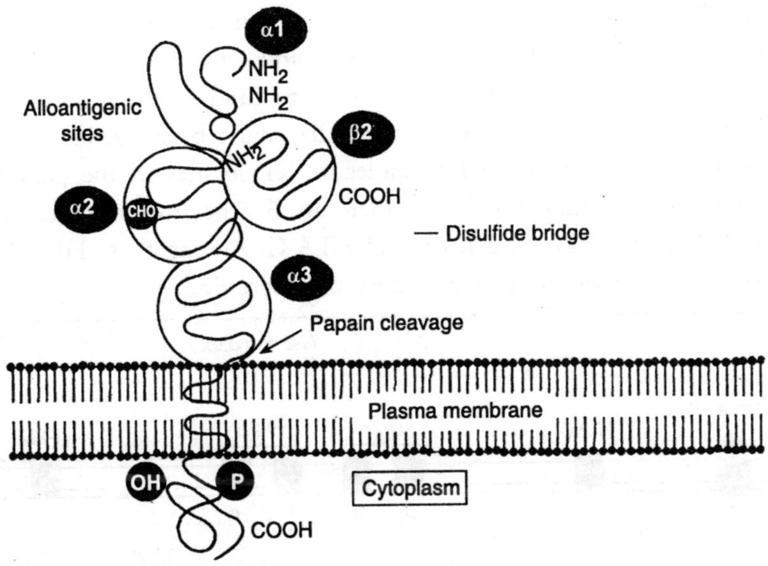

Fig. 17.1: Structure MHC class I molecule

The α-β-chain complex is expressed on the cell surface as the MHC class-I antigen. Without a functional β-2 microglobulin chain, the class I antigen will not be expressed on the cells surface. Individuals with defective a β-2 microglobulin gene do not express any class I antigen and hence they have a deficiency of cytotoxic T-cells.

The MHC class I molecules is organizes into three external domains as each containing 90 amino acid, of which transmembrane domain of about 25 hydrophobic amino acid, then short stretch of charged hydrophilic amino acids and last cytoplasmic anchor segment of 30 amino acids. In size and organization, β2-microglobulin in similar to the external domain, it does not contain a transmembrane region and is non covalently bound to the class I' glycoprotein.

The alpha 3 domain and β2 microglobulin are organized into two β-pleated sheets each formed by antiparallel β strand of amino acids. β2 microglobulin interacts extensively with alpha 3 domain and also alpha 2 and alpha 1 domain. The interaction of β2-microglobulin and a peptide with a class I chain is essential for class I molecule to reach its fully folded conformation. Interaction between alpha chain and β2 microglobulin form dimer is then stabilized by the binding of an appropriate peptide to form the native trimeric class I structure consisting of the class I alpha chain, β2-microglobulin and a peptide. This complete molecular complex is ultimately transported to the cell surface.

The structure of the peptide binding groove, revealed by X-ray crystallography, shows that the groove is composed of two á helices forming a wall on each side and eight β-pleated sheets forming a floor. The peptide is bound in the groove and the residues that line the groove make contact with the peptide. These are the residues that are the most polymorphic. The groove will accommodate peptides of approximately 8-10 amino acids long. Whether a particular peptide will bind to the groove will depend on the amino acids that line the groove. Because class I molecules are polymorphic, different class I molecules will bind different peptides. Each class I molecule will bind only certain peptides and will have a set of criteria that a peptide must have in order to bind to the groove. Thus, for every class I molecule, there are certain amino acids that must be a particular location in the peptide before it will bind to the MHC molecule. These sites in the peptide are referred to as the "anchor sites". Within the MHC there are 6 genes that encode class I molecules HLA-A, HLA-B, HLA-C, HLA-E, HLA-F and HLA-G. Among these HLA-A, HLA-B, and HLA-C are the most important and are most polymorphic.

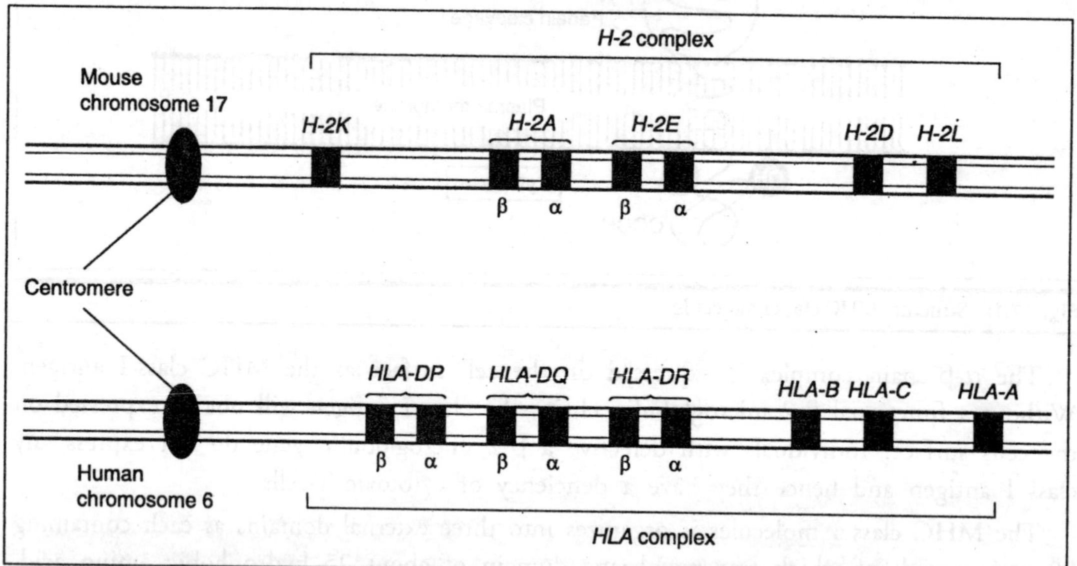

Fig. 17.2: The H-2 and HLA gene complexes

STRUCTURE OF CLASS II MOLECULES

Class II MHC molecules are composed of two polypeptide chains an α and a β chain of approximately equal length. Both chains have four regions: first, a cytoplasmic region containing site phosphorylation and binding to cytoskeletal elements; second, a transmembrane region containing hydrophilic amino acids by which the molecule is anchored in the cell membrane, third, a highly conserved α2 domain and a highly conserved β2 domain to which CD4 binds and fourth, a highly polymorphic peptide binding region formed from the α1 and β1 as with class I MHC molecules, an analysis of which part of the class II MHC molecule is most variable demonstrates that variability is most pronounced in the α1 and β1 domains, which comprise the peptide binding region (Figure 17.3). The structure of the peptide binding groove; revealed by X-ray crystallography, shows that, like class I MHC molecules, the groove is composed of two α helices forming a wall on each side an eight β-pleated sheets forming a floor. Both the α1 and β1 chain contribute to the peptide binding groove. The peptide is bound in the groove and the residues that line the groove make contact with the peptide. These are the residues that are the most polymorphic. The groove of class II molecules is open at one end so that the groove can accommodate longer peptides of approximately 13-25 amino acid long with some of the amino acids located outside of the groove. Whether a particular peptide bind to the groove will depend on the amino acids that line the groove. Because class II molecule are polymorphic, different class II molecules will bind different peptides. Like class I molecules, each class II molecule will bind only certain peptides and will have a set of criteria that a peptide must have in order to bind to the groove within the MHC there are 5 loci that encode class II molecules, each of which contain gene for an α chain and at least one gene for a β chain.

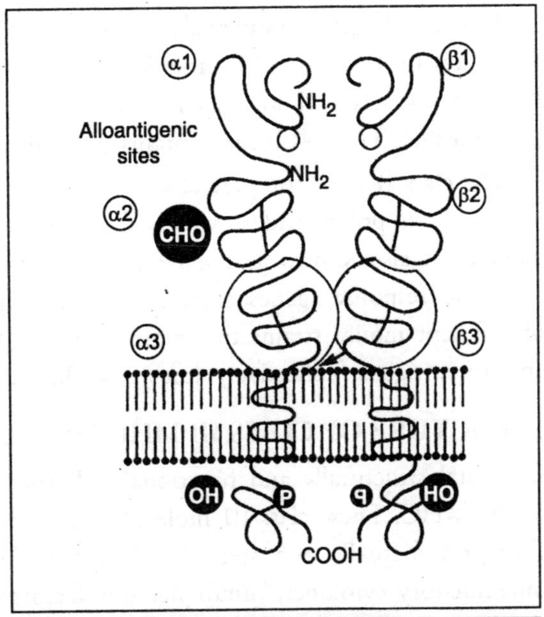

Fig. 17.3: Structure of MHC class II molecule

HLA-DP, HLA-DQ, HLA-DR, HLA-DM, and HLA-DO among these, HLA-DP, HLA-DQ, and HLA-DR are the most important and are most polymorphic.

The class II MHC molecules contain two different polypeptide chains a 33-kDa alpha chain and a 28 kDa â chain, which associate by noncovalent interactions. Like class I MHC molecule, class II MHC molecules are membrane bound glycoproteins that contain external domains, a transmembrane segment and a cytoplasmic anchor segment. Each chain in a class II molecule contains two external domains; alpha 1 and alpha 2 domains in one chain and β1 and β2 domains in the other. The membrane proximal alpha 2 and β2 domains, like the membrane proximal alpha 3 and β2 microglobulin domains of class I MHC molecules bear sequence homology to the immunoglobulin fold structure; for this reason, class II MHC molecules also are classified in the immunoglobulin super family. The membrane-distal portion of a class II MHC molecule is composed of the alpha 1 and β1 domains and forms the antigen-binding cleft for processed antigen.

Class II MHC molecules for human occurred as a dimmer of the alpha beta heterodimer. The dimmer is oriented so that the two peptide-binding clefts face in opposite directions. A recently reported structure of a human class II molecule crystallized in the presence of the super antigen *Staphylococcus aureus* enterotoxin B occurred as a single alpha beta chain pair. Super antigens bind to class II molecules in regions outside of the cleft. This complex included a peptide in the binding site of the class II molecule, suggesting that the alpha beta heterodimer is functional.

The class II gene complex also contains at least three loci, DP, DQ and DR; each of these loci codes for one α- and a variable number of β-chain polypeptides which associate together to form the class II antigens. Like the class I antigens, the class II antigens are also polymorphic. The DR locus contains more than one, possibly four, functional β-chain genes.

The class II gene complex contains two loci, A and E each of which code for one α- and one β-chain polypeptides, which form one class II molecule. The mouse class II gene complex is also known as the **I-region** and the genes in this complex are referred to as **Ir** (immune response) genes since they determine the magnitude of immune responsiveness of different mouse strains to certain antigens. Products of A and E loci are also termed IA and IE antigens, collectively known as Ia antigens.

STRUCTURE OF CLASS III MHC MOLECULES

In several structurally and functionally diverse proteins are encoded within the third region of the MHC. These class III molecules include several complement components as C2, C4a, C4b and foctorB, two steroid, 21 hydroxylase enzyme as 21-OHA and 21-OHB, the inflammatory cytokines, tumor necrosis factors α- and β (TNF-α & TNF-β), and two heat shock proteins. Unlike class I and class II MHC molecules, the class III molecules are not related structurally to class I and class II molecules, and have no role in antigen presentation although most play some role in immune responses. It is not known why the class III genes are situated within the MHC complex. The genetic association of certain MHC alleles with various diseases may in some cases reflect regulatory disorders of genes within the class III region.

MHC Loci

In man and mouse, as in most species, each class of MHC is represented by more than one locus (polygeny), in man these are called HLA for human leucocyte antigen. The class I loci are HLA-A, -B and -C and the class II loci HLA-DR, -DQ and -DP. All the MHC genes map within a single region of the chromosome (hence the term *Complex*); the physical map of the MHC of man and mouse has been determined by cloning, mapping and sequencing.

All MHC loci are expressed **co-dominantly**, that is to say both the set of alleles inherited from one's father and the set inherited from one's mother are expressed on each cell. In about 97% of cases the entire linked MHC complex is inherited intact, i.e. without recombination; the set of linked MHC alleles found on the same chromosome is called a **haplotype.**

ALLELIC POLYMORPHISM

The MHC is highly **polymorphic**. There are a large number of genetic variants (**alleles**) at each genetic **locus.** Crucially many of these **alleles** are represented at significant frequency (> 1%) in the population, and in addition the alleles generally differ from one another by many (up to 30) amino acid substitutions. For example in humans there are more than 200 **alleles** described at some MHC loci. Such a remarkable degree of polymorphism implies a selective pressure to establish and maintain it.

FUNCTION OF MHC

The products of the MHC play a fundamental role in regulating immune responses. T-cells must recognize antigen as a complex with MHC molecules. This requires antigen to be processed by unfolding and proteolytic digestion before it complexes with the MHC molecule. Once formed the complex of antigenic peptide and MHC are generally very stable (half life ~24hrs). Thus the biological role of MHC proteins is to bind small peptides and to "present" these at the cell surface for the inspection of T-cell antigen receptors. The allelic variation of MHC molecules is functionally reflected in the selection of peptides which can bind, each allelic product has a unique set of peptides which it can bind with high affinity.

In a normal animal or cell the majority of MHC molecules will be complexed with self peptides, "empty" MHC molecules are less stable especially in the case of class I products.

EXPRESSION OF MHC MOLECULES

MHC class I molecules are widely expressed, though the level varies between different cell types. MHC class II molecules are constitutively expressed only by certain cells involved in immune responses, though they can be induced on a wider variety of cells.

TABLE 17.1
Expression of MHC in different cell type

Cell Type	MHC I	MHC II
T-cells	+++	Varies, inducible in some species
B-cells	+++	++
Macrophages	+++	+
Dendritic cells	+++ x10	+++ x10
Granulocytes	++	–
Endothelium	++	– (inducible)
Hepatocytes	+	–
Neurons	–	–

MHC RESTRICTION

The term **MHC restriction** refers to the phenomenon whereby T-cells from one individual recognizing antigen fail to recognise cells presenting the same antigen unless the presenting cells express one or more MHC alleles identical to those on the **cortical thymic epithelial cells** of the individual in which those T-cells matured (these alleles will be those expressed by the T-cell in a normal animal). This phenomenon derives in part from the requirement to recognise self MHC and in part from the different peptide binding specificity of different MHC alleles.

By the early 1970's, immunologists had learned to generate cytotoxic T lymphocytes (CTLs) specific for virus-infected target cells. For example, when mice were infected with lymphocyte choriomeningitis (LCM) virus, they would produce CTLs that could lyses LCM—infected target cells in vitro. Yet these some CTLs failed to bind free LCM virus or viral antigens. This finding suggested that antigen recognition by T-cells was fundamentally different from that B-cells.

The CTLs didn't bind the virus or viral antigens directly as immunoglobulin, the answer began to emerge in the experiments of RM Zinkernagel and PC Doherty in 1974. These studies demonstrated that antigen recognition by T-cells is specific not only for viral antigen but also for MHC molecule. T-cells were shown to recognize antigen only when presented on the membrane of a cell by a self-MHC molecule called as **self-MHC restriction,** distinguishes recognition of antigen by T-cells from that by B-cells. In 1996, Doherty and Zinkernagel were awarded the Nobel Prize for this work.

Two models were proposed to explain the MHC restriction of the T-cell receptor. The dual-receptor model envisioned a T-cell with two separate receptors, one for antigen and one for class I or class II MHC molecules. The altered self model proposed that there was a single receptor capable of recognizing foreign antigen bound to a self-MHC molecule. Unlike the dual receptor model, in which an antigen and MHC molecule are recognized separately, the altered-self model predicts that a single receptor recognizes and alteration in self-MHC molecules induced by their association with foreign antigens.

CHAPTER 18

HYBRIDOMA TECHNOLOGY

INTRODUCTION

A hybridoma is a hybrid cell obtained by fusing a B-lymphocyte with usually a tumour cell of antibody forming system. The hybrid cells thus produced possess the ability to produce antibodies due to the B-lymphocytes genome and the capacity for indefinite growth in vitro due to the myeloma cell involved in the fusion. Therefore, hybridomas are either cultured in vitro or passed through mouse peritoneal cavity to obtain monoclonal antibodies this is called **hybridoma technology.**

Antibodies are produced by B-lymphocytes, each B-lymphocyte cell being programmed to respond to a single antigenic determinant. Antigenic determinant denotes that region of an antigen molecule, which reacts with an antibody that is specific to it when an antigen reacts to the cell surface receptor of a B-lymphocyte; it proliferates rapidly to yield a population of B-cells all of which produce antibodies of the same specificity. This is called "clonal selection". Thus B-lymphocytes cell produce antibodies of only one specificity i.e. specific to only one antigenic determinant. In addition an antibody producing B-lymphocyte cell, called plasma cell, is fully differentiated and does not divide when cultured in vitro. These features are critical to hybridoma technology.

B-lymphocytes are isolated from the spleen of an animal e.g. mouse, which had been immunized with the antigen against which monoclonal antibodies are to be raised immunization is achieved by injecting the antigen along with a suitable adjuvant either subcutaneously or in peritoneal cavity followed by booster doses of the antigen. Immunization enhances the population of B-lymphocytes producing antibodies specific to the antigen used, which greatly increases the chances of obtaining the desired hybridoma clone. A large no of these B-lymphocytes are mixed with the cells of selected myeloma and induced to fuse to from hybrid cells.

The idea of a "magic bullet" was first proposed by Paul Ehrlich who at the beginning of the 20th century postulated that if a compound could be made that selectively targeted a disease-causing organism, then a toxin for that organism could be delivered along with the agent of selectivity. In the 1970s the B-cell cancer myeloma was known, and it was understood that these cancerous B-cells all produce a single type of antibody (a paraprotein). This was used to study the structure of antibodies, but it was not yet possible to produce identical antibodies specific to a given antigen.

A process of producing monoclonal antibodies involving human-mouse hybrid cells was described by Jerrold Schwaber in 1973 and remains widely cited among those using human-derived hybridomas. But claims to priority have been controversial. A science history paper on the subject gave some credit to Schwaber for inventing a technique that was widely cited, but stopped short of suggesting that he had been cheated. The invention is generally accredited to Georges Köhler, César Milstein, and Niels Kaj Jerne in 1975; who shared the Nobel Prize in Physiology or Medicine in 1984 for the discovery. The key idea was to use a line of myeloma cells that had lost their ability to secrete antibodies, come up with a technique to fuse these cells with healthy antibody producing B-cells, and be able to select for the successfully fused cells.

In 1988 Greg Winter and his team pioneered the techniques to humanize monoclonal antibodies, removing the reactions that many monoclonal antibodies caused in some patients.

- A tumour cell capable of growing in culture infinitely is fused with an antibody producing cell (lymphocyte) to produce a hybrid cell capable of continuous growth and producing antibodies.
- The antibodies (immunoglobulins) are excreted into the blood serum. They crowd around foreign molecules (antigens) and squeeze them to death.
- The structure of antibody was determined by Oedney Porter at Oxford university in 1959 and since then the structure has become better known. Antibodies are Y shaped. The two upper arms correspond to the antigen binding sites. The specificity of the antibody results from a part of the molecule name as the variable region. The immune system produces several billion different proteins by rearranging a few hundred genetic segment.

Monoclonal antibodies are the unique proteins that recognize and attach to specific foreign substances in the body. Monoclonal antibodies are used in diagnosis of certain diseases and, treatment of cancer. Monoclonal antibodies (mAb or moAb) are antibodies that are identical because they are produced by one type of immune cell that are all clones of a single parent cell. Given (almost) any substance, it is possible to create monoclonal antibodies that specifically bind to that substance; they can then serve to detect or purify that substance. This has become an important tool in biochemistry, molecular biology and medicine. When used as medications, the generic name ends in *-mab* ("Nomenclature of monoclonal antibodies").

Production of monoclonals

Steps involved in the production of monoclonals

- Immunization of a mouse.
- Spleen removal from the mouse.
- Fusion of spleen cells with myeloma cells.
- Spleen cells are mortal but produce antibodies while myeloma cells are immortal but have no ability to produce antibodies.
- The resulting hybrid cell will have both of the above properties.
- Screen the cell population for hybrid cells.
- Clone the isolated hybrid cells.
- Produce these cells in large quantities (Scale-up).

Monoclonal antibodies are made by fusing the spleen cells from a mouse that has been immunized with the desired antigen with myeloma cells. Polyethylene glycol is used to fuse of adjacent plasma membranes, but the success rate is low so a selective medium is used in which only fused cells can grow. This is because myeloma cells have lost the ability to synthesize hypoxanthine-guanine-phosphoribosyl transferase (HGPRT).This enzyme enables cells to synthesize purines using an extracellular source of hypoxanthine as a precursor. Ordinarily, the absence of HGPRT is not a problem for the cell because cells have an alternate biochemical pathway that they can use to synthesize purines. However, when cells are exposed to aminopterin (a folic acid analogue), they are unable to use this other, rescue, pathway and are now fully dependent on HGPRT for survival.

The selective culture medium is called HAT medium because it contains: hypoxanthine, aminopterin, and thymidine. This medium is selective for fused, (hybridoma) cells because unfused myeloma cells cannot grow because they lack HGPRT. Unfused normal spleen cells cannot grow indefinitely because of their limited life span. However, hybridoma cells are able to grow indefinitely because the spleen cell partner supplies HGPRT and the myeloma partner is immortal because it is a cancer cell (Figure 18.1). The fused hybrid cells are called hybridomas, and since they are derived from cancer cells, are immortal and can be grown indefinitely.

This mixture of cells is then diluted and clones are grown from single parent cells. The antibodies secreted by the different clones are then tested for their ability to bind to the antigen (for example with a test such as EIA or Antigen Microarray Assay) or immuno-dot blot, and the most productive and stable clone is then grown in culture medium to a high volume. When the hybridoma cells are injected in mice (in the peritoneal cavity, the gut), they produce tumours containing an antibody-rich fluid called ascites fluid.

The medium must be enriched during selection to further favour hybridoma growth. This can be achieved by the use of a layer of feeder fibrocyte cells or supplement medium such as briclone. Production in cell culture is usually preferred as the ascites technique is painful to the animal and if replacement techniques exist, is considered unethical.

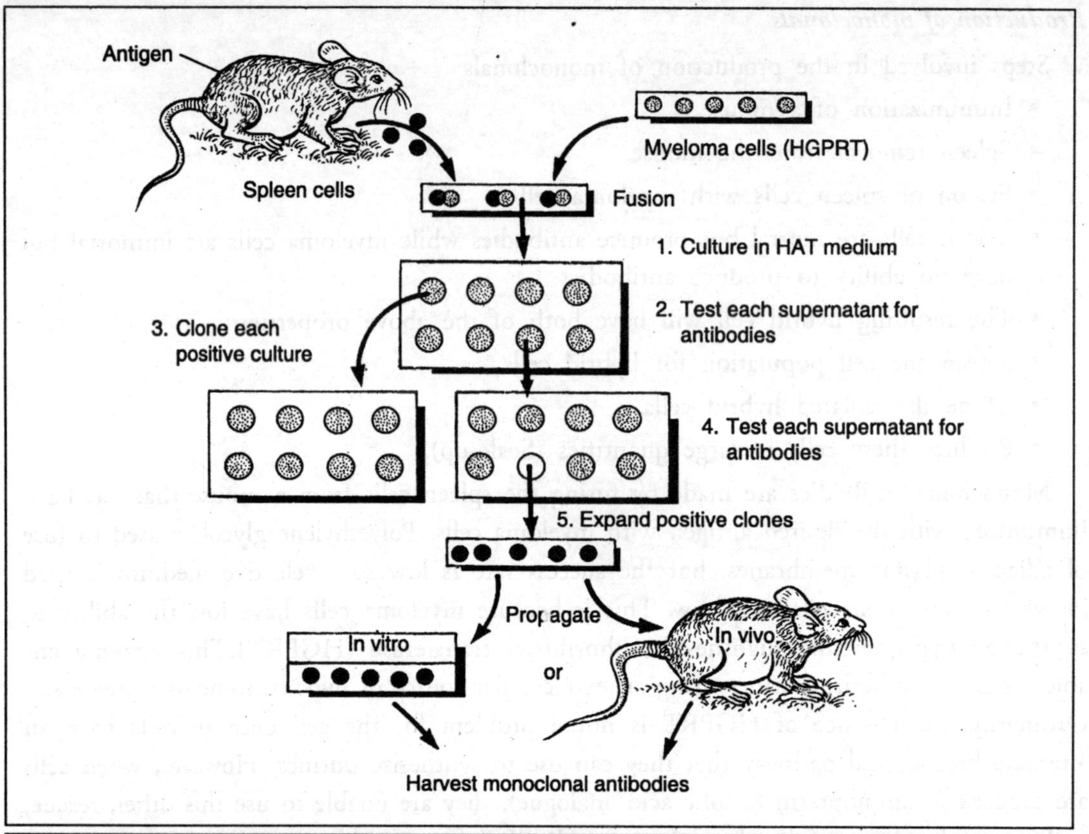

Fig. 18.1: Production of monoclonal antibodies

Recombinant antibody

The production of recombinant monoclonal antibodies involves technologies, referred to as **repertoire cloning** or phage display/yeast display. Recombinant antibody engineering involves the use of viruses or yeast to create antibodies, rather than mice. These techniques rely on rapid cloning of immunoglobulin gene segments to create libraries of antibodies with slightly different amino acid sequences from which antibodies with desired specificities can be selected. These techniques can be used to enhance the specificity with which antibodies recognize antigens, their stability in various environmental conditions, their therapeutic efficacy, and their detectability in diagnostic applications. Fermentation chambers have been used to produce these antibodies on a large scale.

Applications

Once monoclonal antibodies for a given substance have been produced, they can be used to detect the presence and quantity of this substance, for instance in a western blot test (to detect a protein on a membrane) or an immunofluorescence test (to detect a substance in a cell). They are also very useful in immunohistochemistry which detect antigen in fixed tissue sections. Monoclonal antibodies can also be used to purify a substance with techniques called immunoprecipitation and affinity chromatography.

Monoclonal antibody therapy

One possible treatment for cancer involves monoclonal antibodies that bind only to cancer cell-specific antigens and induce an immunological response against the target cancer cell. Such mAb could also be modified for delivery of a toxin, radioisotope, cytokine or other active conjugate; it is also possible to design bispecific antibodies that can bind with their Fab regions both to target antigen and to a conjugate or effector cell. In fact, every intact antibody can bind to cell receptors or other proteins with its Fc region. The illustration below shows all these possibilities.

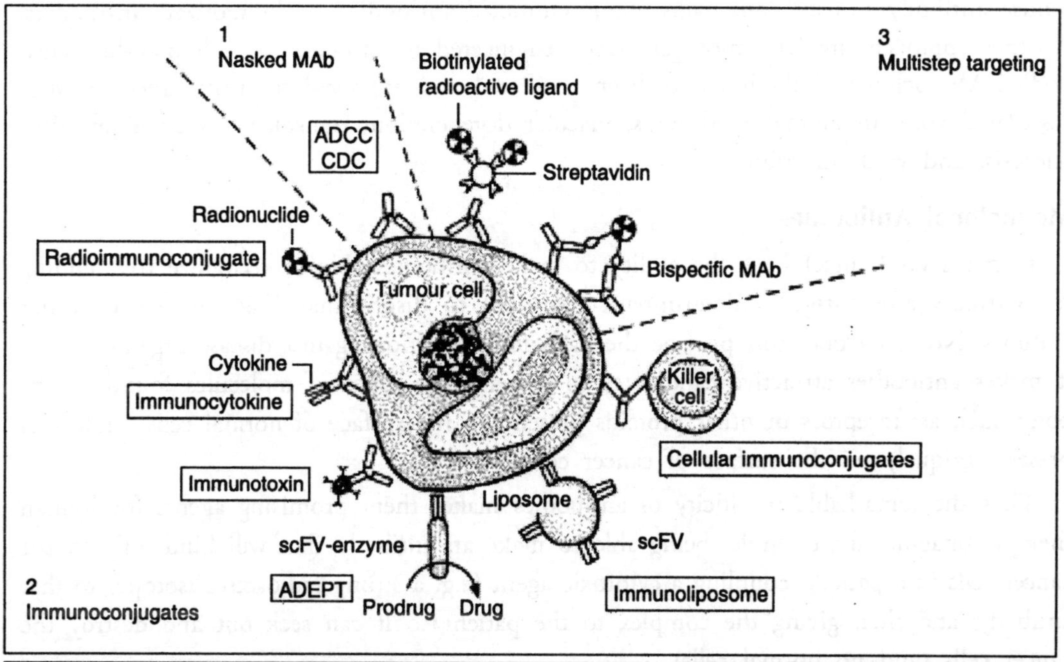

Fig. 18.2: Binding antibodies to the receptor

Monoclonal antibodies for cancer

ADEPT, antibody directed enzyme prodrug therapy; ADCC, antibody dependent cell-mediated cytotoxicity; CDC, complement dependent cytotoxicity; MAb, monoclonal antibody; scFv, single-chain Fv fragment (Fig. 18.2).

Chimeric and humanized antibodies

One problem in medical applications is that the standard procedure of producing monoclonal antibodies yields mouse antibodies. Although murine antibodies are very similar to human ones there are differences. The human immune system hence recognizes mouse antibodies as foreign, rapidly removing them from circulation and causing systemic inflammatory effects.

A solution to this problem would be to generate human antibodies directly from humans. However, this is not easy, primarily because it is generally not seen as ethical to challenge humans with antigen in order to produce antibody; while the ethics of doing the same to non-humans is a matter of debate. Furthermore, it is not easy to generate human antibodies against human tissues.

Various approaches using recombinant DNA technology to overcome this problem have been tried since the late 1980s. In one approach, one takes the DNA that encodes the binding portion of monoclonal mouse antibodies and merges it with human antibody producing DNA. One then uses mammalian cell cultures to express this DNA and produce these half-mouse and half-human antibodies. (Bacteria cannot be used for this purpose, since they cannot produce this kind of glycoprotein). Depending on how big a part of the mouse antibody is used, one talks about **chimeric antibodies** or **humanized antibodies.** Another approach involves mice genetically engineered to produce more human-like antibodies. Monoclonal antibodies have been generated and approved to treat: cancer, cardiovascular disease, inflammatory diseases, macular degeneration, transplant rejection, multiple sclerosis, and viral infection.

Monoclonal Antibodies

Humans (and mice) have the ability to make antibodies able to recognize (by binding to) virtually any antigenic determinant (epitope) to discriminate between even similar epitopes. Not only does this provide the basis for protection against disease organisms, but it makes antibodies attractive candidates to target other types of molecules found in the body, such as: receptors or other proteins present on the surface of normal cells, molecules present uniquely on the surface of cancer cells.

Thus the remarkable specificity of antibodies makes them promising agents for human therapy. Imagine, for example, being able to make an antibody that will bind only to the cancer cells in a patient, coupling a cytotoxic agent (e.g. a strong radioactive isotope) to that antibody, and then giving the complex to the patient so it can seek out and destroy the cancer cells (and no normal cells).

But there are problems to be solved before antibodies can be used in human therapy.

1. The response of the immune system to any antigen, even the simplest, is polyclonal. That is, the system manufactures antibodies of a great range of structures both in their binding regions as well as in their effector regions.
2. Even if one were to isolate a single antibody-secreting cell, and place it in culture, it would die out after a few generations because of the limited growth potential of all normal somatic cells.

What is needed is a way to make "monoclonal antibodies": Antibodies of a single specificity that are all built alike because they are being manufactured by a single clone of plasma cells that can be grown indefinitely.

This problem was solved for mice in 1975 with a technique devised by Köhler and Milstein (for which they were awarded a Nobel Prize). An antibody-secreting B-cell, like any other cell, can become cancerous. The unchecked proliferation of such a cell is called a **myeloma**.

Köhler and Milstein found a way to combine the unlimited growth potential of myeloma cells with the predetermined antibody specificity of normal immune spleen cells.

They did this by literally fusing myeloma cells with antibody-secreting cells from an immunized mouse. The technique is called **somatic cell hybridization.** The result is a **hybridoma.**

Uses for monoclonal antibodies

Monoclonal antibodies are widely used as diagnostic and research reagents. Their introduction into human therapy has been much slower. In some in vivo applications, the antibody itself is sufficient. Once bound to its target, it triggers the normal effector mechanisms of the body. In other cases, the monoclonal antibody is coupled to another molecule, for example a fluorescent molecule to aid in imaging the target; a strongly-radioactive atom, such as Iodine-131 to aid in killing the target.

Monoclonal antibodies in market

Some monoclonal antibodies that have been introduced into human medicine to suppress the immune system are available in market.

- **Muromonab-CD3 (OKT3)** and two humanized anti-CD3 monoclonals. Bind to the CD3 molecule on the surface of T-cells. Used to prevent acute rejection of organ, e.g., kidney, transplants. The humanized versions show promise in inhibiting the autoimmune destruction of beta cells in type 1 diabetes mellitus.
- **Infliximab.** Binds to tumour necrosis factor-alpha (TNF-α). Shows promise against some inflammatory diseases such as rheumatoid arthritis (by blunting the activity of Th1 cells). Side-effects: can convert a latent case of tuberculosis into active disease; can induce the formation of autoantibodies (by promoting the development of Th2 cells).
- **Omalizumab.** Binds to IgE thus preventing IgE from binding to mast cells. Shows promise against allergic asthma.
- **Daclizumab.** Binds to part of the IL-2 receptor produced at the surface of activated T-cells. Used to prevent acute rejection of transplanted kidneys. Has also showed promise against T-cell lymphoma.
- **Rituximab.** Binds to the CD20 molecule found on most B-cells and is used to treat B-cell lymphomas.
- **Zevalin.** This is a monoclonal antibody against the CD20 molecule on B-cells (and lymphomas) conjugated to either the radioactive isotope indium-111 (^{111}In) or the radioactive isotope yttrium-90 (^{90}Y). Both are given to the lymphoma patient, the ^{111}In version first followed by the ^{90}Y version (in each cases supplemented with Rituxan).
- **Bexxar.** This is a conjugate of a monoclonal antibody against CD20 and the radioactive isotope iodine-131 (^{131}I). It, too, is designed as a treatment for lymphoma. Although both Bexxar® and Zevalin® kill normal B-cells, they don't harm the B-

cell precursors because these do not express CD20. So, in time, the precursors can repopulate the body with healthy B-cells.

On 3 February 2005, the New England Journal of Medicine reported that 59% of patients with a B-cell lymphoma were disease-free 5 years after a single treatment with ^{131}I-tositumomab.

- **Herceptin**. Binds HER2, a receptor for **epidermal growth factor** (EGF) that is found on some tumour cells (some breast cancers, lymphomas). The only monoclonal so far that seems to be effective against solid tumors.
- **Erbitux**. Blocks HERl, another epidermal growth factor (**EGF**) receptor.
- **Mylotarg**. A conjugate of a monoclonal antibody that binds **CD33**, a cell-surface molecule expressed by the cancerous cells in **acute myelogenous leukemia** (AML) but not found on the normal stem cells needed to repopulate the bone marrow. **Calicheamicin**, a complex oligosaccharide that makes double-stranded breaks in DNA. Mylotarg is the first **immunotoxin** that shows promise in the fight against cancer.
- **Lymphocide**. Binds to CD22, a molecule found on some B-cell leukemias.
- **Alemtuzumab**. Binds to CD52, a molecule found on white blood cells. Has produced complete remission of chronic lymphocytic leukemia (for 18 months and counting).
- **Lym-1**. Binds to the HLA-DR-encoded histocompatibility antigen that can be expressed at high levels on lymphoma cells.
- **Vitaxin**. Binds to a **vascular integrin (alpha-v/beta-3)** found on the blood vessels of tumours but not on the blood vessels supplying normal tissues. In Phase II clinical trials, Vitaxin has shown some promise in shrinking solid tumours without harmful side effects.
- **Bevacizumab**. Binds to vascular endothelial growth factor (VEGF) preventing it from binding to its receptor. Approved by the US FDA in February 2004 for the treatment of colorectal cancers.
- **Abciximab**. Inhibits the clumping of platelets by binding the receptors on their surface that normally are linked by fibrinogen. Helpful in preventing reclogging of the coronary arteries in patients who have undergone angioplasty.

Problems with monoclonal therapy

Why are there so few monoclonals being used in human therapy a quarter century after their discovery? The main difficulty is that mouse antibodies are "seen" by the human immune system as foreign, and the human patient mounts an immune response against them, producing **HAMA** ("human anti-mouse antibodies"). These not only cause the therapeutic antibodies to be quickly eliminated from the host, but also form immune complexes that cause damage to the kidneys.

Using genetic engineering it is possible to make mouse-human hybrid antibodies in an attempt to reduce the problem of HAMA.

- **Chimeric antibodies.** The antibody combines the antigen-binding parts (variable regions) of the mouse antibody with the effector parts (constant regions) of a human antibody. **Infliximab**, **rituximab**, and **abciximab** are examples.
- **Humanized antibodies.** The antibody combines only the amino acids responsible for making the antigen binding site (the hypervariable regions) of a mouse (or rat) antibody with the rest of a human antibody molecule thus replacing its own hypervariable regions. **Daclizumab, Vitaxin, Mylotarg, Herceptin,** and **Xolair** are examples.

In both cases, the new gene is expressed in mammalian cells grown in tissue culture (E. coli cannot add the sugars that are a necessary part of these glycoproteins).

Man-made antibodies

Monoclonal antibodies can now be genetically engineered and endowed with new properties. In the future, gene technology could enable antigen-binding fragments to be made by exploiting repertoires of variable domain genes derived from immunized animals and expressed in bacteria. How readily can this approach be extended to production of *'in vitro'* repertoires of variable domain genes, and obviate the immunization of animals?

TABLE 18.1
Shows the mabs and their applications

Type	Application	Mechanism	Mode
Infliximab	• rheumatoid arthritis • Crohn's disease	inhibits TNF-α	chimeric
Basiliximab	• Acute rejection of kidney transplants	inhibits IL-2 on activated T-cells	chimeric
Abciximab	• Prevent coagulation in coronary angioplasty	inhibits the receptor GpIIb/IIIa on platelets	chimeric
Daclizumab	• Acute rejection of kidney transplants	inhibits IL-2 on activated T-cells	humanized
Gemtuzumab	• relapsed acute myeloid leukemia	targets an antigen on leukemia cells	humanized
Alemtuzumab	• B-cell leukemia	targets an antigen CD52 on T- and B-lymphocytes	humanized
Rituximab	• non-Hodgkin's lymphoma	targets phosphoprotein CD20 on B-lymphocytes	humanized
Palivizumab	• RSV infections in children	inhibits an RSV protein	humanized
Etanercept	• rheumatoid arthritis	contains TNF receptor	

CHAPTER

19

IMMUNODEFICIENCY

INTRODUCTION

Immunodeficiency diseases are characterized by undue susceptibility to infection. Paradoxically, many immunodeficiency syndromes are also characterized by autoimmune diseases and excessive production of immunoglobulin E (IgE) antibodies. Because of the ability of antibiotics to control many types of infections, autoimmune diseases now account for significant morbidity among patients with immunodeficiency. Finally, there is an increased incidence of malignant disease in patients with immunodeficiency diseases. Whether the latter is the result of increased susceptibility to infection with agents predisposing to malignancy or defective tumour immunosurveillance is unknown.

Defects in one or more component of immune system can leads to serious or often fatal disorders which are called **immunodeficiency.** When system errs by which failing to protect the host from diseases causing agents or from malignant cells, the result is immunodeficiency.

There are two types of immunodeficiency.

1. **Primary or congenital immunodeficiency**

   A condition resulting from genetic or developmental defect in the immune system is called primary immunodeficiency. This is a condition in which defect found at birth or childhood but is sometimes clinically detected later.

   The primary immunodeficiencies are classified into B-cell deficiencies, T-cells deficiencies and common variable deficiencies.

2. **Secondary or acquired immunodeficiency**

   It is loss of immune function and results from exposure to various infectious agents, e.g. AIDS is cause due to HIV infection. Secondary immunodeficiencies may cause similar infections to primary immunodeficiency diseases, but secondarily to some other pathological condition such as malnutrition, age, drugs, tumours, or infections including HIV in AIDS.

Primary or congenital immunodeficiency

Primary immunodeficiency diseases are intrinsic defects of the immune system. Patients with primary immunodeficiency diseases have increased susceptibility to recurrent and persistent infections, but other symptoms are also common. Most primary immunodeficiency diseases are relatively rare disorders. The incidence of primary immunodeficiency diseases varies greatly from about 1:500 births with selective IgA deficiency to only a few known cases for the rarest disorders.

Adaptive immune mechanisms recognize and neutralize foreign molecules or microorganisms in a specific manner. Lymphocytes, B- and T-cells, can respond selectively to thousands of non-self materials. Adaptation is further acquired with memory of previous infections. Immunodeficiencies impair the functioning of the immune system. Deficiencies are highly variable with regard to symptoms, phenotype, genotype, severity, etc, because many cells and molecules are required for both natural and adaptive immunity. However, increased susceptibility to infections is common to all immunodeficiencies.

More than 70 primary immune deficiencies are known, and can be grouped according to the components of the immune system affected. Antibody deficiency disorders are defects in immunoglobulin-producing B-cells. T-cell deficiencies affect the function in killing infected cells or helping other immune cells. T-cell deficiencies result usually in combined immunodeficiencies where both T-cells and antibody production are defective, complement system and phagocytic cells.

Primary immunodeficiency diseases patients have recurrent, serious infections starting early after birth. Children are protected for 6-12 months by the maternal IgG. Life threatening symptoms can arise already within the first few days of life in SCID. Patients with antibody deficiencies are especially susceptible to encapsulated bacteria, which cause pyogenic infections. T-cell immunodeficiencies and severe combined immunodeficiencies (SC-IDs) are marked with opportunistic infections caused by common environmental microorganisms.

Primary immunodeficiency diseases patients require prolonged treatment with high doses antibiotics. Antibody deficiencies are treated with intravenous immunoglobulin substitution therapy. Gammaglobulins are extracted from human blood from donor pools. Leukocytes are produced in stem cells in bone marrow. In SCIDs bone marrow transplantation is the most effective treatment. In certain metabolic disorders (ADA and PNP deficiency) enzyme substitution therapy can be applied.

Primary defects in B lymphocyte or B-cell deficiencies

B-cell immunodeficiencies are antibody deficiency disorders that are restricted to antibody function. Either B lymphocyte development is impaired, or B-cells fail to respond to

T-cell signals, or only selected subsets of immunoglobulins may be deficient. Patients have recurrent pyogenic infections with encapsulated bacteria. Infections require early and vigorous treatment with antibiotics and lifelong immunoglobulin replacement therapy.

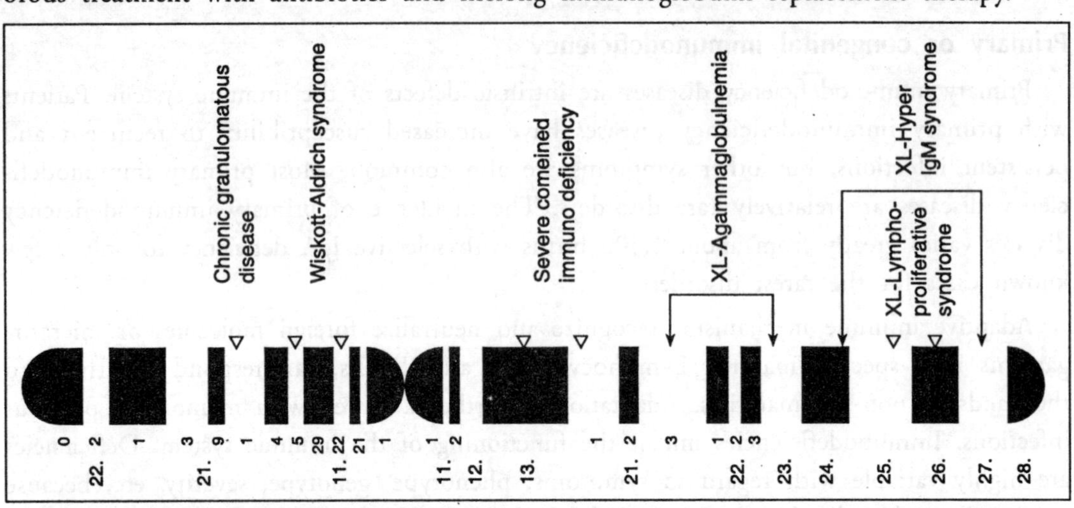

Fig. 19.1: Loci of immunodeficiency on chromosomes

X-linked agammaglobulinemia (XLA)

X-linked agammaglobulinemia (XLA) is one of several immunodeficiency syndromes that have been mapped to the X-chromosome (Figure 19.1). The defect is due to mutations in the **Bruton's tyrosine kinase (*Btk*) gene**, which is expressed early in B-cell development and which is crucial for maturation of the B-cell series from pre-B-cells. Over 100 different *Btk* mutations are now recorded in the XLA mutation registry. Early B-cells are generated in the bone marrow but fail to mature, and the production of immunoglobulin in affected males is grossly depressed. B-cells are reduced or absent in the peripheral blood and there are few lymphoid follicles or plasma cells in the lymph nodes. This disease also called Brutons agammaglobulinaemia, and is characterized by absence of gamma globulin in blood as the name implies. XLA is a typical antibody deficiency in which production of antibodies is prevented due to a block in the B-cell maturation. The prevalence is about 1:200,000. Serum concentrations of IgG, IgA, and IgM are markedly reduced. Levels of circulating B lymphocytes are significantly decreased and plasma cell are absent from lymph nodes and bone marrow, but the number of T-cells is normal or even increased. The clinical phenotype may be variable and even members of the same family can have different symptoms. Patients with XLA have normal response to viral infections and normal V(D)J rearrangement. XLA is caused by a block in the B-cell differentiation. Btk, the affected protein, is a tyrosine kinase that regulates activity of signalling pathways by phosphorylation.

Typically, about one third of X-linked primary immunodeficiency diseases are new, sporadic cases, where the mutation has appeared in the patient. Bruton tyrosine kinase (Btk) is the defective molecule in X-linked agammaglobulinaemia (XLA). Mutations in the gene for *BTK* prevent B-cell maturation since Btk is crucial signalling molecule regulating B-cell development into antibody-producing plasma cells. Btk protein consists of five domains, all of which can be affected by disease-causing alterations. The majority of the known mutations (altogether more than 300) lead to truncation of the protein due to either nonsense or frame shift mutations because of insertions, deletions or splice site defects. Although the mutations are evenly distributed in the *BTK* gene, the most mutable single sites in many genetic disorders, including XLA, are CpG dinucleotides.

In addition to the catalytic tyrosine kinase domain, Btk contains modules for protein-protein interactions, namely Tec homology (TH), Src homology 2 (SH2), and SH3 domains. Pleckstrin homology (PH) domain is responsible for membrane localization of the protein by binding to the phosphatidyl inositol groups of membrane lipids. Btk has several signalling partners. The specificity of the kinase domain may be increased by simultaneous binding of the SH2 and/or SH3 domain into their recognition sites in the substrate. The TH domain binds Zn^{2+}, which is essential for stability. Btk is activated by stepwise phosphorylation, first by Src family kinases and then by autophosphorylation. Although many of the mutations cause truncation of the Btk protein, there are also other types of alterations. For example, a number of structural mutations lead into misfolded proteins. Three-dimensional structures of the Btk domains have been used to describe the consequences of the disease-causing mutations. Of the amino acid-changing missense mutations, many alter structurally of functionally significant conserved residues.

IgG deficiency/Hyper IgM syndrome (HIM)

Hyper-IgM syndrome is usually inherited as X-linked disorder. Hyper-IgM syndrome represents a group of related diseases, the majority of which are X-linked. Affected male children produce only IgM antibodies and therefore susceptible to severe bacterial infection. XHIM is caused by a genetic defect in the gene for the CD40 ligand. The patients have severely reduced IgG, IgA, and IgE serum levels, but normal or even raised IgM levels. In XHIM there is a failure in heavy chain class switch from IgM to IgG and IgA. Interaction between CD40L in T-cells and CD40 on B-cells is a key signal for the generation of memory B-cells and for the formation of germinal centres. Defective CD40L prevents production of certain antibodies. CD40 is also involved in interaction with macrophages and dendritic cells, which induces IL-12 secretion and thereby elicits an immune response to intracellular microorganisms. Infections in patients with HIM are similar to those in XLA, except for higher tendency for persistent or recurrent neutropenia and thrombocytopenia.

CD40 belongs to the tumour necrosis factor (TNF) super family of transmembrane proteins. Signalling through CD40 activates phospholipase Cã 2 and phosphatidylinositol-3 kinase, upregulates CD23, CD54, and CD80, and induces transcription factors such as nuclear factor-ê B. The XHIM causing mutations can appear in any of the domains of CD40L (intracytoplasmic, transmembrane, extracellular unique, and TNF homology regions). However, mutations cluster into the TNHF domain. The most common alterations are point mutations, which typically result in truncation of the polypeptide chain.

Primary defects in T lymphocyte or T-cell immunodeficiencies

There are many major T-cell deficiency are causes, patients with no T-cell or poor T-cell function are susceptible to opportunistic infection. T-cell immunodeficiency also results in humoral immunodeficiency, in other world, T-cell immunodeficiency leads to combine both humoral and cell mediated immunity. There are wide ranges of causes for T-cells deficiency, ranging from absence of lymphocyte to enzyme deficiency. T-cell deficiencies usually affect also other components of the immune system and lead into common immunodeficiency diseases. Patients have a progressive decrease in the number of circulating T cells, whereas numbers of circulating B-cells and serum immunoglobulin are usually normal. Autoimmunity is a frequent complication in T-cell immunodeficiency diseases. In T-cell deficiencies bone marrow transplantation is in many cases the only long-lasting therapeutic option.

Primary T-cell deficiencies are classified as follows

- DiGeorge syndrome,
- Ataxia telangiectasia (AT),
- Purine nucleoside phosphorylase (PNP) deficiency,
- Adenosine deaminase deficiency,
- Wiskott-Aldrich syndrome (WAS),
- Severe combine immunodeficiencies,

DiGeorge syndrome

This is selective T-cell deficiency is due to congenital malformation that results in defective development of third and fourth pharyngeal pouches. It is also called as thymic hyperplasia. DiGeorge syndrome is a congenital immune disorder characterized by lack of embryonic development or underdevelopment. Thymic epithelium is derived from the third and fourth pharyngeal pouches by the sixth week of gestation. The syndrome is associated with several defects and it has been called also for CATCH22, because the gene defect is usually a deletion in chromosome 22, and the symptoms include cardiac abnormalities, abnormal facies, thymic hypoplasia, cleft palate, and Hypocalcinaemia. The degree of thymus problems varies and only about 20% of the patients have the function or levels of T-cells decreased.

Ataxia telangiectasia (AT)

Ataxia telangiectasia is an autosomal recessive disorder characterised by abnormal gait (Ataxia), vascular malformation (telangiectasia), various neurologic deficit, increased incidence of tumour and immunodeficiency. Ataxia telangiectasia (AT) is a rare, autosomal recessive, progressive, neurodegenerative childhood disease that affects the nervous and other body systems. The prevalence of AT is in the order of 1:40,000 to 1:100:000 births. The first signs of the disease usually occur during the first decade of life. Ataxia telangiectasia means ataxia, lack of muscle control, and telangiectasias (tiny, red "spider" veins) first in the corners of the eyes or on ears and cheeks. Immune system is impaired in many patients and they are also predisposed to leukemia and lymphoma and they are extremely sensitive to radiation exposure. About 70% of the patients have IgA deficiency and also some IgG subclasses can be reduced. At the moment there is no cure for AT.

The defective protein, ATM, is a protein kinase that reacts to DNA damage and affects the accumulation of a tumour suppressor p53, which is defective in many cancers. ATM mutations delay the accumulation of p53, allowing cells to replicate without repair of the damaged DNA and thereby increasing the risk of cancer. Large numbers of ataxia telangiectasia causing mutations have been identified from different parts of the molecule.

Purine nucleoside phosphorylase (PNP) deficiency

Purine nucleoside phosphorylase (PNP) deficiency is characterized by accumulation of toxic purine metabolites, primarily dGTP, in cells. PNP catalyses the phosphorolysis of the purine nucleosides, deoxyinosine and deoxyguanosine, to purine bases and ribose-1-phosphate. dGTP is particularly toxic to T-cells by inhibition of ribonucleotide reductase and further DNA synthesis and proliferation. PNP deficiency is often accompanied by neurologic disorder. The enzyme following PNP in the purine catabolism is adenosine deaminase, mutations of which cause SCID.

Adenosine deaminase (ADA) deficiency

Adenosine deaminase (ADA) deficiency accounts for about half of the autosomal recessive forms of SCIDs. ADA follows PNP in purine nucleoside catabolism, but deficiency in this enzyme causes even more severe symptoms than PNP deficiency, which is a T-cell deficiency. ADA degrades toxic adenosine and deoxyadenosine, which accumulate in the cells of patients. Immature lymphoid cells are particularly sensitive to these nucleotides. In addition to immunological defect, most patients with ADA deficiency also have skeletal abnormalities.

ADA deficiency is an ideal disease for development of gene therapy methods, because there is selective pressure on nontransduced T-cells, because they are not viable. ADA protein is tolerated also when overexpressed. The disorder is treated usually with ADA protein substitution therapy or with bone marrow transplantation.

The second most common form of SCID after X_SCID is caused by a defective enzyme, adenosine deaminase (ADA), necessary for the breakdown of purines. Lack of ADA causes accumulation of dATP. This metabolite will inhibit the activity of ribonucleotide diphosphate reductase, the enzyme that reduces ribonucleotides to generate deoxyribonucleotides. The effectiveness of the immune system depends upon lymphocyte proliferation and hence dNTP synthesis. Without functional ribonucleotide reductase, lymphocyte proliferation is inhibited and the immune system is compromised.

Wiskott-Aldrich syndrome (WAS)

This X-linked disorder is an immunodeficiency that increases in severity with age and usually results in fatal infection or lymphoid malignancy. Wiskott-Aldrich syndrome is X-linked diseases include thrombocytopenia (lowered platelet count) which may lead to fatal bleeding. Wiskott-Aldrich syndrome (WAS) is an immunodeficiency disorder of both T- and B-cells characterized by thrombocytopenia, eczema, and recurrent infections. Patients have progressive lymphopenia. WAS is an X-linked immunodeficiency. Female carriers are phenotypically normal since the defective X chromosome is preferentially inactivated. Boys with WAS have increased serum IgA and IgE levels, whereas IgM is decreased. T-cells are defective and they have fewer microvilli on the cell surface than normal cells. WAS patients with autoimmune manifestations are at a high risk of getting malignancies. WAS leads to death within first two decades of life because of viral or bacterial infections without bone marrow transplantation.

The gene for WAS has been identified. WAS protein (WASP) interacts with Cdc42, a GTP-binding protein, that has function in cytoskeleton reorganization. Defective cell polarization due to WASP mutations could affect also B-cell T-cell interactions. WASP consists of several domains, from the N-terminus partly overlapping pleckstrin homology (PH) and WASP homology 1 (WH1) domains, GTPase binding domain, proline rich region and WH2 domain. The proline rich region interacts with adaptor protein Nck. More than 100 WAS-causing mutations scattered along the gene and protein are known. WASP mutations are mainly amino acid substitutions or nonsense mutations leading to truncation of the protein.

WAS generally becomes symptomatic in children. Due to its mode of inheritance, the overwhelming majority are male. It is characterised by bruising caused by thrombocytopenia (low platelet counts), small platelet size on blood film, eczema, recurrent infections, and a propensity for autoimmune disorders and malignancies (mainly lymphoma and leukemia).

Common variable immunodeficiency (CVID)

Common variable immunodeficiency has been attributed to multiple abnormalities including intrinsic B-cell defects. The group of antibody deficiency disorder is usually inherited in an autosomal manner. Common variable immunodeficiency (CVID) is a group of undifferentiated disorders with defective antibody formation. The incidence is approximately 1:25,000. Patients with CVID usually have normal number of circulating B-cells, but low serum levels of IgG and IgA. However, the B-cells are defective. CVID affects both females and males equally and it usually has later age of onset than other antigen immunodeficiency. Patients have an unusually high incidence of lymphoreticular and gastrointestinal malignancies and the incidence of autoimmune disorders is also increased. CVID forms arise from several different genetic defects. Other B-cell deficiencies have been described including, for example μ heavy chain deficiency; λ 5 surrogate light chain deficiency; and κ light chain deficiency.

Severe combined immunodeficiencies (SCIDs)

The term severe combined immunodeficiency disease is given to heterogeneous group of disorder characterised by defective development of function of B and T lymphocyte and deficient humoral and cell mediated immunity. The first example of this disease was discovered in Switzerland in 1950. In combined B- and T-cell immunodeficiencies, the most severe immunodeficiency, all adaptive immune functions are absent. The condition is fatal unless the immune system can be reconstituted, either by transplants of immunocompetent tissue or by enzyme replacement. The immunological, genetic, and enzymatic characteristics of the diseases show great diversity. SCIDs have an average frequency of approximately 1 in 75,000 births.

The infant with SCID develop recurrent infection early in life, They have prolong diarrhea due to rotavirus or bacterial infection of gastrointestinal tract, they develop pneumonia usually due to protozoan. The most common yeast *Candiada albicans* grow luxuriously in their mouth or on their skin. Infant with SCID have very few lymphocyte in their blood(Fewer than 3000/microliter). The SCID are more common male than female infant. This is because over 50% of SCID cases caused by gene on X chromosome. The remaining cases of SCID are due to recessive gene on another chromosome.

Severe combined immunodeficiency, or Boy in the Bubble Syndrome, is a genetic disorder in which both "arms" (B-cells and T-cells) of the adaptive immune system are crippled, due to a defect in one of several possible genes. SCID is a severe form of heritable immunodeficiency. It is also known as the "bubble boy" disease because its victims are extremely vulnerable to infectious diseases. The most famous case is the boy David Vetter.

SCID affects about 1 in 100,000 live births. These babies, if untreated, usually die within 1 year due to severe, recurrent infections. Chronic diarrhea, ear infections, recurrent *Pneumocystis jirovecii* pneumonia, and profuse oral candidiasis commonly occur. Treatment options are much improved since David Vetter, and living in a bubble is no longer necessary.

Recent studies indicate that one in every 2,500 children in the Navajo population inherit **Severe combined immunodeficiency** (SCID). This condition is a significant cause of illness and death among Navajo children. Ongoing research reveals a similar genetic pattern among the related Apache people. Mortan Cowan, MD, director of the Pediatric Bone Marrow Transplant Program at the University of California-San Fransciso, noted that although researchers have identified about a dozen genes that cause SCID, the Navajo and Apache population has the most severe form of the disorder. This is due to the lack of a gene designated "Artemis." Without the gene, children's bodies are unable to repair DNA or develop disease-fighting cells.

Most cases of SCID are due to mutations in the gene encoding the common gamma chain (γ_c), a protein that is shared by the receptors for interleukins IL-2, IL-4, IL-7, IL-9, IL-15 and IL-21. These interleukins and their receptors are involved in the development and differentiation of T- and B-cells. Because the common gamma chain is shared by many interleukin receptors, mutations that result in a non-functional common gamma chain cause widespread defects in interleukin signalling. The result is a near complete failure of the immune system to develop and function, with low or absent T-cells and NK cells and non-functional B-cells.

The common gamma chain is encoded by the gene IL-2 receptor gamma, or IL-2Rã, which is located on the X-chromosome. Therefore, immunodeficiency caused by mutations in IL-2Rγ is known as X-linked severe combined immunodeficiency. The condition is inherited in an X-linked recessive pattern. All males (who have one X and one Y chromosome) who have an X chromosome with the defective gene will have SCID. Females (who have two X chromosomes) with one defective and normal chromosome will be healthy, as one normal IL-2Rγ gene is sufficient to have a normal immune system. However, these females are carriers. Their male children have a 50% chance of having X-linked SCID, and their female children have a 50% chance of being carriers.

The most common treatment for SCID is bone marrow transplantation, which requires matched donors (a sibling is generally best). David Vetter, the original "bubble boy", had one of the first transplantations, and finally died because of an unscreened virus, Epstein-Barr (tests were not available at the time), in his newly transplanted bone marrow from his sister. Today, transplants done in the first three months of life have a high success rate.

More recently gene therapy has proved useful. Transduction of the missing gene to hematopoietic stem cells using viral vectors is being tested in ADA SCID and X-linked SCID. The first gene therapy trials were performed in 1990, with peripheral T-cells. In 2000, the first gene therapy "success" resulted in SCID patients with a functional immune system. These trials were stopped when it was discovered that two of ten patients in one trial had developed leukemia resulting from the insertion of the gene-carrying retrovirus near an oncogene. In 2007, four of the ten patients have developed leukemias. Work is now focusing on correcting the gene without triggering an oncogene. No leukemia cases have yet been seen in trials of ADA-SCID, which does not involve the *gamma c* gene that may be oncogenic when expressed by a retrovirus.

Trial treatments of SCID have been gene therapy's only success; since 1999, gene therapy has restored the immune systems of at least 17 children with two forms (ADA-SCID and X-SCID) of the disorder. SCID mice are used in disease, vaccine, and transplant research, especially as animal models for testing the safety of new vaccines or therapeutic agents in people with weakened immune systems.

An animal variation of the disease, an autosomal recessive gene with clinical signs similar to the human condition, also affects the Arabian horse. In horses, to date, the condition remains a fatal disease, as the animal inevitably succumbs to an opportunistic infection within the first four to six months of life. However, carriers, who themselves are not affected by the disease, can be detected with a DNA test. Thus careful breeding practices can avoid the risk of an affected foal being produced.

Another animal with well-characterized SCID pathology is the dog. There are two known forms, an X-linked SCID in Basset Hounds that has similar ontology to X-SCID in humans, and an autosomal recessive form seen in one line of Jack Russell Terriers that is similar to SCID in Arabian horses and mice.

X-linked SCID

X-linked SCID, accounting for about 50-60% of SCID cases, is caused by IL-2 receptor ã chain mutations, which lead to very low numbers of T-cells and NK cells, whereas B-cells are present in high numbers. However, the B-cells are immature and defective. IL-2 receptor (IL-2R) α-chain (CD25) deficiency has also been reported. The α chain of the receptor forms part of the receptor also for IL-2, -4, -7, -9, and -15, affecting the differentiation and growth of lymphocytes. The ã chain consists of an extracellular domain with WS motif, transmembrane region, as well as intracellular domain with Box1 and Box2 regions. Some 150 different mutations distributed in all the domains have been determined. Patients with X-SCID have extreme susceptibility to infections.

Other combined immunodeficiencies

Combined immunodeficiencies can arise also in association with other diseases or secondarily to other disorders. A great number of disorders are related to immunological deficiencies, although the main defect affects some other functions. **Xeroderma pigmentosum** and DNA ligase 1 defects are photosensitive dermatoses. Sunlight causes skin lesions for patients with these diseases.

Turner syndrome patients have generally normal numbers of T- and B-cells, but serum IgG and IgM levels are reduced. The syndrome is named after Henry Turner, an Oklahoma endocrinologist, who described it in 1938. In Europe, it is often called **Ullrich-Turner syndrome** or even **Bonnevie-Ulrich-Turner syndrome** to acknowledge that earlier cases had also been described by European doctors.

The first published report of a female with a 45, X karyotype was in 1959 by Dr. Charles Ford and colleagues in Harwell, Oxfordshire and Guy's Hospital in London. It was found in a 14-year-old girl with signs of Turner syndrome. **Turner syndrome** or **Ullrich-Turner syndrome** encompasses several chromosomal conditions, of which monosomy X is the most common. It occurs in about 1 out of every 2500 female births. Some sources say 1 out of every 5000. Instead of the normal XX sex chromosomes for a female, only one X chromosome is present and fully functional; in rarer cases a second X chromosome is present but abnormal, while others with the condition have some cells with a second X and other cells without it. A normal female karyotype is labelled 46, XX; individuals with Turner syndrome are 45, X0. In Turner syndrome, female sexual characteristics are present but generally underdeveloped.

Turner syndrome (TS) is a medical problem that affects about one in every 2,500 girls. Although researchers don't know exactly what causes Turner syndrome, they do know that it's the result of a problem with a girl's **chromosomes** (pronounced: **krow-muh-soamz**). Most girls are born with two X chromosomes, but girls with Turner syndrome are born with only one X chromosome or they are missing part of one X chromosome. The effects of the condition vary widely among girls with Turner syndrome. It all depends on how many of the body's cells are affected by the changes to the X chromosome.

Girls with Turner syndrome are usually short in height. Girls with Turner syndrome who aren't treated reach an average height of about 4 feet 7 inches (1.4 meters). The good news is that when Turner syndrome is diagnosed while a girl is still growing, she can be treated with hormones to help her grow taller.

In addition to growth problems, Turner syndrome prevents the ovaries from developing properly, which affects a girl's sexual development. Because the ovaries are responsible for making the hormones that control breast growth and menstruation, most girls with Turner syndrome will not go through all of the changes associated with puberty unless they get treatment for the condition. Nearly all girls with Turner syndrome will be infertile, or unable to become pregnant on their own.

Abnormalities in chromosome 18 may lead to decreased serum IgA levels and also antibody formation can be impaired. Hyper-IgE syndrome, **Job syndrome**, is a disorder of neutrophil oxidative metabolism, typically with extremely high serum IgE levels, chronic eczema, and recurrent skin and sinopulmonary tract infections.

Bloom syndrome

Bloom syndrome is marked by chromosomal breakage disorder affecting multiple systems and characterized primarily by telangiectatic erythema and short stature. The patients have reduced T-cell levels sometimes accompanied by reduced antibody levels. Bloom syndrome is a rare inherited disorder characterized by a high frequency of breaks and rearrangements in an affected person's chromosomes, discovered and first described by dermatologist Dr. David Bloom in 1954.

People with Bloom syndrome are much smaller than average, and often have a high-pitched voice and characteristic facial features including a long, narrow face; small lower jaw; and prominent nose and ears. They tend to develop pigmentation changes and dilated blood vessels in the skin, particularly in response to sun exposure. These changes often appear as a butterfly-shaped patch of reddened skin on the face. The skin changes may also affect the hands and arms.

Other features of the disorder may include learning disabilities, mental retardation, chronic lung problems, diabetes, and immune deficiency that leads to recurrent pneumonia and ear infections. Men with Bloom syndrome usually do not produce sperm and, as a result, are unable to father children. Women with the disorder generally experience menopause earlier than usual.

Relationship to cancer

Chromosomal instability in Bloom syndrome results in a high risk of cancer in affected individuals. Affected individuals develop the full range of cancers found in the general population, but the cancers arise unusually early in life. People with Bloom syndrome may be first diagnosed with cancer at about 25 years old.

Genetics

The BLM gene provides instructions for producing a protein called the Bloom (BLM) syndrome protein, which is a member of the DNA helicase family. Mutations in the BLM gene (locus 15q26.1) cause Bloom syndrome. DNA helicases are enzymes that unwind the two spiral strands of a DNA molecule so that they can be copied. When a cell prepares to divide to form two cells, the chromosomes are duplicated so that each new cell will get a complete set of chromosomes; this replication process involves unwinding the DNA so that it can be copied. The BLM protein is important in maintaining the stability of the DNA during this process. Mutations in the *BLM* gene alter or reduce the BLM protein's DNA helicase activity, which causes errors in the copying process during replication. As a result, people with Bloom syndrome have a higher frequency of chromosome breakage and rearrangement than unaffected people. This increase in chromosome breakage and rearrangement leads to the signs and symptoms of Bloom syndrome. Another suggestion put forward is that people with Bloom syndrome overproduce the superoxide anion, whose volatility damages the affected person's chromosomes.

Bloom syndrome is inherited in an autosomal recessive fashion. Both parents must be carriers in order for a child to be affected. The carrier frequency in individuals of Eastern European ancestry is about 1/100. If both parents are carriers, there is a one in four, or 25%, chance with each pregnancy for an affected child. Genetic counselling and genetic testing is recommended for families who may be carriers of Bloom syndrome.

Down syndrome

In Down syndrome, serum IgM is progressively reduced and thymus can be defective. The patients have facial abnormalities, but also mental retardation. The triplicated chromosome region contains gene for interferon receptor and therefore the patients suffering of Down syndrome are more sensitive to interferons than normal population. Chromosomal defects as Down syndrome, also known as trisomy 21, Turner syndrome and rings or deletions of chromosome 18 lead into immunodeficiencies.

Secondary or Acquired Immunodeficiency syndrome (AIDS)

AIDS

Acquired immunodeficiency syndrome is a disease described in early 1980s, characterized by profound immunosuppression with diverse clinical features, including opportunistic infection, malignancies, and central nervous system degeneration. AIDS is one of group of clinical syndrome caused by retrovirus called HIV. HIV primarily infect CD4 T-cells, including helper T-cells, and macrophages, as well as dendritic cells in lymph node. AIDS is caused by a virus called HIV, the Human Immunodeficiency Virus.

AIDS stands for Acquired Immune Deficiency Syndrome:

- *Acquired* means you can get infected with it;
- *Immune Deficiency* means a weakness in the body's system that fights diseases;
- *Syndrome* means a group of health problems that make up a disease.

Human immunodeficiency virus (HIV) is a retrovirus that can lead to *acquired immunodeficiency syndrome* (AIDS), a condition in humans in which the immune system begins to fail, leading to life-threatening opportunistic infections. Previous names for the virus include human T-lymphotropic virus-III (HTLV-III), lymphadenopathy-associated virus

(LAV), and AIDS-associated retrovirus (ARV). HIV is thought to have originated in non-human primates in sub-Saharan Africa and transferred to humans during the 20th century. Two species of HIV infect humans: HIV-1 and HIV-2. Both species of the virus are believed to have originated in West-Central Africa and jumped species (zoonosis) from a non-human primate to humans. HIV-1 first described and characterized in laboratory Luc mantangnier in Paris. HIV-1 is thought to have originated in southern Cameroon after jumping from wild chimpanzees (*Pan troglodytes troglodytes*) to humans during the twentieth century. It evolved from a Simian Immunodeficiency Virus (SIV_{cpz}). HIV-2, on the other hand, may have originated from the Sooty Mangabey (*Cercocebus atys*), an Old World monkey of Guinea-Bissau, Gabon, and Cameroon.

HIV is a member of the genus Lentivirus, part of the family of Retroviridae. Lentiviruses have many common morphologies and biological properties. Many species are infected by lentiviruses, which are characteristically responsible for long-duration illnesses with a long incubation period. Lentiviruses are transmitted as single-stranded, positive-sense, enveloped RNA viruses. Upon entry of the target cell, the viral RNA genome is converted to double-stranded DNA by a virally encoded reverse transcriptase that is present in the virus particle. This viral DNA is then integrated into the cellular DNA by a virally encoded integrase, along with host cellular co-factors so that the genome can be transcribed. Once the virus has infected the cell, two pathways are possible: either the virus becomes latent and the infected cell continues to function, or the virus becomes active and replicates, and a large number of virus particles are liberated that can then infect other cells.

Two species of HIV infect humans: HIV-1 and HIV-2. HIV-1 is the virus that was initially discovered and termed LAV. It is more virulent, relatively easily transmitted, and is the cause of the majority of HIV infections globally. HIV-2 is less transmittable than HIV-1 and is largely confined to West Africa.

HIV has been found at low concentrations in the saliva, tears and urine of infected individuals, but there are no recorded cases of infection by these secretions and the potential risk of transmission is negligible. The use of physical barriers such as the latex condom is widely advocated to reduce the sexual transmission of HIV. Spermicide, when used alone or with vaginal contraceptives like a diaphragm, actually increases the male to female transmission rate due to inflammation of the vagina; it should not be considered a barrier to infection.

AIDS epidemic

HIV infection in humans is now pandemic. As of January 2006, the Joint United Nations Programme on HIV/AIDS (UNAIDS) and the World Health Organization (WHO) estimate that AIDS has killed more than 25 million people since it was first recognized on December 1, 1981, making it one of the most destructive pandemics in recorded history. It is estimated that about 0.6% of the world's population is infected with HIV. In 2005 alone, AIDS claimed an estimated 2.4–3.3 million lives, of which more than 570,000 were children. A third of these deaths are occurring in sub-Saharan Africa, retarding economic growth and increasing poverty. According to current estimates, HIV is set to infect 90 million people in Africa, resulting in a minimum estimate of 18 million orphans. Antiretroviral treatment reduces both the mortality and the morbidity of HIV infection, but routine access to antiretroviral medication is not available in all countries.

Mode of transmission of HIV

Infection with HIV occurs by the transfer of blood, semen, vaginal fluid, pre-ejaculate, or breast milk. Within these bodily fluids, HIV is present as both free virus particles and virus within infected immune cells. The four major routes of transmission are unprotected sexual intercourse, contaminated needles, breast milk, and transmission from an infected mother to her baby at birth. Screening of blood products for HIV has largely eliminated transmission through blood transfusions or infected blood products in the developed world.

- **Sexual route.** The majority of HIV infections are acquired through unprotected sexual relations. Sexual transmission can occur when infected sexual secretions of one partner come into contact with the genital, oral, or rectal mucous membranes of another.
- **Blood or blood product route.** This transmission route can account for infections in intravenous drug users, hemophiliacs and recipients of blood transfusions (though most transfusions are checked for HIV in the developed world) and blood products. It is also of concern for persons receiving medical care in regions where there is prevalent substandard hygiene in the use of injection equipment, such as the reuse of needles in Third World countries. HIV can also be spread through the sharing of needles. Health care workers such as nurses, laboratory workers, and doctors, have also been infected, although this occurs more rarely. People who give and receive tattoos, piercings, and scarification procedures can also be at risk of infection.
- **Mother-to-child transmission** (MTCT). The transmission of the virus from the mother to the child can occur *in utero* during pregnancy and *intrapartum* at childbirth. In the absence of treatment, the transmission rate between the mother and child is around 25%. However, where combination antiretroviral drug treatment and Cesarian section are available, this risk can be reduced to as low as 1%. Breast feeding also presents a risk of infection for the baby.

Clinical features of HIV infection

Clinical manifestations are quite variable, initial infection may occur, without accompanying symptoms. Many patients experience an acute HIV syndrome within 2-6 weak of exposure of virus. After initial phase latent phase begins it last for 10 years.

Infection with HIV-1 is associated with a progressive decrease of the CD4$^+$ T-cell count and an increase in viral load. The stage of infection can be determined by measuring the patient's CD4$^+$ T-cell count, and the level of HIV in the blood. HIV infection has basically four stages: incubation period, acute infection, latency stage and AIDS. The initial incubation period upon infection is asymptomatic and usually lasts between two and four weeks. The second stage, acute infection, which lasts an average of 28 days and can include symptoms such as fever, lymphadenopathy (swollen lymph nodes), pharyngitis (sore throat), rash, myalgia (muscle pain), malaise, and mouth and esophageal sores. The latency stage, which occurs third, shows few or no symptoms and can last anywhere from two weeks to twenty years and beyond. AIDS, the fourth and final stage of HIV infection shows as symptoms of various opportunistic infections.

HIV primarily infects vital cells in the human immune system such as helper T-cells (specifically $CD4^+$ T-cells), macrophages and dendritic cells. HIV infection leads to low levels of $CD4^+$ T-cells through three main mechanisms: firstly, direct viral killing of infected cells; secondly, increased rates of apoptosis in infected cells; and thirdly, killing of infected $CD4^+$ T-cells by CD8 cytotoxic lymphocytes that recognize infected cells. When $CD4^+$ T-cell numbers decline below a critical level, cell-mediated immunity is lost, and the body becomes progressively more susceptible to opportunistic infections. Eventually most HIV-infected individuals develop AIDS (Acquired Immunodeficiency Syndrome). These individuals mostly die from opportunistic infections or malignancies associated with the progressive failure of the immune system. About 9 out of every 10 persons with HIV will progress to AIDS after 10-15 years. Treatment with anti-retrovirals, where available, increases the life expectancy of people infected with HIV. After the diagnosis of AIDS is made, the current average survival time with antiretroviral therapy (as of 2005) is estimated to be more than 5 years. Without antiretroviral therapy, death normally occurs within a year. It is hoped that current and future treatments may allow HIV-infected individuals to achieve a life expectancy approaching that of the general public.

Symptom of HIV infection

Many people do not develop any symptoms when they first become infected with HIV. Some people, however, get a flu-like illness within three to six weeks after exposure to the virus. This illness, called Acute HIV Syndrome, may include fever, headache, tiredness, nausea, diarrhea and enlarged lymph nodes (organs of the immune system that can be felt in the neck, armpits and groin). These symptoms usually disappear within a week to a month and are often mistaken for another viral infection. During this period, the quantity of the virus in the body will be high and it spreads to different parts, particularly the lymphoid tissue. At this stage, the infected person is more likely to pass on the infection to others. The viral quantity then drops as the body's immune system launches an orchestrated fight. More persistent or severe symptoms may not surface for several years, even a decade or more, after HIV first enters the body in adults, or within two years in children born with the virus. This period of "asymptomatic" infection varies from individual to individual. Some people may begin to have symptoms as soon as a few months, while others may be symptom-free for more than 10 years. However, during the "asymptomatic" period, the virus will be actively multiplying, infecting, and killing cells of the immune system.

Later symptoms of HIV infection

- Lack of energy
- Weight loss
- Frequent fevers and sweats
- A thick, whitish coating of the tongue or mouth (thrush) that is caused by a yeast infection and sometimes accompanied by a sore throat
- Severe or recurring vaginal yeast infections
- Chronic pelvic inflammatory disease or severe and frequent infections like herpes zoster
- Periods of extreme and unexplained fatigue that may be combined with headaches, lightheadedness, and/or dizziness
- Rapid loss of more than 10 pounds of weight that is not due to increased physical exercise or dieting
- Bruising more easily than normal
- Long-lasting bouts of diarrhea
- Swelling or hardening of glands located in the throat, armpit, or groin
- Periods of continued, deep, dry coughing
- Increasing shortness of breath
- The appearance of discoloured or purplish growths on the skin or inside the mouth
- Unexplained bleeding from growths on the skin, from mucous membranes, or from any opening in the body
- Recurring or unusual skin rashes
- Severe numbness or pain in the hands or feet, the loss of muscle control and reflex, paralysis or loss of muscular strength
- An altered state of consciousness, personality change, or mental deterioration.

CD4-HIV Interaction

Once HIV enters the human body, it attaches itself to a white blood cell (WBC) called CD4. Also, called T4 cells, they are the main disease fighters of the body. Whenever there is an infection, CD4 cells lead the infection-fighting army of the body to protect it from falling sick. Damage of these cells, hence can affect a person's disease-fighting capability and general health. After making a foothold on the CD4 cell, the virus injects its RNA into the cell. The RNA then gets attached to the DNA of the host cell and thus becomes part of the cell's genetic material. It is a virtual takeover of the cell. Using the cell's division mechanism, the virus now replicates and churns out hundreds of thousands of its own copies. These cells then enter the blood stream, get attached to other CD4 cells and continue replicating. As a result, the number of the virus in the blood rises and that of the CD4 cells declines. Because of this process, immediately after infection, the viral load of an infected individual will be very high and the number of CD4, low. But, after a while, the body's immune system responds vigorously by producing more and more CD4 cells to fight the virus. Much of the virus gets removed from the blood. To fight the fast-replicating virus, as many as a billion CD4 cells are produced every day, but the virus too increases on a similar scale. The battle between the virus and the CD4 cells continues even as the infected person remains symptom-free. But after a few years, which can last up to a decade or even more, when the number of the virus in the body rises to very high levels, the body's immune mechanism finds it difficult to carry on with the battle. The balance shifts in favour of the virus and the person becomes more susceptible to various infections. These infections are called opportunistic infections because they swarm the body using the opportunity of its low immunity. At this stage, the number of CD4 cells per millilitre of blood (called CD4 Count), which ranges between 500 to 1,500 in a healthy individual, falls below 200. The viral load, the quantity of the virus in the blood, will be very high at this stage.

Structure of HIV

The primary cause of AIDS is called HIV; It has been isolated from bone marrow, blood, lymph node, spleen, saliva, and semen. HIV-1 is RNA virus that employs the enzyme reverse transcriptase it is therefore called retrovirus. In retroviruses, Viral RNA is first transcribe into DNA using reverse transcriptase. HIV is spherical virus containing dense core surrounded by lipoprotein envelope. It is about 120 nm in diameter and roughly spherical. On envelope each virion expresses 72 spikes composed of gp120 and gp41 (Fig. 19.2). The gp41 molecule is transmembrane molecule that crosses the lipid bilayer of viral envelope. The gp120 is associated with gp41 and serve as viral receptor for CD4 on host cells. HIV genome composed of two copies of positive single-stranded RNA that codes for the virus's nine genes enclosed by a conical capsid composed of 2,000 copies of the viral protein p24. The single-stranded RNA is tightly bound to nucleocapsid proteins, p17 and enzymes needed for the development of the virion

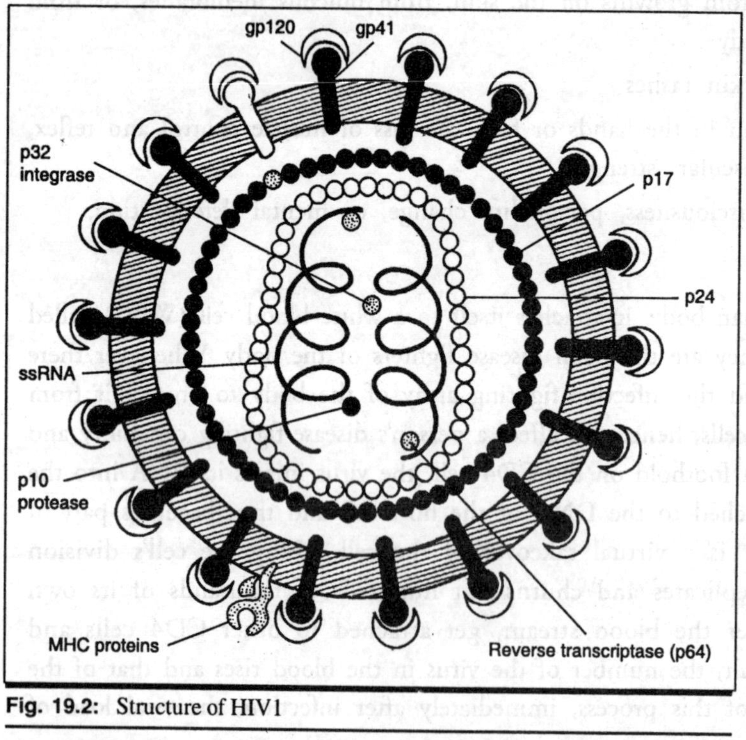

Fig. 19.2: Structure of HIV1

such as reverse transcriptase (p64), proteases, ribonuclease and integrase. A matrix composed of the viral protein p17 surrounds the capsid ensuring the integrity of the virion particle. This is, in turn, surrounded by the viral envelope which is composed of two layers of fatty molecules called phospholipids taken from the membrane of a human cell when a newly formed virus particle buds from the cell. Embedded in the viral envelope are proteins from the host cell and about 70 copies of a complex HIV protein that protrudes through the surface of the virus particle.

Genome of HIV

Genetic organization of HIV of the nine genes that are encoded within the RNA genome, three of these genes, *gag*, *pol*, and *env*, contain information needed to make the structural proteins for new virus particles. *gag* encode core structural protein For example, *env* codes for a envelope glycoprotein called gp160 that is broken down by a viral enzyme to form gp120 and gp41. The *pol* gene codes for reveres transcriptase, endonuclease, and viral protease required for viral replication. The other six are regulatory genes. The six remaining genes, *tat*, *rev*, *nef*, *vif*, *vpr*, and *vpu* (or *vpx* in the case of HIV-2), are regulatory genes for proteins that control the ability of HIV to infect cells, produce new copies of virus (replicate), or cause disease. The tat(transactivator) has been likened to a sort of fast forward switch (Fig. 19.3). It stimulates the production of viral protein. The protein encoded by *nef*, for instance, appears necessary for the virus to replicate efficiently, and the *vpu*-encoded protein influences the release of new virus particles from infected cells. The ends of each strand of HIV RNA contain an RNA sequence called the long terminal repeat (LTR). Regions in the LTR act as switches to control production of new viruses and can be triggered by proteins from either HIV or the host cell.

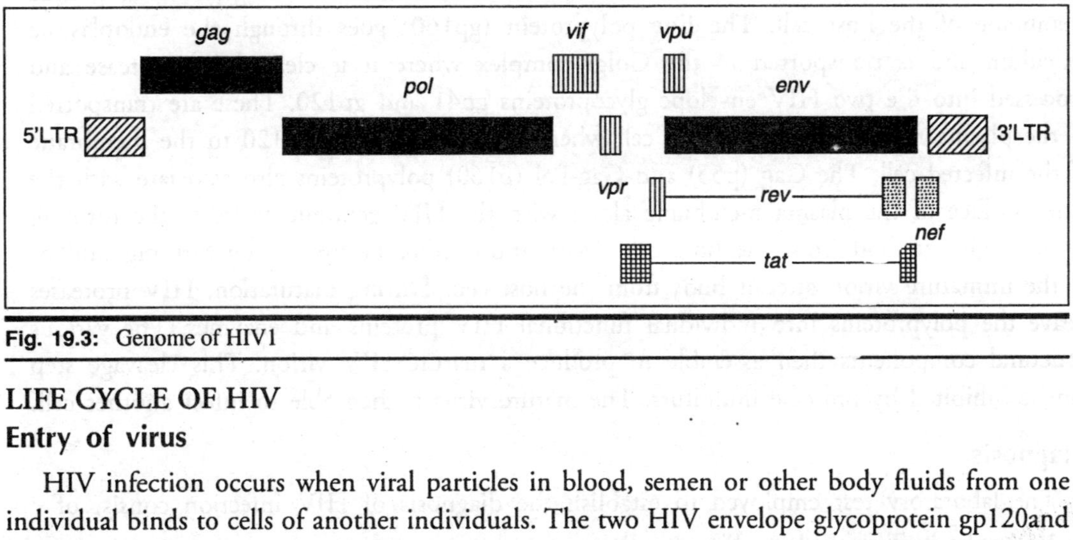

Fig. 19.3: Genome of HIV1

LIFE CYCLE OF HIV

Entry of virus

HIV infection occurs when viral particles in blood, semen or other body fluids from one individual binds to cells of another individuals. The two HIV envelope glycoprotein gp120and gp40 are critical for HIV infection. The first step of HIV infection is high affinity binding of gp120 to CD4 molecule on the surface T-cells. Entry to the HIV begins through interaction of the trimeric envelope complex (gp160 spike, and both CD4 and a chemokine receptor (generally either CCR5 or CXCR4, but others are known to interact) on the cell surface. gp120 binds to integrin $\alpha_4\beta_7$ activating LFA-1 the central integrin involved in the establishment of virological synapses, which facilitate efficient cell-to-cell spreading of HIV-1. The gp160 spike contains binding domains for both CD4 and chemokine receptors. The first step in fusion involves the high-affinity attachment of the CD4 binding domains of gp120 to CD4. Once gp120 is bound with the CD4 protein, the envelope complex undergoes a structural change, exposing the chemokine binding domains of gp120 and allowing them to interact with the target chemokine receptor (Fig.19.4). This allows for a more stable two-pronged attachment, which allows the N-terminal fusion peptide gp41 to penetrate the cell membrane. Repeat sequences in gp41, HR1 and HR2 then interact, causing the collapse of the extracellular portion of gp41 into a hairpin. This loop structure brings the virus and cell membranes close together, allowing fusion of the membranes and subsequent entry of the viral capsid.

Internalization and transcription

Once HIV has bound to the target cell, the HIV RNA and various enzymes, including reverse transcriptase, integrase, ribonuclease and protease, are injected into the cell. All enzymes within nucleocapsid complex become active and begins the viral reproductive cycle. Nucleoprotein core of virus becomes disrupted. RNA genome of HIV is transcribe into dsDNA from viral reverse transcriptase and viral DNA enters the nucleus. The viral integrase also enter the nucleus and catalyze the integration of viral genome into the host cell genome. The integrated DNA from HIV is called provirus.

Assembly and release

The final step of the viral cycle, assembly of new HIV-1 virons, begins at the plasma membrane of the host cell. The Env polyprotein (gp160) goes through the endoplasmic reticulum and is transported to the Golgi complex where it is cleaved by protease and processed into the two HIV envelope glycoproteins gp41 and gp120. These are transported to the plasma membrane of the host cell where gp41 anchors the gp120 to the membrane of the infected cell. The Gag (p55) and Gag-Pol (p160) polyproteins also associate with the inner surface of the plasma membrane along with the HIV genomic RNA as the forming virion begins to bud from the host cell. Maturation either occurs in the forming bud or in the immature virion after it buds from the host cell. During maturation, HIV proteases cleave the polyproteins into individual functional HIV proteins and enzymes. The various structural components then assemble to produce a mature HIV virion. This cleavage step can be inhibited by protease inhibitors. The mature virus is then able to infect another cell.

Diagnosis

The laboratory test employed to establish the diagnosis of HIV infection consist of

1. Antibody Test: ELISA, Western Blot
2. Test for virus and viral antigen
3. Test for proviral DNA
4. Test for immune defect

1. **Binding and Fusion:** HIV begins its life cycle when it binds to a CD4 receptor and one of two co-receptors on the surface of a CD4+ T-lymphocyte. The virus then fuses with the host cell. After fusion, the virus releases RNA, its genetic material, into the host cell.
2. **Reverse Transcription:** An HIV enzyme called reverse transcriptase converts the single-stranded HIV RNA to double-standed HIV DNA.
3. **Integration:** The newly formed HIV DNA enters the host cell's nucleus, where an HIV enzyme called integrase "hides" the HIV DNA within the host cell's own DNA. The integrated HIV DNA is called provirus. The provirus may remain inactive for several years, producing few or no new copies of HIV.
4. **Transcription:** When the host cell receives a signal to become active, the provirus uses a host enzyme called RNA polymerase to create copies of the HIV genomic material, as well as shorter strands of RNA called messenger RNA (mRNA). The mRNA is used as a blueprint to make long chains of HIV proteins.
5. **Assembly:** An HIV enzyme called protease cuts the long chains of HIV proteins into smaller individual proteins. As the smaller HIV proteins come together with copies of HIV's RNA genetic material, a new virus particle is assembled.
6. **Budding:** The newly assembled virus pushes out ("buds") from the host cell. During budding, the new virus steals part of the cell's outer envelope. This envelope, which acts as a covering, is studded with protein/sugar combinations called HIV glycoproteins. These HIV glycoproteins are necessary for the virus to bind CD4 and coreceptors. The new copies of HIV can now move on to infect other cells.

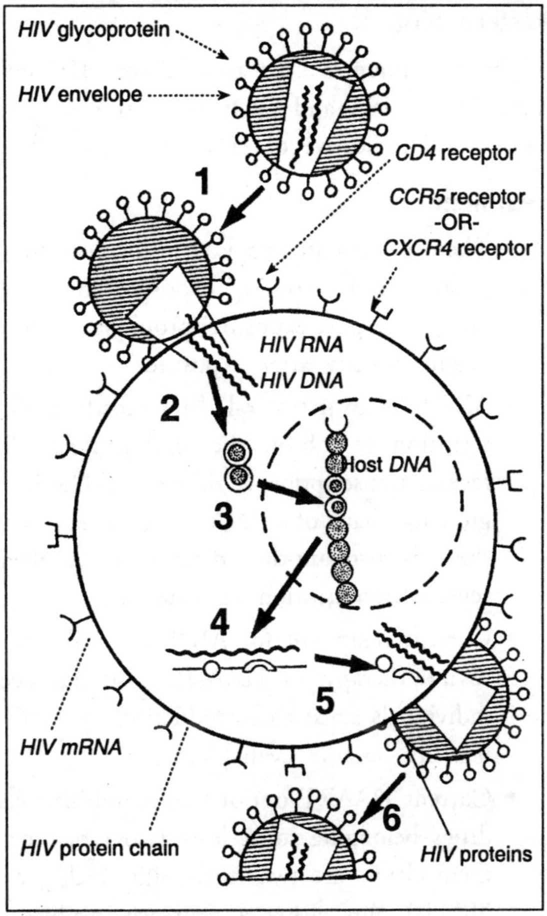

Fig. 19.4: Life cycle of HIV

ELISA

For initial and screening test of HIV serology the ELISA test is preferred. The specificity and sensitivity of this test is generally 90-99%. HIV-1 testing consists of initial screening with an enzyme-linked immunosorbent assay (ELISA) to detect antibodies to HIV-1. Specimens with a nonreactive result from the initial ELISA are considered HIV-negative unless new exposure to an infected partner or partner of unknown HIV status has occurred. Specimens with a reactive ELISA result are retested in duplicate. If the result of either duplicate test is reactive, the specimen is reported as repeatedly reactive and undergoes confirmatory testing with a more specific supplemental test (e.g. Western blot)Sera which test positively twice, consecutively by ELISA are subject to confirmation by western blot.

Western blot

This technique is used to confirm HIV infection. Only specimens that are repeatedly reactive by ELISA and positive by IFA or reactive by western blot are considered HIV-positive and indicative of HIV infection.

Treatment

- There is currently no vaccine or cure for HIV or AIDS. The only known method of prevention is avoiding exposure to the virus. However, an antiretroviral treatment, known as post-exposure prophylaxis, is believed to reduce the risk of infection if begun directly after exposure.
- AZT-It is antiretroviral drug mostly useful in AIDS that interface with reverse transcription or inhibit the viral protease. The prototype of drug that interface with reverse transcription is zidovudin. The introduction of AZT, a nucleoside analog, into growing chain of cDNA of retrovirus causes chain termination. In addition to AZT, there is also another drug such as Nevirapine which inhibit the action of that reverse transcription enzyme.
- Current treatment for AIDS is combination therapy consists of highly active antiretroviral therapy, or HAART. This has been highly beneficial to many HIV-infected individuals since its introduction in 1996, when the protease inhibitor-based HAART initially became available.
- Current HAART options are combinations (or "cocktails") consisting of at least three drugs belonging to at least two types, or "classes," of antiretroviral agents. Typically, these classes are two nucleoside analogue reverse transcriptase inhibitors (NARTIs or NRTIs) plus either a protease inhibitor or a non-nucleoside reverse transcriptase inhibitor (NNRTI). Because AIDS progression in children is more rapid and less predictable than in adults, particularly in young infants, more aggressive treatment is recommended for children than adults.
- There is no question that treatment should be started before the patient's CD4 count falls below 200, and most national guidelines say to start treatment once the CD4 count falls below 350; but there is some evidence from cohort studies that treatment should be started before the CD4 count falls below 350.
- Anti-retroviral drugs are expensive, and the majority of the world's infected individuals do not have access to medications and treatments for HIV and AIDS. This is because a vaccine would cost less, thus being affordable for developing countries, and would not require daily treatment.

Promising new treatments include Cre recombinase and the enzyme Tre recombinase both of which are able to remove HIV from an infected cell. These enzymes promise a treatment in which a patient's stem cells are extracted, cured, and reinjected to promulgate the enzyme into the body. The carried enzyme then finds and removes the virus.

CHAPTER 20

AUTOIMMUNITY

INTRODUCTION

The term "autoimmune disease" applies to conditions where the autoimmune process contributes to the pathogenesis of the disease. This is different to situations where apparently harmless autoantibodies are formed following tissue damage, for example heart antibodies appearing after a myocardial infarction.

Although the immune system has an elaborate system of checks and balances to ensure self tolerance, occasionally this system breaks down. When the immune system attacks host components causing pathological change, this is called **autoimmunity**. Many people experience an autoimmune reaction during their lifetime. Mostly these are short-lived, self-resolving sequelae of infection. However in some 5% of individuals, the reaction is chronic, debilitating and even (rarely) life-threatening. It is these latter conditions where serious immunopathology occurs which are usually considered autoimmune disease. We shall consider the following aspects:

1. The characteristics of autoimmune diseases
2. Which immune mechanisms are involved in bringing about the pathogenic change?
3. What factors initiate the autoreactivity?

Autoimmune diseases form a spectrum ranging from organ-specific conditions in which one organ only is affected to systemic diseases in which the pathology is diffused throughout the body (Fig. 20.1). The extremes of this spectrum result from quite distinct underlying mechanisms, but there are many conditions in which there are components of both organ-specific and systemic damage.

Examples of autoimmune disease

Grave's Disease (thyrotoxicosis)
Hashimoto's thyroiditis
pernicious anaemia
Addison's disease
insulin dependent diabetes mellitus
Goodpasture's syndrome
myasthenia gravis
multiple sclerosis(?)
autoimmune haemolytic anaemia
idiopathic thrombocytopenic purpura
rheumatoid arthritis
scleroderma
systemic lupus erythematosis (SLE)

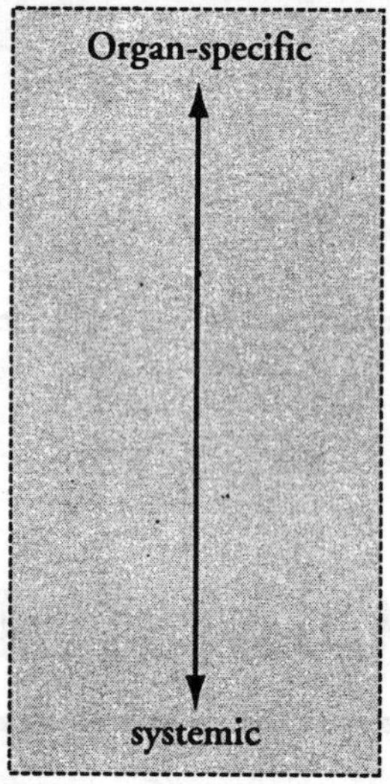

Fig. 20.1: Shows the various types of autoimmune diseases

Organ-specific autoimmune diseases

In an organ-specific autoimmune disease, the immune response is directed to a target antigen unique to a single organ or gland, so that the manifestations are largely limited to that organ. The cells of the target organs may be damaged directly by humoral or cell-mediated effector mechanisms. Alternatively, the antibodies may overstimulate or block the normal function of the target organ. Autoimmune diseases involving direct cellular damage occur when lymphocytes or antibodies bind to cell-membrane antigens causing cellular lysis and/or an inflammatory response in the affected organ. Gradually, the damaged cellular structure is replaced by connective tissue (scar tissue), and the function of the organ declines.

Graves' disease

This is a rare example of an autoimmune disease which can be transferred with IgG antibodies. Firstly passive transfer of IgG from patients to rats often produces similar symptoms transiently in the animals. Secondly babies born to mothers with Graves' have shown transient symptoms of hyperthyroidism which disappear with catabolism of the maternal IgG (transferred via the placenta) and are relieved by plasma exchange. The disease causing antibodies can be shown to recognise the thyroid stimulating hormone (TSH) receptor and to stimulate thyrocytes *in vitro.*

The production of thyroid hormones is carefully regulated by thyroid-stimulating hormone (TSH), which is produced by the pituitary gland. Binding of TSH to a receptor on thyroid cells activates adenylate cyclase and stimulates the synthesis of two thyroid hormones, thyroxine and triiodothyronine. A patient with **Graves' disease** produces auto-antibodies that bind the receptor for TSH and mimic the normal action of TSH, activating adenylate cyclase and resulting in production of the thyroid hormones (Figure 20.2). Unlike TSH, however, the autoantibodies are not regulated, and consequently they over-stimulate the thyroid. For this reason these auto-antibodies are called long-acting thyroid-stimulating (LATS) antibodies.

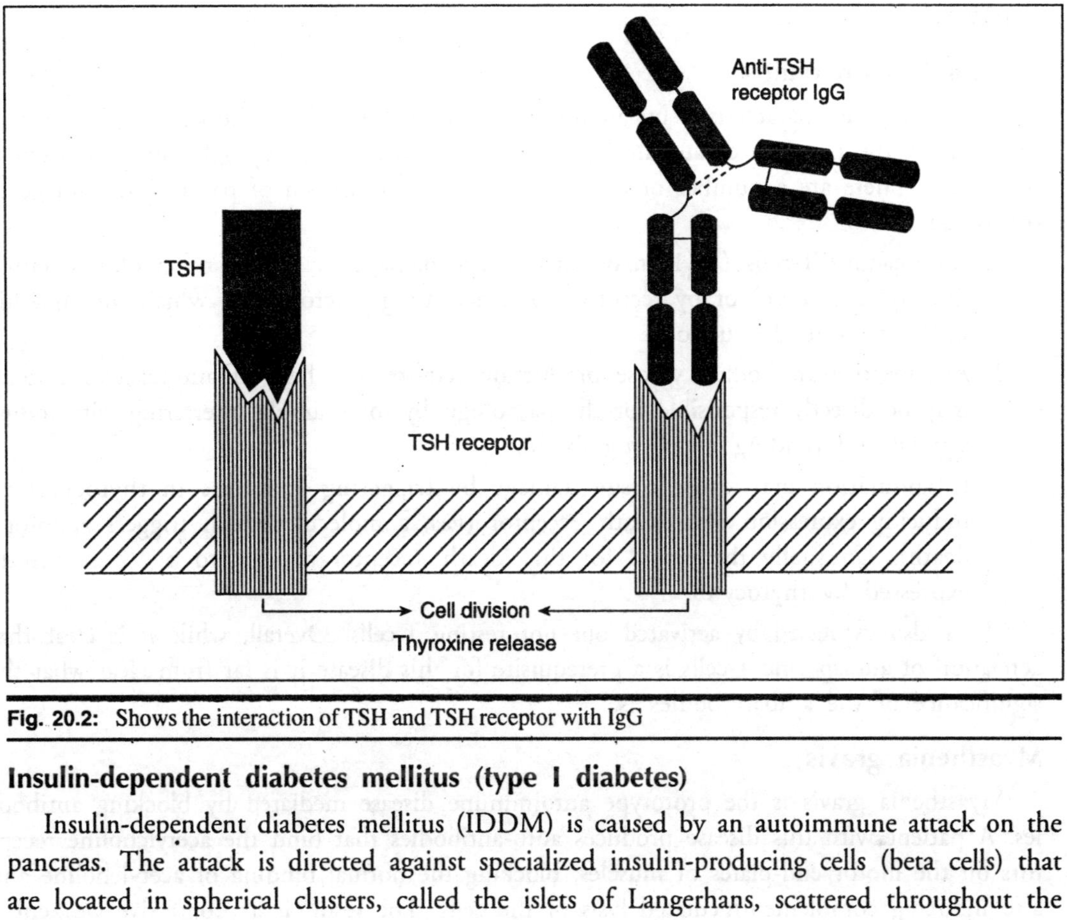

Fig. 20.2: Shows the interaction of TSH and TSH receptor with IgG

Insulin-dependent diabetes mellitus (type I diabetes)

Insulin-dependent diabetes mellitus (IDDM) is caused by an autoimmune attack on the pancreas. The attack is directed against specialized insulin-producing cells (beta cells) that are located in spherical clusters, called the islets of Langerhans, scattered throughout the pancreas. The autoimmune attack destroys beta cells, resulting in decreased production of insulin and consequently increased levels of blood glucose. Several factors are important in the destruction of beta cells. Auto-antibodies to beta cells may contribute to cell destruction by facilitating either antibody-plus-complement lysis or antibody-dependent cell-mediated cytotoxicity (ADCC). A disease afflicting 0.2% of the population.

Insulin-dependent diabetes mellitus (type I diabetes) is a multifactorial autoimmune disease the major feature of which is a T-cell infiltration of the islets of Langerhans and progressive **T-cell-mediated destruction** of insulin-producing B-cells. Auto-reactive T-cells specific for B-cell antigens such as glutamic acid decarboxylase (GAD) and insulin can be identified in the blood of newly diagnosed diabetic patients. Autoantibodies specific for the same antigens can also be detected in the serum of patients and can be used as a sensitive marker to predict the risk of developing the disease. It is not clear if these antibodies are the result of T-cell-mediated injury or if they play some role in the causation of the disease.

Delay in onset of disease can be achieved by early treatment with cyclosporin, since this drug targets T-cell cytokine synthesis so specifically. The disease has a strong genetic basis and is associated with genes within the MHC, particularly the *DR3.DQ2* and the *DR4.DQ8* haplotypes, and with polymorphisms within the insulin gene region on chromosome 11p15.5.

Hashimoto's thyroiditis

This disease is characterised by an intense mononuclear cellular infiltrate into the thyroid and by the presence of autoantibodies primarily directed at thyroglobulin and thyroid peroxidase. There are a number of theories about the mechanism of pathogenic damage to the tissue.

1. Autoreactive T-cells (T_H1) may cause tissue damage by release of cytokines, either directly (e.g. TNF) or by recruiting and activating macrophages, which subsequently mediate tissue destruction.
2. Autoreactive antibodies, whose production requires the help of autoreactive T cells, may be directly responsible for the pathology, by for example interfering with iodine uptake and binding by thyroglobulin.
3. Inflammation may cause tissue damage by triggering apoptosis in thyrocytes by inducing expression of a 'death' receptor (Fas, a molecule which triggers apoptotic death). Unusually the ligand for this 'death' receptor appears to be constitutively expressed by thyrocytes.

It is also expressed by activated but not resting T-cells. Overall, while it is clear that activation of autospecific T-cells is a prerequisite for this disease it is far from clear what the significance of the autoantibodies is.

Myasthenia gravis

Myasthenia gravis is the prototype autoimmune disease mediated by blocking antibodies. A patient with this disease produces auto-antibodies that bind the acetylcholine receptors on the motor end-plates of muscles, blocking the normal binding of acetylcholine and also inducing complementmediated lysis of the cells. The result is a progressive weakening of the skeletal muscles. Ultimately, the antibodies destroy the cells bearing the receptors. The early signs of this disease include drooping eyelids and inability to retract the corners of the mouth, which gives the appearance of snarling. Without treatment, progressive weakening of the muscles can lead to severe impairment of eating as well as problems with movement. However, with appropriate treatment, this disease can be managed quite well and afflicted individuals can lead a normal life.

Systemic autoimmune diseases

This is second type of autoimmune diseases called as systemic autoimmune diseases. In systemic autoimmune diseases, the response is directed toward a broad range of target antigens and involves a number of organs and tissues. These diseases reflect a general defect in immune regulation that results in hyperactive T-cells and B-cells. Tissue damage is widespread, both from cell-mediated immune responses and from direct cellular damage caused by auto-antibodies or by accumulation of immune complexes.

Systemic lupus erythematosus(SLE)

Systemic lupus erythematosus attacks many tissues. One of the best examples of a systemic autoimmune disease is **systemic lupus erythematosus** (SLE), which typically appears in women between 20 and 40 years of age; the ratio of female to male patients is 10:1. SLE is characterized by fever, weakness, arthritis, skin rashes, pleurisy, and kidney dysfunction. Lupus is more frequent in African-American and Hispanic women than in Caucasians, although it is not known why this is so. Affected individuals may produce auto-antibodies to a vast array of tissue antigens, such as DNA, histones, RBCs, platelets, leukocytes, and clotting factors; interaction of these auto-antibodies with their specific antigens produces various symptoms. Auto-antibody specific for RBCs and platelets, for example, can lead to complement-mediated lysis, resulting in hemolytic anaemia and thrombocytopenia, respectively. When immune complexes of auto-antibodies with various nuclear antigens are deposited along the walls of small blood vessels, a type III hypersensitive reaction develops. The complexes activate the complement system and generate membrane-attack complexes and complement split products that damage the wall of the blood vessel, resulting in vasculitis and glomerulonephritis.

Systemic lupus erythematosus is a relapsing and remitting multisystem disease where auto-antibodies are formed against soluble components to which they have continual access. The complexes that are generated give rise to lesions similar to those occurring in serum sickness. A variety of different autoantigens are present in lupus, many of them within the nucleus, with the most pathognomonic being **double-stranded DNA(dsDNA)**. The detection of antinuclear antibodies (ANAs) and particularly the measurement of anti-DNAantibodies are valuable tests for the diagnosis of SLE. Deposition of complexes is widespread, as the name implies, and lesions are often present in the kidney, skin, joints, muscle, lung and brain. About 40% of patients eventually develop kidney involvement and eluates of renal tissue from patients with lupus nephritis will show anti-dsDNA. Immunofluorescent staining of kidney biopsies from patients with evidence of renal dysfunction will show the presence of antigen-antibody complexes and complement. The staining pattern with a fluorescent anti-IgG or anti-C3 is punctate or "lumpy-bumpy", in marked contrast with the linear pattern caused by the g.b.m. antibodies in Goodpasture's syndrome. Patients with SLE produce antibodies to a variety of tissue antigens including those on red cells, platelets, lymphocytes and various clotting factors. During the active phase of the disease, serum complement levels fall as components are binding to immune aggregates in the kidney and circulation.

Rheumatoid arthritis

Rheumatoid arthritis is a common autoimmune disorder, most often affecting women from 40 to 60 years old. The major symptom is chronic inflammation of the joints, although the hematologic, cardiovascular, and respiratory systems are also frequently affected. Many individuals with rheumatoid arthritis produce a group of auto-antibodies called rheumatoid factors that are reactive with determinants in the Fc region of IgG. The classic rheumatoid factor is an IgM antibody with that reactivity. Such auto-antibodies bind to normal circulating IgG, forming IgM-IgG complexes that are deposited in the joints. These immune complexes can activate the complement cascade, resulting in a type III hypersensitive reaction, which leads to chronic inflammation of the joints.

Rheumatoid arthritis is a common destructive arthropathy of unknown etiology, strongly linked to the MHC class II proteins HLA-DRB1*0404 and *0401. The joint changes are produced by the hyperplasia of the synovial cells associated with increased vascularity and infiltration of inflammatory cells forming a pannus overlaying and destroying cartilage and bone (Figure 26.3). The infiltrating cells are primarily CD4+T-cells, which stimulate monocytes, macrophages and mast cells to secrete IL-1, IL-6, TNF and a variety of chemokines, which recruit neutrophils into the joints. The IL-1 and TNF in the synovium stimulate fibroblasts and chondrocytes to release tissue-destroying proteolytic enzymes which lead to joint damage, and bone destruction follows the stimulation of osteoclasts by these cytokines. As the malign pannus (cover) grows over the cartilage, tissue breakdown can be seen at the margin, almost certainly as a result of the release of enzymes, ROIs and especially IL-1, IL-6 and TNF. B-cells are also activated and plasma cells are frequently observed. Secondary lymphoid follicles with germinal centers may be present in the synovium. Immune complexes of rheumatoid factor and IgG may initiate an Arthus reaction in the joint space leading to an influx of polymorphs.

Treatment

The central role of cytokines in RA is appreciated, a number of anticytokine therapies have been developed. These include a soluble TNF receptor-IgG1 fusion protein (Etanercept) which neutralizes free TNF, and a chimeric monoclonal antibody against TNF itself (Infliximab). By blocking TNF activity, its ability to activate the cytokine cascade of IL-1, IL-6, IL-8 and other inflammatory cytokines is impeded, making this a valuable adjunct to the treatment of RA. Another approach is to block the activation of CD28 on T-cells using a fusion protein made up of CTLA-4 and IgG1.This competes with CD28 for binding to B7.1 (CD80) and B7.2 (CD86) and prevents T-cell activation. By acting so early in the inflammatory cascade CTLA4Ig inhibits the secondary activation of macrophages and B-cells and shows considerable promise in treating patients with RA.

Factors initiate the auto reactivity

Endocrine factors

Most autoimmune disease do not occur with equal frequency in males and females. For example Graves' and Hashimoto's are 4-5 times, and SLE 10 times, more common in females while Ankylosing Spondylitis is 3-4 × more frequent in males. These differences are believed to be the result of hormonal influences.

A second well documented hormonal effect is the marked reduction in disease severity seen in many autoimmune conditions during pregnancy. Rheumatoid arthritis is perhaps the classic example of this effect. In some cases there is also a rapid exacerbation (rebound) after giving birth.

Environment

However, it is clear that environmental factors also play a role in autoimmune disease. If you examine how frequently identical twins both develop a disease (the **concordance** rate), it is only about 20-40% for common autoimmune diseases such as diabetes, SLE and rheumatoid arthritis. This makes it highly likely that environmental factors must also be important. While we might expect factors such as diet to play a role, we can postulate that infectious organisms are the most significant environmental factor.

Mechanisms of autoimmune pathology

Although in many cases the precise combination of pathogenetic mechanisms is not understood, there are three general types of mechanism which it is important to distinguish.

- Direct antibody mediated effects
- T-cell mediated effects (cellular immune)
- Immune complex mediated effects

Direct antibody mediated pathology

We have already seen one example in Graves' disease where stimulatory antibodies dysregulate thyroid function. A counter effect can also be seen as in Myasthenia Gravis. In this disease, auto-antibodies to the acetylcholine receptor block neuromuscular transmission from cholinergic neurons by blocking the binding of acetylcholine and by causing down-regulation (degradation) of its' receptor. Rheumatic fever is an example of direct tissue pathology following antibody binding.

T-cell mediated damage

This term implies that the recognition of autoantigen by T-cells leads to tissue destruction without requiring the production of auto-antibody. There are a number of ways this can come about:

1. Direct T-cell cytotoxicity via $CD8^{+}$ CTL
2. Self-destruction of tissue cells induced by cytokines, e.g. TNFa
3. Recruitment and activation of macrophages leading to bystander tissue destruction
4. Induction of target tissue apoptosis by the T-cell membrane protein **FasL**

In most cases we do not know what the relative contribution of these factors is, though mechanism 4 would have to be confined to those tissues which constitutively express the 'death' protein Fas or which express it as a consequence of inflammation.

Immune-complex mediated pathology

Immune-complexes are frequently implicated in autoimmune pathology. Systemic diseases such as SLE and vasculitis almost certainly result from autoantibody-antigen complexes and their consequences. Certain organs are especially sensitive to immune complex deposition particularly the kidney. SLE patients possess a wide variety of auto-antibodies to both cytoplasmic and nuclear antigens. The presence of IgG anti double-stranded DNA is characteristic of this condition (Note: IgM anti-ds DNA is NOT pathogenic). Two significant facts point to the role of immune complexes in SLE. First, patients demonstrate significant depletion of complement (C3) and neutrophils resulting from activation by the complexes. Second, complement deficiencies which impair IC clearance (C1, C2 or C4, see lecture 10) are very strong predisposing factors for SLE.

TABLE 20.1
Autoimmune diseases and antibody

Diseases	Antibody	Comments
Primary myxedema	Thyroid	Tests +ve in 99% of cases.
Hashimoto's thyroiditis	Thyroid	Distinction from colloid goiter, thyroid cancer and subacute thyroiditis Thyroidectomy usually unnecessary in Hashimoto goiter.
Thyrotoxicosis	Thyroid	High titers of cytoplasmic Ab indicate active thyroiditis and tendency to post-operative myxedema: anti-thyroid drugs are the treatment of choice although HLA-B8 patients have high chance of relapse.
Pernicious anaemia	Stomach	Help in diagnosis of latent PA.
Insulin-dependent diabetes mellitus (IDDM)	Pancreas	Insulin Ab early in disease. GAD Ab standard test for IDDM. Two or more autoAb seen in 80% of new onset children or prediabetic relatives but not in controls.
Idiopathic adrenal atrophy	Adrenal	Distinction from tuberculous form
Myasthenia gravis	Muscle ACh receptor	When positive suggests associated thymoma (more likely if HLA-B12), positive in >80%.
Pemphigus vulgaris and pemphigoid	Skin	Different fluorescent patterns in the two diseases.
Autoimmune hemolytic anaemia	Erythrocyte (Coombs' test)	Distinction from other forms of anaemia.
Rheumatoid arthritis	Antiglobulin,e.g. SCAT and latex fixation Antiglobulin + raised agalacto-Ig	High titer indicative of bad prognosis Prognosis of rheumatoid arthritis.
SLE	High titer antinuclear, DNA	DNA antibodies present in active phase. Ab to double-stranded DNA characteristic; highaffinity complement-fixing Ab give kidney damage, low affinity CNS lesions.

CHAPTER 21

TRANSPLANTATION

EARLY HISTORY

The earliest known records of tissue transplantation are those of the Hindu surgeon Sushrutu who reported the use of a flap from a patient's forehead to repair an amputated nose. This procedure was probably practiced by Hindu surgeons as early as 700 B.C. In the 15th century, Italian surgeons began to practice rhinoplasty by means of flaps and extended the donor site to the patient's arm. In 1503, one such surgeon reported the first allograft—the grafting of skin from a slave for the reconstruction of the master's nose. A sizable legend grew out of such reports, although obviously unfounded in fact. Skin grafting became an accepted practice in the late 1800s. Many workers, however, did not distinguish between autografts (donor and recipient the same individual) and allografts (donor and recipient of the same species) or even, sometimes, xenografts (donor and recipient of different species). The last of these formed the basis for an extensive practice known as zoografting in which patients were subjected to grafts from animals ranging from pigs to frogs. Billingham points out that no one apparently cared whether the grafted skin "took" or merely promoted healing of the wound. The results of these efforts led to a period of confusion in transplantation. Without any clear understanding of the processes involved, surgeons embarked on all sorts of transplants, and a series of operations were reported that we know, from our present understanding of the laws of transplantation, could not possibly have been successful. Dr. Serge Voronoff, for example, attained considerable fame (and fortune) in Europe by developing an unusual grafting procedure in which testicles were transplanted from ape to human in order to restore men to youth and vitality. In 1936, Voronoy, a Russian surgeon, reported the first clinical renal allograft. There was apparently a mismatch of blood types and the patient died having demonstrated only minimal renal function. The early postwar years saw reports of attempts at clinical renal homotransplantation from various locales around the world. One of the important contributions to the understanding of transplantation in this era was the work of Medawar. In 1943, Gibson and Medawar reported their experience with autologous and allogeneic skin grafts on a woman who had suffered extensive third-degree burns. The allografts in this case were taken from the patient's sibling, and for clinical reasons were transplanted in two different stages separated by about a week. The authors observed accelerated rejection of the second grafts. Appreciating the possible significance of these observations, Medawar followed them with a series of grafting experiments in rabbits and mice. By 1945, he was able to conclude that "resistance to homologous grafted skin therefore belongs to the general category of actively acquired immune reactions", thereby establishing the relationship of clinical transplantation to the field of immunology. In 1952, the first successful renal transplant was performed in Boston using the kidney of an identical twin.

The transplantation of organs and cells between individuals saves or prolongs thousands of lives each year. The growing list of organs transplanted includes corneas, kidneys, livers, hearts, lungs, small intestines, pancreata, and even hands. Currently, clinical cellular transplantation includes islets of Langerhans and hematopoietic cells, but the list is likely to expand in the future to include other cell types, such as hepatocytes and myoblasts, which are currently under investigation in experimental models. Success of all types of transplants depends on the ability to avoid rejection due to a host-versus-graft immune response. Hematopoietic cell transplantation is, in addition, associated with some special considerations, since the administration of a donor graft that contains mature T-cells to a conditioned, and consequently immunoincompetent recipient, is associated with the risk of rejection in the graft-versus-host (GVH) direction, that is, graft-versus-host disease (GVHD). Because all of these transplants are performed between different humans (i.e. members of the same species), they are referred to as allotransplants. However, improvements in immunosuppressive therapies have increased the success of allotransplantation to the point that a new limitation has been encountered—the insufficient supply of human organs. While this insufficiency can be partly overcome by the increasing use of living donors as sources of kidneys, liver, and lungs, such procedures are associated with significant risks to the donor, and some organs (e.g. the heart) can, obviously only be obtained from cadaveric donors. This organ shortage has led to considerable interest in alternative sources of organs, such as artificial organs, tissue grafts engineered in vitro from stem cells, and donors from other species. Transplants from other species, referred to as xenografts, are felt by many to be a promising solution to the organ shortage, but present even greater immunologic challenges than allografts, as well as potentially novel infectious risks, and these problems must be addressed before xenotransplantation can be effectively used in the clinic. This chapter presents an overview of the immunology of organ and cellular allotransplantation, as well as xenotransplantation.

Transplantation is the introduction of biological material—organs, tissue, cells, and fluids—into an organism. It can be distinguished into three critical relationships between the transplanted material and the recipient. **Syngeneic** transplants—from genetically identical individuals, usually the same individual *(these are similar to grafts between identical twins* *or isogenic strains of experimental animals).* **Allogeneic** transplants—from one individual to another of the same species. **Xenogeneic** transplants—between individuals of different species. Unsurprisingly syngeneic transplants do not usually generate any immunological problems, but allogeneic and xenogeneic transplants are almost always destroyed by immunological processes unless some action is taken to impair the immunological process. Basically therefore transplantation presents two key problems.

1. Genetic variation between donor and recipient.
2. Immunological recognition of the variation.

1. Genetic variation between donor and recipient

Genetic variation between individuals that results in protein sequence differences is at the heart of the transplant problem. In an essentially outbred species like *Homo sapiens* the extent of this allelic polymorphism is considerable. We do not yet have an accurate estimate of the average number of proteins whose sequence varies from one individual to another but it must be greater than a few hundred, possibly as high as several thousand. Obviously the variation is even greater between individuals from different species.

2. Immunological recognition of the variation

However genetic differences between donor and recipient are only of significance in transplantation if they cause incompatibility. Almost ubiquitous in allogeneic transplants is immunological rejection. Early experimental work on allotransplantation in mice identified a very clear distinction between one chromosomal region and the remainder of the genome. Non-identity at this special region always led to very rapid rejection of the transplanted tissue, even if this was the only genetic difference between the donor and recipient. This region was therefore termed the **Major Histocompatibility Complex (MHC).** We now know that this region exists in all vertebrates and that it is highly polymorphic, so that in outbred populations two individuals will almost certainly differ at this region unless they are monozygotic. However the corollary is NOT true, that is even if identical at the MHC, transplants between individuals are likely to be rejected due to **minor histocompatibility loci.** In the mouse there are about 50 such loci, in humans probably more. The key distinction is that **individually** these minor H loci are less 'strong' and in particular the strength varies between allelic differences.

For example, mice differing only at the locus H-1 will reject skin grafts in 25 days if H-1^b skin is transplanted onto H-1^c recipients but in 100 days if the transplant is done in the reverse direction (by comparison MHC-disparate grafts will be rejected in 11 days in all cases).

Recognition and rejection mechanisms

There are three basic types of 'recognition' which allows the host to know that the transplanted tissue is foreign.

1. Recognition by antibody
2. Direct recognition-recognition of foreign MHC by T-cells
3. Indirect recognition-recognition of minor H loci by T-cells

As implied above these three recognitions may lead to very different time scales of destruction of the transplanted cells/tissue and trigger distinct effector mechanisms.

Mechanisms of graft rejection

There are at least four distinct mechanisms that can cause graft rejection that have been identified at this time and it is likely that additional mechanisms will be characterized in the future. It is convenient to describe these mechanisms according to the time frame in which they tend to occur in clinical practice, especially as their names (hyperacute rejection, accelerated rejection, acute rejection, and chronic rejection) have a clear temporal distinction. However, it is increasingly possible to characterize these mechanisms according to the cell types and processes involved and, in some cases, they may occur at uncharacteristic times.

Early or accelerated rejection

A second mechanism of rejection, usually caused by antibodies, is almost as infrequent as hyperacute rejection. It occurs as a result of antibodies that are induced very rapidly after a transplant is performed. This type of rejection has sometimes been called "accelerated" rejection because it typically occurs within the first 5 days, but there is no consensus regarding this name. The process is characterized by fibrinoid necrosis of donor arterioles with intravascular thrombosis.

Hyperacute rejection

This type of rejection occurs very rapidly, resulting in necrosis of the transplanted tissue within minutes or a few hours of contact. It always results from the reactivity of the donor cells with **pre-existing** antibody. The most common situation in which this occurs is in ABO blood group incompatible transplants, therefore we will briefly review the ABO system.

Blood groups

Blood groups arise from genetic variations within a species. Blood transfusion is the oldest form of transplantation and today we should be careful to distinguish what component of blood we are transplanting; what we call blood groups are genetically variable (polymorphic) structures present on red blood cells. It is important to note that some of these structures are only present on red blood cells and thus incompatibility at these loci only affects the transplantation of RBC (blood transfusion), whereas other are present on many tissue cells and therefore affect other types of transplantation.

ABO is special in that it is the only histocompatibility alloantigen for which pre-existing antibody is present in naive, previously untransplanted (untransfused) recipients. Actual mechanism of hyperacute rejection involves, complement activation and secondary activation of coagulation.

It is rejection caused by preformed antibodies. Hyperacute rejection is said to occur when a vascularized organ suffers from rejection within minutes to hours after transplantation. The phenomenon is visible and dramatic. Transplanted kidneys that have initially perfused well turn blue and mottled shortly after vascularization is established. Urine output ceases and recovery does not occur. Microscopically, organs show evidence of extensive vascular thrombosis and hemorrhage with little evidence of a mononuclear cell infiltrate. There are several important components involved in the mechanism of hyperacute rejection. First, there are donor endothelial MHC antigens or carbohydrate determinants as described above. Second, there are preformed antibodies that can bind these antigens. Third, the complement and coagulation cascades are activated by the binding of preformed antibodies to the donor antigens. Finally, there are complement regulatory proteins that can modify complement activation, and anticoagulants that can modify the coagulation pathway. The target of the hyperacute rejection process is the donor vascular endothelium.

The crucial event in the process is the formation of the membrane attack complex (MAC) made up of C5-9 of the complement cascade. In allogeneic combinations, this is always initiated by antibody/antigen binding, which activates complement through the classical pathway. In some xenogeneic combinations, complement activation can also occur through the alternative pathway, and thus does not require antibody binding. Complement activation is controlled by several regulatory molecules, including complement receptor 1, decay accelerating factor (DAF, CD55), membrane cofactor protein (MCP, CD46), and CD59, which act at different stages along the cascade. Many of these molecules are produced by the vascular endothelial cells. Since these regulatory proteins prevent unwanted complement activation in the face of low levels of perturbation to the system, the initial stimulus for activation must be strong enough to overcome these down-regulating molecules. Thus, the titer and avidity of the preformed antibodies must be relatively high. Preformed antibodies directed at MHC antigens almost always accomplish this activation, whereas the lower-affinity blood group antibodies lead to hyperacute rejection in only about 25% of cases. One of the reasons that hyperacute rejection is such an important feature in xenogeneic transplantation is that the complement regulatory proteins produced by the donor vascular endothelium of one species do not always function effectively with complement molecules derived from a different species. Because of this homologous restriction, lower levels of an initial triggering signal lead to explosive complement activation. Although the membrane attack complex is often thought of as a lytic molecule, its effect on the donor vascular endothelium, even before cell lysis, is to cause endothelial activation. This occurs rapidly, before there is time for new gene transcription or protein synthesis, and has been referred to as type I endothelial activation. The two principal manifestations of this activation are cell retraction, leading to gaps between endothelial cells, and the loss of antithrombotic molecules from the endothelium. Thus, type I endothelial activation is responsible for the two principal pathologic findings in hyperacute rejection: extravascular hemorrhage and edema, and intravascular thrombosis.

Rejection caused by T-Cells (Acute Rejection)

Although in clinical practice few allogeneic organs suffer either hyperacute or accelerated rejection after careful crossmatching, rejection episodes occurring toward the end of the first week after transplantation are not infrequent, despite the use of immunosuppression. These episodes are separable from the humoral rejection processes by the later timing of their occurrence, by the absence in many cases of antidonor antibodies in the recipient, and by the cellular infiltrate usually present in the biopsy called "acute rejection" episodes, most rejection treated by clinicians is of this type. Acute rejection occurs with decreasing frequency after the first 3 months, but rejection by apparently similar mechanisms may occur much later after transplantation, especially if immunosuppressive medication is withdrawn. Acute rejection of organ allografts is T-cell–mediated. Therefore, treatment is usually with increased doses of standard immunosuppressive drugs or with anti lymphocyte antibodies. These strategies are so likely to be successful that the diagnosis of acute rejection is doubtful if they are not.

Chronic rejection

Chronic rejection is mediated by B- or T-cell. Most experimental studies of rejection are performed without immunosuppression and, therefore, graft destruction usually occurs within the first several weeks by one of the mechanisms described above. In clinical practice, however, the use of immunosuppression usually allows graft survival for much longer periods of time. Nonetheless, clinical survival statistics reveal that even when 1-year graft survival has been achieved, the loss of transplanted organs continues to occur at a rate of about 3% to 5% per year and a significant portion of this loss appears to be due to immunologic mechanisms. The term "chroniç rejection" has been used to describe this late process of graft destruction. As immunosuppressive reagents have become more effective at controlling acute rejection, chronic rejection has emerged as one of the most important problems in clinical practice. Indeed, while there has been ongoing improvement over the past 30 years in the 1-year graft survival rates for kidney transplants, the half-life for organs that have survived for 1 year has not changed significantly over that entire period of time. As a result of this ongoing loss, only about 50% of transplants are still functioning 10 years later. In stark contrast to the considerable progress in the management of acute rejection essentially no improvement in treatment in past 25 years has made a significant impact on the long-term loss of transplanted organs through chronic rejection. Possibly in part this is because the mechanism(s) of chronic rejection are still obscure.

Particular transplant situations

Transplants at certain anatomical sites are generally accepted without any immune rejection. The most important of these is the **cornea**. The absence of lymphatic drainage is probably the critical common factor

1. Vascularised solid organs are kidney, lung, liver, heart, pancreas, combined
2. Haemopoietic stem cell transplants often previously called bone marrow transplants, now renamed as source is frequently blood
3. Sources of stem cells are used, listed in order of decreasing mature T-cell contamination
   - peripheral blood (enriched by cytokine administration)
   - bone marrow
   - cord blood

A high percentage of stem cell transplants are autologous, given to restore haemopoiesis after lethal, high-dose chemo- or radiotherapy. These do not give rise to immunological problems.

Due to the preconditioning of the recipients which greatly impairs their immunological function, the major immunological problem associated with allogeneic stem cell transplants is **Graft-versus-Host disease (GvHD)**, the converse to the usual graft rejection problem. This is because the mature T-cells (and also NK cells) of the transplanted innoculum are immunocompetent, reactive cells.

In fact the GvH may be clinically useful in leukaemic patients, where it has been found to help eliminate residual (host) tumour cells. Thus while removing T-cells from the innoculum reduces GvHD, it increases tumour relapse in leukaemic patients; to the point where some centres are adding back measured doses of T-cells to T-cell depleted stem cells (typically from marrow).

Influence of HLA-matching on allograft survival

For haemopoietic stem cell (HSC) transplants, the degree of HLA (=human MHC) matching is critical in determining the probability of Graft-versus-Host disease (GvHD). This is the major form of rejection in clinical HSC transplants. Most centres will only transplant across 1 MHC class I (HLA-A or B locus) difference. The difference between HLA-genotyped and zero mismatches reflects the imperfection of the HLA typing process. HLA genotype matched means HLA-identical sibling, thus all alleles are truly identical; whereas zero mismatched unrelated donors may be mismatched due to the typing not distinguishing between very closely related alleles.

In addition, of course there may be some effect of non-HLA loci since HLA identical sibs will share 50% of their minor histocompatibility loci.

In the UK the majority view seems to be that you try to get a match for 4/6 loci—especially class II which has a slightly larger effect—and if a suitably matched recipient is not available then a less favourable match is accepted. This is argued to be the optimal use of organs since favourable matching gives the best longevity of organ success and since most recipients will need a second kidney at some point, the greater mismatch the more difficult finding a second graft is likely to be (the existance of prior reactive antibodies is very unfavourable to graft survival).

Immunosuppression

Immunosuppression is essential to clinical transplantation "standard" regime varies for different tissue/organs. For kidney transplants involves combination therapy with three different sorts of drugs, Steroids—given for systemic immunosuppressive effects, complex mechanism of action e.g. prednisolone. Cytotoxic-lead to cell death, espcially on entry into cell cycle e.g. azathioprine. Immunosuppressive-targeting cell signalling pathways in lymphocytes e.g. cyclosporin A; others include FK506, rapamycin.

The immunosuppressive drug

Allogeneic transplantation requires some degree of immunosuppression if the transplant is to survive. Most of the immunosuppressive treatments that have been developed have the disadvantage of being nonspecific; that is, they result in generalized immunosuppression of responses to all antigens, not just those of the allograft, which places the recipient at increased risk of infection.

Many novel treatments are in development, particularly antibodies blocking surface molecules critical in T-cell activation e.g. B7-1,2 (CTLA4-Ig); anti-CD40L. Currently licenced biotherapeutic treatments used to treat graft rejection include Anti-CD3 monoclonal antibodies (CD3 is a component of the T-cell receptor). Anti-CD25 monoclonal antibodies (CD25 is a component of the receptor for IL2, a major growth factor for T-cells). A major future goal in transplant research is donor specific tolerance. Because tolerance induction requires a signal from the T-cell receptor (and possibly other co-receptors), this conflicts with the use of some forms of immunosuppression. For example the major immunosuppressant cyclosporin A is known to block tolerance induction.

TRANSPLANTATION OF SPECIFIC ORGANS AND TISSUES

Skin grafting

Although allogeneic skin grafting represents a frequently used experimental model, its application in humans is unusual. Most skin transplantation in humans is done with autologous tissue. Recently, however, "artificial" skin grafts have been created that consist of stromal elements and cultured cells of allogeneic and even xenogeneic origin. Evidence suggests that these grafts are not rejected, although some of their components may be replaced by recipient tissues over time.

Skin grafts are frequently used experimentally on small animals to examine rejection because they can be performed rapidly in large numbers. On the one hand, the use of skin grafts has the disadvantage that they are not primarily vascularized and, thus, may not be susceptible to precisely the same mechanisms of rejection as are solid organs. On the other hand, the difficulty in prolonging skin graft survival in rodents more accurately reflects the difficulty in prolonging transplantation of solid organ survival in larger animals than does the transplantation of solid organs in rodents.

Kidney transplantation

Kidneys have been the most frequently transplanted organs for many years. At present, approximately 10,000 kidney transplants are performed annually in the United States. The likelihood that a renal allograft will survive with good function for at least 1 year has slowly been rising. Patient survival after 1 year is expected to be better than 90% and the current likelihood of graft function, at 1 year, now exceeds 85% in many units, even when organs from totally unrelated donors are used. Nonetheless, even well-matched recipients of renal transplants must, for the most part, continue to take immunosuppressive medications for the rest of their lives. These patients are susceptible to the complications of their immunosuppressive medications, including increased risks of infection, cancer, hypertension, and metabolic bone disease. Thus, they pay a price for their new organ stemming from our inability to provide specific immunosuppression. The success of clinical transplantation is a double-edged sword. With such good patient and graft survival rates initially, it is difficult to justify risky clinical trials of new approaches to immunosuppression which, by achieving antigen-specific tolerance, might avoid the long-term need for immunosuppression altogether. Because of the success of modern nonspecific immunosuppression, the major obstacle to achieving a successful kidney transplant is no longer the early rejection of the organ after transplantation. Instead, the three major obstacles are now the shortage of organs, the problem of sensitization, and the chronic rejection of kidneys that may occur after many years. Partly because of the increasing success of renal transplantation, the number of candidates for the procedure has continued to grow and now far exceeds the supply of available organs. Unlike hearts and livers, where an inadequate supply of organs leads to the death of many candidates, those waiting for renal transplants are instead faced with long periods on dialysis. This waiting time is often 4 or more years even for unsensitized candidates seeking kidneys from cadaver donors. The second major obstacle in obtaining a successful renal transplant stems from the problem of sensitized candidates with broadly reactive antibodies resulting from prior antigen exposure. These highly sensitized individuals may wait many years to obtain a kidney that is crossmatch negative, and some never receive a transplant at all. The late failure of kidney transplants many years after the procedure probably involves a component of immunologic rejection but may also be due to other factors, including the effects of the early ischemic injury and the ongoing effects of drugs and metabolic abnormalities in the recipient. Although the 1-year survival rate for kidney transplants has improved from roughly 40% to nearly 90% over the past 40 years, the conditional half-life for kidney transplants (defined as the number of years at that half of the kidneys have been lost among those recipients who had functioning kidneys at 1 year) has barely increased during that time period and is still less than 10 years. One of the important changes in clinical kidney transplantation in recent years is the greater use of living unrelated donors. In the past, living-donor kidney transplantation was performed almost only when a related donor was available to the recipient with some, or complete, matching of the HLA antigens. The use of better immunosuppressive drugs has reduced the importance of genetic matching and the growing waiting list for cadaver donors has made the importance of finding a living donor much greater. Thus, it is now common for the living donor for a kidney transplant to be a completely unrelated and genetically unmatched individual, such as a spouse or best friend. The results of transplants from unrelated donors have been better than those from cadaver donors and as good as those from all but the most closely matched living related donors.

Liver transplantation

Transplantation of the liver represents a major technical challenge. For this reason the organ and patient survival rates are less good than those for renal transplantation. However, successful liver transplantation can now be achieved with survival of about three-quarters of recipients at 1 year. From an immunologist's point of view, liver transplantation is of interest first because the organ is apparently quite resistant to immediate antibody-mediated rejection. Transplantation across blood group barriers and in the face of a positive crossmatch (by retrospective analysis) has generally been successful in the short term. There is evidence, however, that long-term organ survival is diminished in blood-group incompatible patients. Second, the long-term survival of liver transplants has never seemed to depend on better HLA matching between donor and recipient, even before the use of new immunosuppressive agents. In fact, some data have suggested the opposite correlation. The possibility that poorly matched livers may survive better than well-matched ones might be due to an inability of recipient T-cells, sensitized to viral pathogens in association with self-MHC antigens, to recognize those pathogens in the donor liver presented in association with donor HLA antigens. Thus, poorly matched livers might escape injury caused by immunologic responses to hepatotrophic viruses. Third, transplantation of the liver carries with it large numbers of donor lymphoid cells, thus creating the setting for GVHD. Donor lymphocytes can mediate an antibody-dependent hemolysis of recipient red blood cells in the case of recipient blood-group incompatibility with the donor. Thus, "A" or "B" recipients of "O" livers have been subject to an immune hemolytic anaemia during the early post-transplant period.

There may also be other manifestations of GVH disease even in blood-group compatible recipients.

Heart and lung transplantation

Heart transplantation is also a relatively recent component of standard clinical transplantation, with survival rates frequently in excess of 80% at 1 year. One of the immunologic issues of particular importance in heart transplantation is the high rate of new atherosclerotic disease in the coronary arteries of the donor organ. This atherosclerotic disease is a manifestation of chronic "rejection," but the causes appear to include more than just immunologic responses. Lung transplantation, either in conjunction with heart transplantation or alone, is a still more recent addition to clinical transplantation. Recipients of lung transplants have demonstrated a tendency to develop pathologic changes of "bronchiolitis obliterans" that is also thought to be a manifestation of chronic rejection.

Pancreas and islet transplantation

Transplantation of the whole pancreas provides, of course, more tissue than is needed to treat diabetes mellitus. The intriguing aspect of pancreas transplantation, therefore, is the potential that useful results might be accomplished by transplantation of insulin-producing islet cells alone. Although islet cell transplants have been very successful in animal models, success in humans was unusual until very recently. Transplantation of the whole pancreas was almost without success until about 1980, largely for technical reasons. More recently, successful pancreas transplantation to treat diabetes mellitus has been achieved using new technical approaches, and with success rates approaching those for kidney transplantation, as long as the two organs are transplanted together. The lower survival rates achieved when pancreas transplantation is performed alone probably reflect the difficulty in diagnosing rejection episodes involving this organ. By the time blood sugar levels begin to rise, destruction of the pancreas is generally so far advanced that it cannot be reversed by immunotherapy. Measurement of the serum creatinine, reflecting early dysfunction of a simultaneous kidney transplant, allows much earlier detection of rejection activity and, thus, better outcomes. On the other hand, simultaneous transplantation of both a kidney and a pancreas from a single donor has demonstrated the interesting phenomenon that rejection activity in one organ is not always associated with rejection activity in the other. According to the 1999 report of the Islet Transplant Registry, only 8% of over 200 type 1 diabetic patients had achieved insulin independence after an islet transplant.

Hematopoietic cell transplantation

Allogeneic bone marrow transplantation may also be thought of as immunotherapy leading to an attack on the residual leukemic cells that remain in cytoablated hosts. Bone marrow transplants, and more recently, transplants of hematopoietic stem cells and progenitors mobilized from the marrow into peripheral blood by treatment with G-CSF, are used most commonly for the treatment of otherwise incurable leukemias and lymphomas, aplastic anemia, and congenital immunodeficiency states. One fundamental difference between hematopoietic cell transplantation and the transplantation of all other organs is that the recipient's treatment for his or her malignancy usually results in ablation of the immune and hematopoietic systems prior to transplantation—that is, the "conditioning" for transplantation is myeloablative. Originally, hematopoietic cell allografts were administered only as a means of replacing these ablated host functions. However, clinical experience soon showed that one of the main therapeutic benefits of allogeneic HCT is due to the graft-versus-leukemia (GVL) effect of donor lymphocytes.

Additional applications include hemoglobinopathies and inborn errors of metabolism. Autologous hematopoietic cell transplants have been used quite widely for hematologic rescue following high-dose chemo/radiotherapy for the treatment of malignancies. Additionally, autologous hematopoietic cell transplants (HCT) are currently being explored as a treatment for autoimmune diseases. However, autologous HCT will not be considered further here because it does not involve the broaching of any immunologic barriers.

The manifestation of the rejection

In clinical organ transplantation, the most obvious manifestation of the rejection process is diminished function of the transplanted organ. Other causes of graft dysfunction exist, however, and it is obviously important to confirm the immunologic origin of the event before increasing immunosuppression. The clinical pattern of dysfunction often helps to suggest the diagnosis of rejection. However, no clinical sign can definitively diagnose rejection. It would be useful, therefore, to determine a means of identifying rejection episodes based on systemic manifestations of the immunologic mechanisms involved. Unfortunately, a well-established assay to measure rejection activity does not yet exist. Two approaches include the measurement of antidonor antibody production and the sequential measurement of cell-mediated responses to donor antigens. Antibody responses have frequently been documented following graft rejection, but they tend to appear after rejection is complete. *In vitro* cell-mediated responses, both proliferative and cytotoxic, may or may not be present while a graft is in place and are not well correlated with clinical rejection episodes. Assays of humoral and cellular responses to donor antigens both suffer from the possibility that donor-specific elements of the response may be absorbed by the antigens of the graft, at least until the late stages of rejection. Furthermore, the existence of tissue-specific peptides may allow T-cell responses to occur *in vivo* that cannot be measured *in vitro* when donor lymphocytes are used as stimulators. A recent report suggests that urinary perforin and granzyme B levels may be useful in the diagnosis of renal allograft rejection, but did not rule out the possibility that other conditions associated with renal allograft dysfunction, such as CMV infection, might produce a similar profile.

The "gold standard" in the diagnosis of allograft rejection has always been the biopsy of the transplanted organ itself. Pathologists have been able to identify the abnormal lymphocytic infiltrate within grafts, to grade the intensity of the infiltrate, and, for some organs, to describe histologic findings characterizing the effects of immunologic injury. Some pathologic changes, including a lymphocytic infiltrate of the vascular wall, seem to be well correlated with rejection activity. In addition, pathologic changes suggesting non-immunologic causes of renal dysfunction may be helpful in patient management. Despite the widespread reliance on the biopsy to define episodes of rejection, however, differentiation of rejection from its absence is often difficult, particularly when cyclosporine immunosuppression has been used. Since most clinical allograft biopsies are performed when the organ is not functioning well, and only after mechanical causes of dysfunction have been excluded, most organs that are biopsied, by selection, are undergoing rejection. Therefore, inability to detect nonrejection events in a few cases pathologically will still leave an excellent correlation between the diagnosis of rejection and the response to therapy. When routine biopsies of transplanted organs have been done, regardless of organ dysfunction, they have revealed a poor correlation between histologic findings and clinical evidence of rejection. Experimental studies of skin grafts, as discussed above, have found that the degree of lymphocytic infiltrate in an allograft correlated better with the nature of the antigenic disparity than with the intensity of the rejection process. Furthermore, several experimental models of tolerance induction have shown intense lymphocytic infiltrates in organs that go on to survive indefinitely in recipients who develop tolerance to the donor antigens. These studies suggest that the amount of lymphocytic infiltrate detected pathologically may not be helpful in diagnosing rejection episodes and determining the need for treatment.

CHAPTER

22

TUMOUR IMMUNOLOGY

INTRODUCTION

The tumour immunology involves the study of antigens on tumour cells and the immune response to these antigens. The existence of an immune response against a tumour is based on changes in the surface component of malignant cell. Malignant tumour or cancer grow in an uncontrolled manner, invade a normal tissue and often metastasize and grow at site distant from tissue of origin. Cancer can arise from any tissue in body. Although tumour are derived from self tissue the process of a malignant transformation may be associated with expression of molecule on tumour cells that are recognized as a foreign by specific immune system, such molecule are called as tumour antigen.

Tumour antigen is a substance produced in tumour cells that triggers an immune response in the host. Tumour antigens are useful in identifying tumour cells and are potential candidates for use in cancer therapy. Tumours of the immune system are classified as lymphomas or leukemias. Lymphomas proliferate as solid tumours within a lymphoid tissue such as the bone marrow, lymph nodes, or thymus; they include Hodgkin's and non-Hodgkin's lymphomas. Leukemias tend to proliferate as single cells and are detected by increased cell numbers in the blood or lymph. Leukemia can develop in lymphoid or myeloid lineages. Historically, the leukemias were classified as acute or chronic according to the clinical progression of the disease. The acute leukemias appeared suddenly and progressed rapidly, whereas the chronic leukemias were much less aggressive and developed slowly as mild, barely symptomatic diseases. These clinical distinctions apply to untreated leukemias; with current treatments, the acute leukemias often have a good prognosis, and permanent remission can often be achieved. Now the major distinction between acute and chronic leukemias is the maturity of the cell involved. Acute leukemias tend to arise in less mature cells, whereas chronic leukemias arise in mature cells. The acute leukemias include acute lymphocytic leukemia and acute myelogenous leukemia; these diseases can develop at any age and have a rapid onset. The chronic leukemias include chronic lymphocytic leukemia and chronic myelogenous leukemia; these diseases develop slowly and are seen in adults.

Mechanism of tumour antigenesis

Normal proteins in the body are not antigenic because of self-tolerance, a process in which self-reacting cytotoxic T lymphocytes (CTLs) and autoantibody-producing B lymphocytes are pulled in the thymus. Thus any protein that is not exposed to the immune system triggers an immune response. This may include normal proteins that are well sequestered from the immune system, proteins that are normally produced in extremely small quantities, proteins that are normally produced only in certain stages of development, or proteins whose structure is modified due to mutation.

Types of tumour antigens

There are two types of tumour antigen first is tumour specific antigen and tumour-*associated* antigens. Any protein produced in a tumour cell that has an abnormal structure due to mutation can act as a tumour antigen. Such abnormal proteins are produced due to mutation of the concerned gene. Mutation of protooncogenes and tumour suppressors which lead to abnormal protein production are the cause of the tumour and thus such abnormal proteins are called tumour-*specific* antigens. Tumour-specific antigens have been identified on tumours induced with chemical or physical carcinogens and on some virally induced tumours. Demonstrating the presence of tumourspecific antigens on spontaneously occurring tumours is particularly difficult because the immune response to such tumours eliminates all of the tumour cells bearing sufficient numbers of the antigens and in this way selects for cells bearing low levels of the antigens. Examples of tumour-specific antigens include the abnormal products of ras and p53 genes. In contrast, mutation of other genes unrelated to the tumour formation may lead to synthesis of abnormal proteins which are called tumour-*associated* antigens.

Several tumour-associated transplantation antigens have been identified on human melanomas. Five of these—MAGE-1,MAGE-3, BAGE, GAGE-1, GAGE-2—are oncofetal-type antigens. Each of these antigens is expressed on a significant proportion of human melanoma tumours, as well as on a number of other human tumours, but not on normal differentiated tissues except for the testis, where it is expressed on germ-line cells.

Proteins that are normally produced in very low quantities but whose production is dramatically increased in tumour cells, trigger an immune response. An example of such a protein is the enzyme tyrosinase, which is required for melanin production. Normally tyrosinase is produced in minute quantities but its levels are very much elevated in melanoma cells.

Oncofetal antigens are another important class of tumour antigens. **Oncofetal tumour antigens,** as the name implies, are found not only on cancerous cells but also on normal fetal cells. Examples are alphafetoprotein (AFP) and carcinoembryonic antigen (CEA). These proteins are normally produced in the early stages of embryonic development and disappear by the time the immune system is fully developed. Thus self-tolerance does not develop against these antigens if these antigens appear later on cancer cells, they are recognized as nonself and induce an immunologic response. Abnormal proteins are also produced by cells infected with oncoviruses, e.g. EBV and HPV. Cells infected by these viruses contain latent viral DNA which is transcribed and the resulting protein produces an immune response. Tumours may express proteins normally expressed on fetal but not adult tissue. Such **oncofetal antigens** include a-fetoprotein (AFP), found on primary hepatocellular carcinoma cells, and carcinoembryonic antigen (CEA) expressed by gastrointestinal and breast carcinomas. These oncofetal antigens may also be released in various inflammatory conditions but measurement of their levels in the blood may be helpful in the diagnosis of malignant disease and in the monitoring of progression of the disease. The majority of tumour-specific antigens are either completely abnormal peptides or mutated forms of normal cellular proteins produced by tumour cells and complexed to class I MHC products. These peptides, which are not normally destined to be positioned in the surface plasma membrane, can still signal their presence to T-cells in the outer world by a processed peptide/MHC mechanism. In addition to proteins, other substances like cell surface glycolipids and glycoproteins may also have an abnormal structure in tumour cells and could thus be targets of the immune system.

Mutant antigens are encoded by genes that are expressed in normal cells but are mutated in tumour cells. Such mutations may change as few as one amino acid. Single point mutations in oncogenes or tumour suppressor genes can account for the large diversity of antigens found on carcinogen-induced tumours. The gene encoding the p53 cell cycle inhibitor is a hotspot for mutation in cancer, while the oncogenic human *ras* genes differ from their normal counterpart by point mutations usually leading to single amino acid substitutions. Such mutations have been recorded in 40% of human colorectal cancers and their preneoplastic lesions, in more than 90% of pancreatic carcinomas, in acute myelogenous leukemia and in preleukemic syndromes. The oncogene *HER-2/neu* encodes a membrane protein that is present on various human tumours, especially ovarian and breast carcinomas, where it could serve as a target for either monoclonal antibodies or Tc.

DNA tumour viruses

The cancer-inducing DNA viruses that encode functionally similar but antigenically distinct proteins required for the induction and maintenance of malignant transformation are SV40 and polyoma viruses. The investigator(1980) analyzed how the SV40-specific CTL could recognize the nuclear T antigen on the surface of tumour cells.These virally encoded proteins are shared by all tumours induced by the same virus regardless of tissue origin or animal species. These proteins appear intracellularly as predominantly nuclear antigens, but the same genes that encode these proteins are also required for the expression of the CTL-recognized, virus-specific, MHC-presented peptides acting as transplantation antigens on the tumour cell surface. The rejection antigens are tumour-specific in the sense that they are not expressed by normal host cells and are not encoded by genes of the host or presented on the virion particles themselves, which carry viral capsid antigens.

Thus, **DNA Tumour Viruses** e.g. SV40, polyoma virus, HPV, and adenovirus, show stable but random integration of at least the early-region genes of the virus into the genome of the cells they transform. It is the protein product of the early region rather than the site of integration of the virus into the genome that seems to be important for inducing and maintaining the transformed phenotype.

Thus, the early-region proteins are virally encoded, tumour-specific antigens that are clearly related to the transformed phenotype and the establishment of malignant behavior.

RNA tumour viruses

Ribonucleic acid (RNA) tumour viruses, which can be of exogenous or endogenous type, integrate into host cell DNA; however, for most of the viral life cycle, exogenous viruses exist and replicate as infectious particles capable of infecting other cells. In contrast, endogenous viruses remain, most of the time, as proviruses integrated into the host's genome and produce infectious particles only when "induced" by ionizing radiation, chemical carcinogens, mutagens, or protein synthesis inhibitors. Genomic DNA of inbred mice contains about 50 copies of normal murine leukemia virus (MuLV)—related proviral sequences. These proviruses can be categorized into three major groups, which are determined by the range of the host cells that can be infected by the infectious particles derived from these proviruses: ecotropic viruses, which replicate in mouse cells but not in nonmouse cells; xenotropic viruses, which replicate in nonmouse cells but not in mouse cells; and polytropic (amphotropic or dual-tropic) viruses, which replicate in both types of cells. The differences in host range of the viruses are determined by differences in their viral envelope (env) genes, which encode a 70-kDa glycoprotein (gp70). Thus, oligonucleotide probes specific for the three types of env genes or antibody probes specific for the three antigenic types of gp70 viral envelope proteins can be used to group these proviruses into one of the three categories just listed.

Importance of tumour antigens

Tumour antigens, because of their relative abundance in tumour cells are useful in identifying specific tumour cells. Certain tumours have certain tumour antigens in abundance. Two types of tumour antigens have been identified on tumour cells: tumour-specific transplantation antigens and tumour-associated transplantation antigens. Tumour-specific antigens are unique to tumour cells and do not occur on normal cells in the body. They may result from mutations in tumour cells that generate altered cellular proteins; cytosolic processing of these proteins would give rise to novel peptides that are presented with class I MHC molecules, inducing a cell-mediated response by tumour-specific CTLs. Tumour-associated antigens, which are not unique to tumour cells, may be proteins that are expressed on normal cells during fetal development when the immune system is immature and unable to respond but that normally are not expressed in the adult. Reactivation of the embryonic genes that encode these proteins in tumour cells results in their expression on the fully differentiated tumour cells. Tumour-associated antigens may also be proteins that are normally expressed at extremely low levels on normal cells but are expressed at much higher levels on tumour cells.

TABLE 22.1
Tumour antigen and their location

Tumour antigen	Tumour in which it is found
Alphafetoprotein (AFP)	Germ cell tumours
Carcinoembryonic antigen (CEA)	—
CA-125	Ovarian cancer
MUC-1	Breast cancer
Epithelial tumour antigen (ETA)	Breast cancer
Tyrosinase	Malignant melanoma
Melanoma-associated antigen (MAGE)	Malignant melanoma
Abnormal products of ras, p53	Various tumours

Certain tumour antigens are thus used as tumour markers. More importantly, tumour antigens can be used in cancer therapy as tumour antigen vaccines.

Cancer

About 100 and 350 of each 100,000 people die of cancer each year worldwide. Cancer is due to failures of the mechanisms that usually control the growth and proliferation of cells. During normal development and throughout adult life, intricate genetic control systems regulate the balance between cell birth and death in response to growth signals, growth-inhibiting signals, and death signals. Cell birth and death rates determine adult body size, and the rate of growth in reaching that size. In some adult tissues, cell proliferation occurs continuously as a constant tissue-renewal strategy. Intestinal epithelial cells, for instance, live for just a few days before they die and are replaced; certain white blood cells are replaced as rapidly, and skin cells commonly survive for only 2–4 weeks before being shed. The cells in many adult tissues, however, normally do not proliferate except during healing processes. Such stable cells (e.g. hepatocytes, heart muscle cells, neurons) can remain functional for long periods or even the entire lifetime of an organism.

CHAPTER 23

VACCINES

INTRODUCTION

Vaccines constitute cost-effective measures for preventing disease. A vaccination is a cost-effective means of disease prevention that has already prevented millions from death and suffering caused by the many diseases that afflict humankind. Since the advent of wide-spread vaccinations, the incidence of many diseases such as measles, mumps, diphtheria, pertussis, poliomyelitis, and rubella has dropped significantly. One of the great triumphs for vaccines has been the eradication of smallpox during the 1970s. While during the mid-1960s some ten to fifteen million cases of smallpox occurred, today smallpox is an affliction of the past. Advances in biotechnology and an understanding of the inductive and effector components of immune responses have ushered in a 'golden age' of vaccine development and implementation.

Even though vaccines have been helpful in the prevention of disease, there are still many diseases that have no effective vaccine against them. Fortunately, our understanding of the human immune system has increased significantly since the days of Edward Jenner's small-pox vaccine. With this increased knowledge, vaccine design can be focused on the specific areas of the immune system which would be most effective against the pathogen in question.

Immunization

Immunity to infectious microorganisms can be achieved by active or passive immunization. In each case, immunity can be acquired either by natural processes (usually by transfer from mother to foetus or by previous infection by the organism) or by artificial means such as injection of antibodies or vaccines. The agents used for inducing passive immunity include antibodies from humans or animals, whereas active immunization is achieved by inoculation with microbial pathogens that induce immunity but do not cause disease or with antigenic components from the pathogens. This section describes current usage of passive and active immunization techniques.

Jenner and Pasteur are recognized as the pioneers of vaccination, or induction of active immunity, but similar recognition is due to Emil von Behring and Hidesaburo Kitasato for their contributions to passive immunity. These investigators were the first to show that immunity elicited in one animal can be transferred to another by injecting it with serum from the first.

Passive immunization

The aim of passive immunization is transient protection or alleviation of an existing condition. Passive immunization, in which preformed antibodies are transferred to a recipient, occurs naturally by transfer of maternal antibodies across the placenta to the developing foetus. Maternal antibodies to diphtheria, tetanus, streptococci, rubeola, rubella, mumps, and poliovirus all afford passively acquired protection to the developing foetus. Passive immunization can also be achieved by injecting a recipient with preformed antibodies. In the past, before vaccines and antibiotics became available, passive immunization provided a major defense against various infectious diseases. Despite the risks incurred by injecting animal sera, usually horse serum, this was the only effective therapy for otherwise fatal diseases. Currently, there are several conditions that warrant the use of passive immunization. Passive immunization is routinely administered to individuals exposed to botulism, tetanus, diphtheria, hepatitis, measles, and rabies. Passively administered antiserum is also used to provide protection from poisonous snake and insect bites. Passive immunization can provide immediate protection to travelers or health-care workers who will soon be exposed to an infectious organism and lack active immunity to it. Because passive immunization does not activate the immune system, it generates no memory response and the protection provided is transient.

Active immunization

The goal of active immunization is to elicit protective immunity and immunologic memory. Active immunization can be achieved by natural infection with a microorganism, or it can be acquired artificially by administration of a **vaccine**. In active immunization, as the name implies, the immune system plays an active role—proliferation of antigen-reactive T- and B-cells results in the formation of memory cells. Active immunization with various types of vaccines has played an important role in the reduction of deaths from infectious diseases, especially among children. When active immunization is successful, a subsequent exposure to the pathogenic agent elicits a heightened immune response that successfully eliminates the pathogen or prevents disease mediated by its products.

The goal of vaccines is to prime the recipient's immune response in order to generate memory cells, so a heightened immune response will be elicited upon exposure to the specified pathogen. It is important that the response will occur in the correct pathways and physiological regions. Many intravenous vaccines focus on bloodborne immunity, but this would be inappropriate in a vaccine against a gut parasite, or a pathogen which entered via mucus membranes. Intracellular pathogens are not effectively guarded against by vaccines which limit themselves to the humoral pathway.

The vaccine itself should produce only limited, if any, undesirable side effects. Furthermore, it should be relatively inexpensive and mass producible for population vaccinations. These two factors are as important as efficacy when regulatory bodies and pharmaceutical companies consider possible vaccines. The cost to administer a vaccine must not outweigh its benefits if it is to be mass produced for public health. It also must not endanger its recipients, either through danger of infection and adverse reaction, or by encouraging unwarranted risky behavior. Effective vaccines (e.g. TB in the U.S.) are occasionally not incorporated into public health scheme if their use would obscure the tracking of a disease by producing "false positive" results.

The various types of vaccines are:

1. Pneumococcal conjugate vaccine (PCV);
2. Inactivated (Salk) polio vaccine (IPV);
3. Hepatitis B vaccine
4. Haemophilus influenzae (Hib) vaccine
5. Diphtheria-pertussis (acellular)-tetanus (DPaT) combined vaccine
6. Measles-mumps-rubella (MMR) combined vaccine
7. Varicella zoster (Var) vaccine for chickenpox

Strategies for vaccine development

Modern technologies offer rational strategies for the development of new and improved vaccines against diseases of public health importance. Attributes of various vaccine development strategies and methods of administration of vaccines are as follows. The major strategies in developing vaccines are:

- Whole-Organism
  - ❑ Attenuated (Live)
  - ❑ Inactivated (Killed)
- Purified (Subunit) Antigen
- Recombinant Antigen
- Synthetic Peptide
- Recombinant Vector
- DNA

Whole-organism vaccines

The use of whole organism vaccination is the oldest strategy of vaccine development. Jenner's smallpox vaccine used live cowpox virus (vaccinia) to generate protective immunity. There are two classes of a whole organism vaccine, attenuated (live) and inactivated (killed). Many of the common vaccines currently in use consist of inactivated (killed) or live but attenuated (avirulent) bacterial cells or viral particles. Attenuated vaccines have advantages and disadvantages. A major disadvantage of attenuated vaccines is the possibility that they will revert to a virulent form. Because of their capacity for transient growth, such vaccines provide prolonged immune-system exposure to the individual epitopes on the attenuated organisms, resulting in increased immunogenicity and production of memory cells. As a consequence, these vaccines often require only a single immunization, eliminating the need for repeated boosters.

Examples of attenuated vaccines include: an attenuated strain of *Mycobacterium bovis* called Bacillus Calmette-Guerin (BCG) for tuberculosis, the Sabin vaccine for polio, the measles vaccine, the mumps vaccine, and the rubella vaccine. After 13 years, this strain had adapted to growth in strong bile and had become sufficiently attenuated that it was suitable as a vaccine for tuberculosis. The Sabin polio vaccine and the measles vaccine both consist of attenuated viral strains. The poliovirus used in the Sabin vaccine was attenuated by growth in monkey kidney epithelial cells. In some cases, microorganisms can be attenuated so that they lose their ability to cause significant disease (pathogenicity) but retain their capacity for transient growth within an inoculated host. Attenuation often can be achieved by growing a pathogenic bacterium or virus for prolonged periods under abnormal culture conditions.

Advantage and disadvantage

The advantage of the attenuated vaccine is that the attenuated pathogen stimulates an infection without conferring the disease. Since the microorganism is still living, it provides continual antigenic stimulation giving sufficient time for memory cell production. Also, in the case of viruses or intracellular microorganisms where cell-mediated immunity is usually desired, attenuated pathogens are capable of replicating within host cells. However, since the vaccines contain living organisms, there is a degree of unpredictability. Attenuated pathogens have the possibility of reverting to its pathogenic form, or some attenuated pathogens which are weak enough to be contained by a normal immune system, may still be pathogenic to those with compromised immune systems. Currently, genetic engineering techniques are being used to bypass these disadvantages by removing the genes that cause virulence.

Inactivated (Killed) organisms as vaccines

The simplest way to destroy the ability of microbes to cause disease yet maintain their antigenicity is to prevent their replication by killing in an appropriate manner, such as with formaldehyde treatment. An inactivated whole organism vaccine uses pathogens which are killed and are no longer capable of replicating within the host. The pathogens are inactivated by heat or chemical means while assuring that the surface antigens are intact. Inactivated vaccines are generally safe, but are not entirely risk free.

Surface endotoxins on inactivated pertussis vaccine occasionally induce DTH responses, and influenza virus has been linked to similar reactions, though this may be due more to the immunogenicity of the egg whites in which the virions are raised. Also, inactivated vaccines do not always induce protective immunity. Multiple boosters are usually necessary in order to generate continual antigen exposure, as the dead organism is incapable of sustaining itself in the host, and is quickly cleared by the immune system. Parasitic worms and, to a lesser extent, protozoa are extremely difficult to grow up in bulk to manufacture killed vaccines. This problem does not arise for many bacteria and viruses and, in these cases, the inactivated microorganisms have generally provided safe antigens for immunization. Examples are typhoid, cholera and killed poliomyelitis (Salk) vaccines. Care has to be taken to ensure that important protective antigens are not destroyed in the inactivation process.

Examples of inactivated vaccines include: vaccines for cholera, pertussis, influenza, rabies, and the Salk vaccine for polio. Another common approach in vaccine production is inactivation of the pathogen by heat or by chemical means so that it is no longer capable of replication in the host. It is critically important to maintain the structure of epitopes on surface antigens during inactivation. Heat inactivation is generally unsatisfactory because it causes extensive denaturation of proteins; thus, any epitopes that depend on higher orders of protein structure are likely to be altered significantly. Chemical inactivation with formaldehyde or various alkylating agents has been successful. The Salk polio vaccine is produced by formaldehyde inactivation.

Attenuated vaccines generally require only one dose to induce long-lasting immunity. Killed vaccines, on the other hand, often require repeated boosters to maintain the immune status of the host. In addition, killed vaccines induce a predominantly humoral antibody response; they are less effective than attenuated vaccines in inducing cell-mediated immunity and in eliciting a secretory IgA response.

Recombinant antigen vaccine

The risks and financial burden of extracting subunit antigens from pathogens can be high. With the development of recombinant DNA techniques in the past few decades, other organisms can be recruited and altered to produce the desired antigens more quickly, efficiently and safely. The first such recombinant antigen vaccine approved for human use is the hepatitis B vaccine. This vaccine was developed by cloning the gene for the major surface antigen of hepatitis B virus (HBsAg) and expressing it in yeast cells. The recombinant yeast cells are grown in large fermenters, and HBsAg accumulates in the cells. The yeast cells are harvested and disrupted by high pressure, releasing the recombinant HBsAg, which is then purified by conventional biochemical techniques.

This recombinant hepatitis B vaccine has been shown to induce the production of protective antibodies. A number of genes encoding surface antigens from viral, bacterial, and protozoan pathogens have been successfully cloned into bacterial, yeast, insect, or mammalian expression systems, and the expressed antigens used for vaccine development. Using recombinant techniques, non-pathogenic bacteria or yeasts were altered to produce the HBV-soluble antigen for bulk harvesting. The dangers of viral infection were eliminated, and the mass production of the vaccine brought costs down to a level where it can now be a part of the regular childhood immunization schedule. Examples of recombinant antigen vaccines include: hepatitis B, foot-and-mouth disease.

Peptide vaccine

The synthetic peptide approach to vaccine development arose in response to rapid DNA cloning and sequencing technology. This made it possible to quickly obtain primary sequences and construct various peptides. Peptides are not as immunogenic as proteins, and it is difficult to elicit both humoral and cellular immunity to them. The use of conjugates and adjuvants can assist in raising protective immunity to peptides, but barriers to the widespread use of peptide vaccines remain and pose an interesting problem for immunologists. Most importantly, advances in techniques to produce recombinant proteins or fragments of proteins in transfected cell culture have removed the impetus to develop vaccines based on synthetic peptides. Nonetheless, there remains theoretical interest in immunity to them, and studies of peptide immunity may generate insights leading to new vaccines. Experiments by E. Sercarz have identified nonoverlapping amino acid sequences within hen egg-white lysozyme that induce a strong helper response to an antigen and other peptides that induce immunologic suppression.

The synthetic peptide vaccine has had an impact on three major fronts:

1. **Bacteria:** synthetic peptide vaccines for bacterial diseases have focused primarily on trying to neutralize toxins. For example, synthetic peptide vaccines have been developed for both the diphtheria and cholera toxins.
2. **Viruses:** Synthetic peptide vaccines for HIV proteins and glycoproteins are currently under development. Currently, invariant regions amino acid sequence highly conserved are the focus of synthetic peptide vaccine design for viruses.
3. **Parasites:** Scientists hope to achieve stage specific anti-malarial immunity by constructing synthetic peptide vaccines for key malarial sporozoite epitopes.

DNA vaccines

DNA vaccines seem to be a safer alternative to immune deficient recipients who may not be able to be immunized with live viral or bacterial vaccines. DNA vaccines also induce both humoral and cell-mediated immunity; to stimulate both arms of the immune response with non-DNA vaccines normally requires immunization with a live attenuated preparation, which introduces additional elements of risk. Injection of plasmid DNA directly into recipient cells has been shown to elicit a strong immune response. This method induces a prolonged expression of the DNA encoded peptides which has been shown to promote antibodies, CD8+ and CD4+ T-cells that differentiate towards the Th1 response. Of course, when adding foreign DNA to cells one runs the risk of sensitizing the immune cells to the DNA itself. Cross reactivity with human DNA could lead to devastating autoimmunity. A recently developed vaccination strategy, plasmid DNA encoding antigenic proteins is injected directly into the muscle of the recipient. Muscle cells take up the DNA and the encoded protein antigen is expressed, leading to both a humoral antibody response and a cell-mediated response. What is most surprising about this finding is that the injected DNA is taken up and expressed by the muscle cells with much greater efficiency than in tissue culture. The DNA appears either to integrate into the chromosomal DNA or to be maintained for long periods in an episomal form.

The viral antigen is expressed not only by the muscle cells but also by dendritic cells in the area that take up the plasmid DNA and express the viral antigen. The fact that muscle cells express low levels of class I MHC molecules and do not express costimulatory molecules suggests that local dendritic cells may be crucial to the development of antigenic responses to DNA vaccines.

Advantages

DNA vaccines offer advantages over many of the existing vaccines. For example, the encoded protein is expressed in the host in its natural form—there is no denaturation or modification. The immune response is therefore directed to the antigen exactly as it is expressed by the pathogen. Finally, DNA vaccines cause prolonged expression of the antigen, which generates significant immunological memory. The practical aspects of DNA vaccines are also very promising. Refrigeration is not required for the handling and storage of the plasmid DNA, a feature that greatly lowers the cost and complexity of delivery. The same plasmid vector can be custom tailored to make a variety of proteins, so that the same manufacturing techniques can be used for different DNA vaccines, each encoding an antigen from a different pathogen.

Characteristics of ideal vaccines

Many licensed vaccines have one or more ideal characteristics, but none manifests them all.

1. Although vaccine safety is an issue worldwide, this concern is particularly conspicuous in industrialized countries where the very success of vaccines has led the public to forget the dangers of previously common infectious diseases and instead to dwell on rare adverse events attributed to vaccines.
2. Another chief consideration for vaccines is that they confer long-lived efficacy, an important determinant of cost-effectiveness after implementation.
3. Ideally, just a single dose of vaccine should confer robust, long-lived immunity.
4. Vaccine efficacy depends on the immune responses of individuals. Thus, immunizing very young infants, the elderly and immunocompromised persons poses a challenge.
5. Vaccines that have greater immunogenicity and powerful yet safe adjuvants are needed to successfully immunize the very young and the elderly. For immunocompromised subjects, less immunogenicity is generally acceptable, as long as the vaccine is safe.
6. Ideally, vaccines should be administrable without needles and syringes (orally, nasally or transcutaneously, or with needle-free injection devices).

7. Another trait of an ideal vaccine is its feasibility to be administered in combination with other vaccines. This allows fewer healthcare visits, minimizes inconvenience and trauma and diminishes risks of needle sticks.
8. Combination formulations are particularly desirable so that infants are not given multiple injections at the same visit.
9. Ideally, vaccines should have uncomplicated, economical large-scale manufacture processes, because simplicity of manufacture has long-term implications for vaccine supply and cost.
10. An ideal vaccine would be formulated to resist high and low temperatures, to facilitate distribution.

CHAPTER 24

TOLERANCE

INTRODUCTION

The immune system is often thought of as a protective device for responding to the dangers of pathogenic invaders and injury. However, one of the most serious threats to the organism is the immune system itself. Without the various phenomena referred to as tolerance, the system would surely self-destruct. Tolerance refers to the specific immunological non-reactivity to an antigen resulting from a previous exposure to the same antigen. While the most important form of tolerance is non-reactivity to self antigens, it is possible to induce tolerance to non-self antigens. When an antigen induces tolerance, it is termed tolerogen.

Our own bodies produce some 100,000 different proteins and one of the longstanding conundrums of immunology has been to understand how the immune system produces a virtually universal repertoire while at the same time avoiding reacting to self. We can consider two different definitions of immunological tolerance. The strict definition might be that immunological tolerance occurs when an immunocompetent host fails to respond to an immunogenic challenge (i.e. one that would produce a measurable response in some other, nontolerant host) with a specific antigen. A more operational definition might be that immunological tolerance occurs when an immunocompetent host fails to respond to the presence of a specific antigen.

Induction of tolerance to non-self

Tolerance can be induced to antigenic components on both soluble proteins and cells (tissues) by injecting these materials into animals. Induction of such a tolerance depends on a number of variables.

Tolerance to tissues and cells

Tolerance to tissue and cell antigens can be induced by injection of hemopoietic (stem) cells in neonatal or severely immunocompromised (by lethal irradiation or drug treatment) animals. Also, grafting of allogeneic bone marrow or thymus in early life results in tolerance to the donor type cells and tissues. Such animals are known as **chimeras**. These findings are of significant practical application in bone marrow grafting.

Immunologic features of tolerance

Tolerance is different from non-specific immunosuppression and immunodeficiency. It is an active antigen-dependent process in response to the antigen. Like immune response, tolerance is specific and like immunological memory, it can exist in T-cells, B-cells or both and like immunological memory, tolerance at the T-cell level is longer lasting than tolerance at the B-cell level.

Induction of tolerance in T-cells is easier and requires relatively smaller amounts of tolerogen than tolerance in B-cells. Maintenance of immunological tolerance requires persistence of antigen. Tolerance can be broken naturally (as in autoimmune diseases) or artificially (as shown in experimental animals, by x-irradiation, certain drug treatments and by exposure to cross reactive antigens).

Tolerance may be induced to all epitopes or only some epitopes on an antigen and tolerance to a single antigen may exist at the B-cell level or T-cell level or at both levels.

Mechanism of tolerance induction

The exact mechanism of induction and maintenance of tolerance is not fully understood. Experimental data, however, point to several possibilities.

Clonal deletion: Functionally immature cells of a clone encountering antigen undergo a programmed cell death; for example, auto-reactive T-cells are eliminated in the thymus following interaction with self antigen during their differentiation (negative selection). Recent studies have shown that a variety of antigens are expressed in thymic epithelial cells. Likewise, differentiating early B-cells become tolerant when they encounter cell-associated or soluble self antigen. B-cells expressing only IgM (no IgD) on their surface when exposed to antigen are more prone to tolerance induction than immune response. Clonal deletion has been shown to occur also in the periphery.

Clonal anergy: Auto-reactive T-cells, when exposed to antigenic peptides which do not possess co-stimulatory molecules (B7-1 or B7-2), become anergic to the antigen. Also, B-cells when exposed to large amounts of soluble antigen down regulate their surface IgM and become anergic. These cells also up-regulate the *Fas* molecules on their surface. An interaction of these B-cells with Fas-ligand-bearing cells results in their death via apoptosis.

Clonal ignorance: T-cells reactive to self antigen not represented in the thymus will mature and migrate to the periphery, but they may never encounter the appropriate antigen because it is sequestered in inaccessible tissues. Such cells may die out for lack of stimulus. Auto-reactive B-cells that escape deletion may not find the antigen or the specific helper T-cells and hence not be activated and die out.

Receptor editing: B-cells which encounter large amounts of soluble antigen, as they do in the body, and bind to this antigen with very low affinity become activated to re-express their RAG1 and RAG2 genes. These genes cause them to undergo DNA recombination and change their specificity.

Anti-idiotype antibody: Anti-idiotype antibodies produced experimentally have been demonstrated to inhibit immune response to specific antigens. Anti-idiotype antibodies are produced during the process of tolerization and such antibodies have been demonstrated in tolerant animals. These antibodies prevent the receptor from combining with antigen.

Suppressor cells: Both low and high doses of antigen may induce suppressor T-cells which can specifically suppress immune responses of both B- and T-cells, either directly or by production of cytokines, most importantly, TGF-beta and IL-10.

Termination of tolerance

Experimentally induced tolerance can be terminated by prolonged absence of exposure to the tolerogen, by treatments which severely damage the immune system (x-irradiation) or by immunization with cross reactive antigens. These observations are of significance in the conceptualization of autoimmune diseases.

Types of tolerance

Tolerance divides into two main types:

- Central Tolerance—this occurs during lymphocyte development.
- Peripheral Tolerance—occurs after lymphocytes leave the primary organs.

Central Tolerance are of two types T-cell tolerance and B-cell tolerance.

T-cells tolerance

During T-cell development in the thymus the process of **negative selection** leads to the deletion of thymocytes whose T-cell receptors have 'high affinity' for self. One aspect we did not consider was what other parameters influenced this selection event. The only one we need to discuss is antigen concentration. Remember that in the context of T-cells, antigen means the complex of specific peptide and MHC so the density of any given peptide-MHC complex is determined by the abundance of the peptide and it's affinity for the particular MHC allele. The number of molecules of any given peptide-MHC complex on the surface of a thymic antigen presenting cell probably varies from 1-50,000. We can imagine that if there is only a mean of 1 complex/cell then even the highest affinity T-cells might fail to be deleted whereas for the most abundant peptide-MHC complexes quite low affinity T-cells would be deleted. Thus we would predict an inverse relationship between the threshold affinity of TCR for peptide-MHC required to produce deletion and the abundance of that peptide-MHC complex. Using transgenic mice and organ culture techniques it has been possible to provide evidence for this hypothesis.

B-cells tolerance

During B-cell development in the bone marrow when the complete antigen receptor (IgM) is first expressed on 'immature' B-cells if those cells encounter their target antigen in a form which can **cross-link** their sIgM then such cells are programmed to die (deleted from the repertoire). This was discovered many years ago when it was shown that injection of a polyclonal anti-IgM from birth prevented the development of B-cells, resulting in a 'B-less' mouse. The requirement for crosslinking means that the antigen has to be **polyvalent**, the most obvious example of this being cell-surface molecules. This has been directly demonstrated by using transgenic mice expressing rearranged immunoglobulin genes specific for natural or artificial membrane bound molecules. Presumably other multivalent self antigens to which immature B-cells are exposed also induce deletion of self reactive cells.

Peripheral tolerance

Because many of the proteins which the body uses are not expressed in thymus or serum, and in some cases not expressed until after the immune system has matured. There need to be some mechanisms to prevent autoreactivity of lymphocytes after they have emigrated from the thymus/bone marrow. In fact there are several such mechanisms: **Ignorance, Suppression, Anergy and Split Tolerance.** There are a number of factors such as timing, location, costimulation and dose which affects which of these mechanisms operates.

Ignorance

It can be shown that there are in fact both T-cells and B-cells specific for autoantigens present in circulation. In some cases these cells are quite capable of making a response but are unaware of the presence of their autoantigen. This arises for two reasons. The first is that the antigen may simply be present in too low concentration. Since all lymphocytes have a threshold for receptor occupancy which is required to trigger a response then very low concentrations of antigen (in the case of T-cells these are very low) will not be sensed.

The second possibility is a more interesting one. Some antigens are sequestered from the immune system in locations which are not freely exposed to surveillance. These are termed **immunologically privileged sites.** Examples of such sites are the eye, CNS and testis.

Split tolerance

This term simply means that some part of the immune system is tolerant and some other is not. The most frequent situation is that where T-cell tolerance has been established but autoreactive B-cells are still present. This arises because T- and B-cells have different thresholds for activation and therefore tolerance. In this situation the B-cells are 'helpless'. That is to say, as we have already discussed for most antigens B-cells require help from an antigen-specific T-cell in order to make a response. Thus autoreactive B-cells can be present without being able to become activated provided that there is no T-cell help available. If T-cell help is provided, for example by injecting the autoantigen chemically coupled to an immunogenic foreign carrier, then these B-cells will mount a response. Experiments of this kind indicate that it takes 100-1000 × more antigen to tolerise B-cells that T-cells. As a result this type of split tolerance situation is reasonably common for self serum proteins.

Anergy-T-cells

The naive T-cells need co-stimulatory signals to become activated. The expression of these co-stimulatory molecules is restricted so that most tissue cells lack either B7.1/B7.2 or CD40 or both. Such cells also normally lack class II MHC molecules. Thus tissue cells normally present a spectrum of peptides from their endogenously synthesised proteins on self MHC class I in the absence of co-stimulation. Interaction of such cells with autoreactive T-cells leads to the T-cell becoming refractive to later encounter with the same antigen even when co-stimulation is present. This refractory state is termed **anergy**.

B-Cell anergy

The early studies of Katz and colleagues convincingly demonstrated that some forms of tolerance induction in adult B-cells could be reversed by trypsination of the cells to remove bound antigen and the receptors, followed by receptor reexpression. Unresponsiveness in these cases was likely caused by receptor blockade without signalling. Tolerance induced by other antigens, however, could not be reversed by simply removing the surface molecules, indicating a requirement for active metabolic processes during induction. For example, the D co-polymers induced unresponsiveness at 37°C even if the B-cells were subsequently trypsinized to clear the bound immunoglobulin receptors from the cell surface. In contrast, exposure of the cells to antigen at 4°C did not induce tolerance. The low temperature presumably prevented the necessary signalling to the cell required for the tolerance. In several other systems, stimulation with mitogens such as LPS was also required to reverse B-cell tolerance. These observations suggested that there might exist a stable but reversible unresponsive state for mature B-cells in the short-term absence of antigen.

Because the total number of B-cells capable of binding labelled antigen was not diminished in these models, the tolerant state was referred to by Nossal and Pike as clonal anergy rather than clonal deletion or abortion. This was difficult to prove, however, because in a normal mouse only 1% to 3% of the antigen-binding cells (assayed at limiting dilution) were responsive to antigen or produced specific antibodies when stimulated with LPS. Thus, a small fraction of functionally important cell could have been deleted, but their absence may not have been detected amidst the mass of low-affinity antigen-binding cells. Subsequent limiting dilution experiments, with a more potent mitogenic mixture of dextran sulfate and LPS as a stimulant, revealed that some tolerized B-cells could be stimulated to differentiate into antibody-forming cells when the BCR was bypassed. This reversal suggested that at least a portion of the cells had been rendered functionally unresponsive (anergized) rather than deleted.

As implied from the section on split tolerance, there is a mechanism for tolerising B-cells to soluble antigens if they are present at sufficiently high concentration. The general rules for this tolerance mechanism were worked out using mice transgenic for rearranged immunoglobulin molecules and a transgenic soluble protein whose concentration in serum could be regulated. It is apparent that the critical parameter is receptor (surface Ig) occupancy. When more than 5%* of the sIgM molecules are normally occupied by monomeric soluble antigen the B-cell becomes **anergic**. This anergic state can be recognised in the case of B-cells by the **downregulation of surface IgM.** Note the level of surface IgD remains unaffected and the precise explanation for why such B-cells are refractory to stimulation even when T-cell help is available is not known.

Suppression

It is also clear that in some cases there are autoreactive T-cells present which are capable of reacting to their cognate antigen *as presented within the host* [i.e. they are not anergic or ignorant], yet do not express this reactivity in the normal intact animal. These cells appear to be prevented from reacting by the presence of other T-cells, a phenomenon which has been termed **dominant regulation** or **suppression**. The mechanism is again obscure, but the following experiments make it likely that this form of tolerance is at least as important as anergy.

Timing

Timing is an important parameter on the tolerisability of the immune system. Medewar subsequently investigated the effects of transferring haemopoietic cells from histoincompatible mice at different times after birth. He found that if the cells were transferred in the first few days of life (but not later) the recipient mouse **acquired** lifelong tolerance to the antigens of the donor. Some 50 years ago Owen observed two types of non-identical twin cattle, those that had shared a haemopoietic system *in utero* were tolerant of blood cells from each other and those who had not, were not cross-tolerant. Burnet postulated that there was a temporal window of tolerance such that antigens encountered while the immune system was immature tolerised the relevant lymphocytes. We now understand a good deal about the mechanism of this acquired tolerance. In essence bone marrow stem cells establish chimaerism of the host.

Some of these cells differentiate into antigen-presenting cells and migrate to the thymus where they tolerise developing thymocytes by deletion (**central tolerance**). Lifelong chimaerism is needed to maintain tolerance and even a low level of chimaerism is sufficient. For the first few days of the mouse's life there are too few post-thymic T-cells to sustain a response and these are tolerised either by peripheral deletion or some other mechanism. If the transfer is done later the number and maturity of the peripheral T-cell pool is sufficient to destroy the donor stem cells before they can engraft. Even antigens which cannot establish chimaerism can establish a state of tolerance in neonatal mice, this is less permanent however.

Experimental tolerance

Indeed a number of regimes have been successfully applied under experimental conditions and with particular antigens. A clinically usable protocol is still elusive however. If we could induce tolerance in adults, this would have obvious benefits in transplantation and the treatment of autoimmunity. Much effort has therefore been expended towards this goal. It will suffice here to give just two examples of tolerising conditions which have worked.

Peptide sniffing (inhalation tolerance)

If immunogenic (that is MHC binding) peptides are aerosolised and administered via the nostrils to mice, the animals can be rendered tolerant to immunogenic challenge with the same peptide (i.e. given subcutaneously in adjuvant). **Co-receptor blockade:** The use of monoclonal antibodies to any of the important co-receptors, CD4/8, B7, CD40 where a particular response is dependent on the blockaded co-receptor can lead to the development of tolerance.

The danger hypothesis

Matzinger has proposed that there is not a special window for tolerance during neonatal life but that whether encounter with an antigen results in tolerance or an immune response is determined by whether the prevailing host environment promotes a response via **nonspecific cues 'sensing' danger**. She has further suggested that the controlled death process of apoptosis is critical in preventing autoimmunity when old or surplus cells are disposed of. The notion that the normal, default pathway of the immune system is tolerance rather than response is not a new idea to immunologists - antigens usually fail to elicit a response unless given with adjuvants, whose purpose is probably to generate stimulatory cues (cytokines). Recent experiments have shown that not only can adults be tolerised under certain circumstances, but that neonates can make effective immune responses if the antigen is presented in sufficiently immunogenic form. I believe that the supposed conflict between Matzinger and Medewar is rather 'hyped up' and essentially a matter of detail. Neonatal T-cells are not intrinsically tolerisable but the systemic neonatal environment does predispose to tolerance. Nevertheless, I think that her hypothesis has drawn the attention of a wider audience to current ideas about tolerance induction and the factors determining immune responsiveness.

Oral and nasal tolerance

Intravenous administration generally favors induction of tolerance, whereas subcutaneous administration favors immunity. Intravenous injection might allow antigen presentation by co-stimulatory molecule-deficient naive B-cells in the spleen, whereas subcutaneous injection would favor uptake and presentation by Langerhans cells, which following activation are very effective at initiating immune responses in the draining lymph nodes. The route of antigen introduction is a critical variable in determining the outcome of an immune response. Oral and nasal administration of antigen has also been shown to favor tolerance induction. From the earliest studies of Wells in 1911, it was clear that the oral route of administration induces some form of immunoregulation. Orally immunized animals usually make an initial systemic antibody response that subsequently diminishes. The tolerance state is often associated with large amounts of IgA production in the gut. In other cases (e.g. for myelin basic protein), where the antigen is administered in a form (peptides) preferentially recognized by T-cells rather than B-cells, the immunoregulation has been reported to be mediated either by T-cells that secrete transforming growth factor β (TGF-β) on antigen stimulation or by induction of clonal anergy and deletion. The mechanism observed depends on the antigen dose, with high doses inducing direct inactivation of the antigen-specific T-cells and low doses eliciting TGF-β–mediated immunoregulation. A role for T-cells secreting IL-10 has also been suggested by Hanninen and Harrison. Nasal administration of soluble proteins or peptides prevents and reduces ongoing Th1 IFN-α responses by similar mechanisms.

CHAPTER

25

TECHNIQUES IN IMMUNOLOGY

INTRODUCTION

The fundamental reaction of immunology involves the interaction of antibodies (Ab) and antigens (Ag). These interactions are useful in the defense of the body against bacterial and viral infections and toxins. The defense capabilities are dependent upon the recognition of antigens by humoral components of the immune system. Specific antibodies are then produced in response to exposure to the antigen. The formation of antigen-antibody complexes is the first step in removing infectious agents from the body. Because each antibody can bind more than one antigen and each antigen can be bound by more than one antibody molecule, very large macromolecular complexes can form. These complexes form precipitates which can be cleared from the body through various means. These precipitates are also useful for laboratory and diagnostic tests. When antibodies and antigens are inserted into different areas of an agarose gel, they diffuse toward each other and form opaque bands of precipitate at the interface of their diffusion fronts. Precipitation reactions of antibodies and antigens in agarose gels provide a method of analyzing the various antibody-antigen reactions in a system.

Precipitation

The immunoprecipitation technique detects soluble antigen that react with antibodies called **precipitins**. A precipitation reaction occurs when bivalent or multivalent antibodies and antigen are mixed in proper proportion. The antibody linked the antigen to form antibody-antigen network or lattice that settled out solution when it becomes sufficiently large. The immunoprecipitation reaction occurs only at equivalence zone when their is optimal ration of antigen to antibody so that a lattice form (see Figure 25.1). A precipitation reaction occurs as a result of the combination of antibodies in solution with soluble substances with which the antibodies react. If such a precipitation reaction occurs *in vivo* in the joints or the kidney, inflammation results because the immune complexes are filtered out in those areas, causing irritation.

In a serum precipitin test, dilutions of antigen and antibody in aqueous solution are combined until a precipitation reaction occurs and visible particles accumulate. The amount of "flocculation" can be measured by an instrument designed to measure the light scattering properties of particulate aggregations (nephelometer). The rapid plasma reagin test remains the "Gold Standard" as a screening test for syphilis and has an immunological flocculation reaction as its basis.

The precise "Zone of Equivalence" where there is a perfect concentration of both antibody and antigen in solution, is difficult to achieve. Many tubes of various antibody and antigen concentrations would need to be prepared in order to get just the right amount; the thin ring of precipitate visible only after centrifugation is much more likely than the tube full of precipitate shown at the top of the page. In practice, precipitation reactions are generally prepared as agar diffusion reactions, to make precipitates more visible.

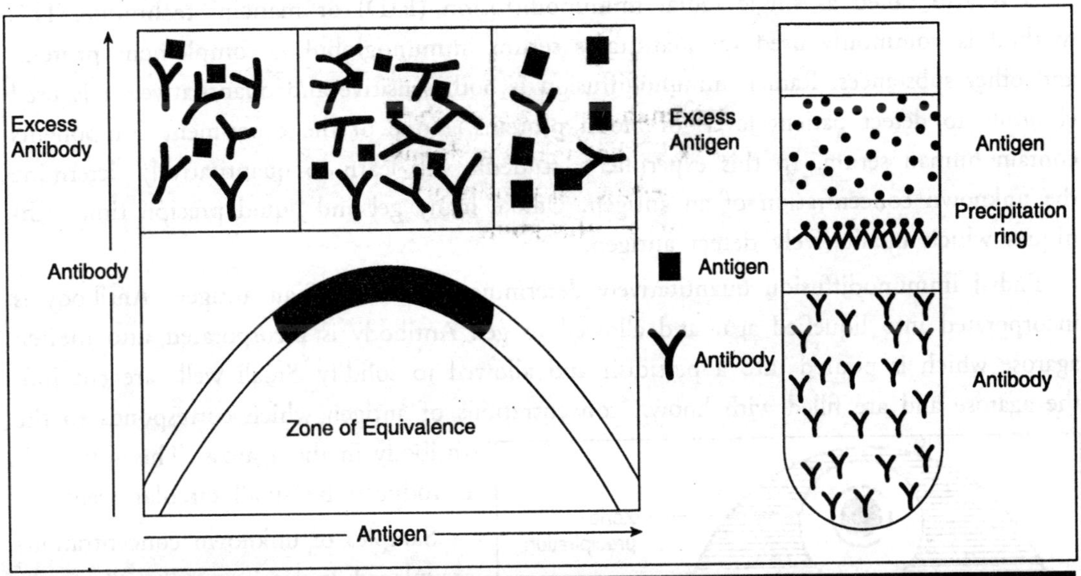

Fig. 25.1: Precipitation reaction (a) Graph showing precipitation curve based on ratio of antigen and antibody, zone of equivalence shows the optimal ratio of for precipitation. (b) Precipitation ring test, antigen and antibody diffuse towards each other and form a zone of equivalence

Immunodiffusion

Immunodiffusion refers to precipitation reaction that occurs between an antibody and antigen in an agarose gel medium. Two techniques are routinely used, single radial immunodiffusion and double immunodiffusion in agar.

Immunodiffusion in agar gels

An agarose gel is used as a matrix for combining diffusion with precipitation. The reactants diffuse through the gel towards each other and precipitation results when the equivalence points have been reached. A single antigen will give rise to a single line of precipitation in the presence of its homologous antibody. When two antigens are present in a system, each behaves independently of the other. Therefore, if several bands of precipitation are detected, there are at least as many antigen-antibody combinations present.

Serum antibodies are not homogeneous (you have antibodies that recognize all kinds of infectious organisms and allergens). It is difficult to obtain a single antigen even in the purest preparations, so there are several antigen-antibody reactions possible in a single mixture. Gel diffusion permits the examination of such multiple systems since the technique allows the separation of the reactions. The individual components in a system react independently of each other so precipitation tests in agar are termed "double immunodiffusion" tests or simply immunodiffusion. (Ouchterlony is used in the older terminology). The hazy white background of the solid agar matrix allows visualization of the precipitation reaction. The method is used clinically in immunology/rheumatology evaluations.

Radial immunodiffusion

It is also called as single radial immunodiffusion (RID) or mancini technique. This method is commonly used to quantitates serum immunoglobulin, complement protein, and other substances. Radial immunodiffusion is both sensitive and quantitative. It is used clinically to detect patient levels of blood proteins. None of the experiment components contain human serum. In this experiment, students will learn to quantitatively determine the unknown concentration of an antigen. Unlike many gel and liquid precipitation techniques which qualitatively detect antigen.

Radial immunodiffusion quantitatively determines the level of an antigen. Antibody is incorporated into liquefied agar and allowed to gel. Antibody is incorporated into molten agarose which is poured into a petridish and allowed to solidify. Small wells are cut into the agarose and are filled with known concentrations of antigen which corresponds to the antibody in the agarose. The antigen is introduced by small circular wells.

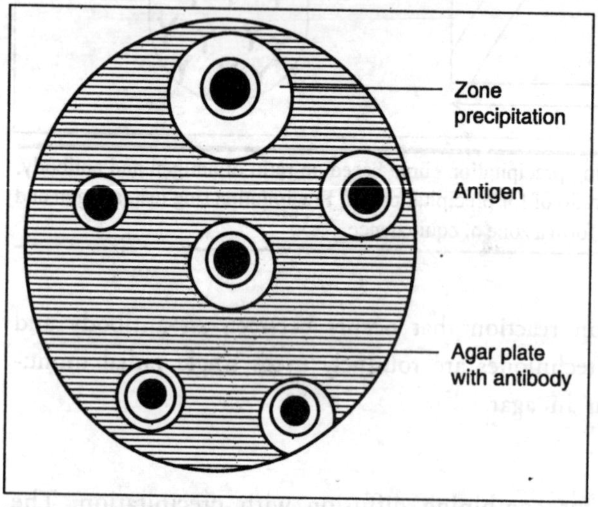

Fig. 25.2: Radial immunodiffusion

Samples of unknown concentrations are placed in similar wells. The antigens in solution then diffuse outwards from the well in a circular pattern surrounding the well. Antibody is present in excess and diffusion of the antigen will continue until a stable ring of antigen antibody precipitate forms (see Figure 25.2).

There are antigen-antibody complexes throughout the zone surrounding the well within the precipitin line. At the precipitin line is where the greatest number of complexes can be found because the antigen and antibody are present in roughly equal proportions. This is known as the equivalence zone or equivalence point. Generally, it takes 24 to 48 hours for optimal diffusion to occur and precipitation to become apparent. For each antigen, an endpoint precipitation ring of a certain diameter will form. From the known standard concentrations, a standard curve can be drawn by plotting antigen concentration versus the diameter squared measurements of the rings. From this linear calibration curve the concentration of the unknown antigen samples may be determined.

Ouchterlony double diffusion

Double diffusion in two dimensions is a simple procedure invented by the Swedish scientist, Ouchterlony. Antigen and antibody solutions are placed in separate wells cut in an agarose plate. The reactants diffuse from the wells toward each other and precipitate where they meet at equivalent proportions. A single antigen will combine with its homologous antibody to form a single precipitation line.

Ouchterlony double immunodiffusion (or agar gel immunodiffusion) is a simple, rather dated method which is still considered to be the gold standard for detection of antigens. A gel plate is cut to form a series of holes in the gel (The geometry is critical). An extract of human cells (harvested from tonsil tissue) is placed in the centre well. This extract contains a cocktail of natural human antigens that we wish to look for. The patients serum is then placed in one of the outer wells and the plate left for 48 hours to develop. During this time the antigens in the tonsil extract migrate in a radial fashion out of the centre well and the antibodies in the patients serum migrate in a radial fashion out of the outer well. When the two meet if there are antibodies in the patients serum to any of the antigens in the tonsil extract the antibodies will bind to the antigens and form what is known as an immune complex. This immune complex precipitates in the gel to give a thin white line.

Precipitation occurs with most antigens because the antigen is multivalent i.e., has several antigenic determinants per molecule to which antibodies can bind. Antibodies have at least two antigen binding sites, thus large aggregates or lattices of antigen and antibody are formed. Experimentally, an increasing amount of antigen is added to a constant amount of antibody in solution, initially at low antigen concentration, all of the antigen is contained in the precipitate this is called the antibody-excess zone (Prozone phenomenon). As more antigen is added, the amount protein precipitated increases until the antigen/antibody molecules are at an optimal ratio. This is known as the equivalence zone or equivalence point. When the amount of antigen in solution exceeds the amount of antibody, the amount of precipitation will decrease. This is known as the antigen excess zone.

The line will give a full identity (continuous line), partial identity (continuous line with a spur coming off one end—like a branch off the main line) or a non identity where the two lines cross completely (See Figure 25.3).

The binding of an antibody to an antigen is a fundamental reaction of immunology. Antibodies and antigens form complexes that precipitate, making it possible to assay antibody-antigen systems. The objective of this experiment is to introduce the principles of

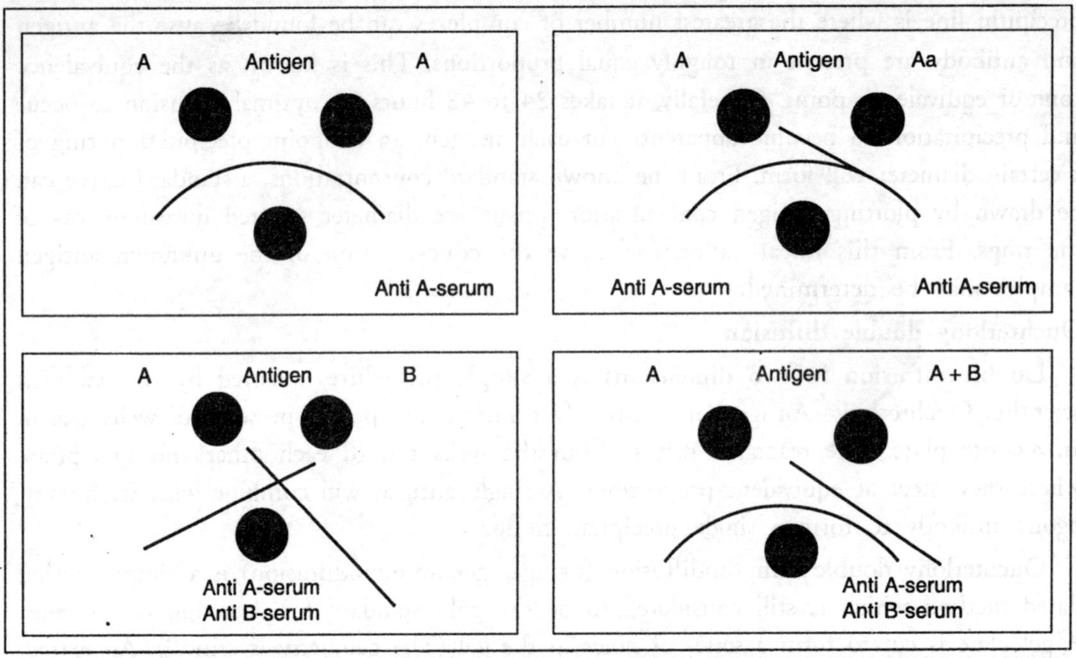

Fig. 25.3: Immuno-double diffusion assay showing characters of identity (top left), reaction of nonidentity (bottom left) partial identity (bottom right) and complex pattern (bottom right)

antigen-antibody interactions by using the Ouchterlony procedure. This binding interaction results in the formation of a white precipitate after diffusion in agarose.

The interactions of an antibody (Ab) with an antigen (Ag) is the fundamental reaction of immunology $Ab + Ag \rightarrow Ab - Ag$.

Most antigens are proteins. The exact identity of the groups that react with the antibody are usually not known. Macromolecular antigens and antibodies form complexes that become insoluble and precipitate from solution. This property makes it possible to perform qualitative and quantitative assays on the antibody-antigen system.

Precipitation occurs with most antigens because the antigen is multivalent, i.e., has several antigenic determinants per molecule to which antibodies can bind. Antibodies have at least two antigen binding sites, thus large aggregates or lattices of antigen and antibody are formed. Experimentally, an increasing amount of antigen is added to a constant amount of antibody in solution. Initially at low antigen concentration, all of the antigen is contained in the precipitate. This is called the antibody-excess zone. As more antigen is added, the amount of protein precipitated increases until the antigen and antibody molecules are at an optimal ratio. This is called the equivalence zone, or equivalence point, where maximum precipitation occurs. When the amount of antigen in solution exceeds the amount of antibody, the amount of precipitation will decrease. This is known as the antigen-excess zone.

When antibodies and antigens are inserted into different areas of an agarose gel, they diffuse toward each other and form opaque bands of precipitate at the interface of their diffusion fronts. Precipitation reactions of antibodies and antigens in agarose gels provide a method of analyzing various antibody-antigen reactions.

THE OUCHTERLONY PROCEDURE

Double diffusion in two dimensions is a simple procedure invented by and named after the Swedish scientist, Orjan Ouchterlony. Antigen and antibody solutions are placed in separate wells cut in an agarose plate.

The reactants diffuse from the wells toward each other and precipitate where they meet at equivalent proportions. A single antigen will combine with its homologous antibody to form a single precipitation line. When two antigens are present, each behaves independently of each other. Thus, the number of precipitin bands indicates there are at least that many antibody-antigen pairs present (see Figure 25.3). Arrows indicate diffusion patterns of antigens and antibodies. Double diffusion in two dimensions is a useful technique for comparing antigens for the number of identical or cross-reacting determinants. If a solution of antigen is placed in two adjacent wells and the homologous antibody is placed in the center well, the two precipitin bands that form will join at their closest ends and fuse. This is known as a reaction of identity.

Immunoelectrophoresis

Immunoelectrophoresis identifies proteins based on their combined electrophoretic and immunological properties. This method is useful to monitor antigen and antigen-antibody purity and to identify a single antigen in a mixture of antigens. In this experiment, serum proteins are separated by agarose gel electrophoresis and the point of equivalence is observed by the maximum antigen-antibody complex formation. None of the experiment components contain human serum.

Immunoelectrophoresis (IEP) is a procedure combining separation of antigens by zone electrophoresis with diffusion of precipitating antibodies at right angles to the direction of antigen migration. The extent of migration of any protein during electrophoresis is dependent upon the electrophoresis buffer, time, voltage, and the isoelectric point of the protein. Particularly, pH determines the separation of one protein species from another.

The net difference between the isoelectric point of a given protein and the pH of the electrophoresis buffer determines the extent of migration toward the cathode or the anode. This immunological method is used to detect abnormalities in specific serum proteins and to identify specific immunoglobulins using anti-human antibody reagents. The anti-human antibody reagents have been prepared against purified fractions of human globulins and are commercially available for research or biomedical purposes.

The precipitation appears as a continuous line in the form of an angle between those two wells and the C well. There are no spurs at the angle and this type of reaction is termed a band of identity.

Immunoelectrophoresis is used in both clinical and research laboratories for separating and identifying proteins on the basis of their electrophoretic behavior and their immunological properties. Proteins such as rabbit serum proteins, which are antigens, when injected into another animal such as a goat (the host), elicit the production of antibodies in the host. The interaction between the antigen and its antibody, which is also a protein, is both strong and highly specific. If solutions of antigen and antibody are mixed in different ratios, it is found that at a specific ratio, known as the equivalence point, the binding is maximized and a precipitate of antigen-antibody complex is formed. In the clinical laboratory, immunoelectrophoresis is used diagnostically in examining certain serum abnormalities, especially those involving immunoglobulins, urine protein, cerebrospinal fluid, pleural fluids and other body fluids. In research, this procedure may be used to monitor antigen and/or antibody purifications, to detect impurities, analyze soluble antigens from plant and animal tissues, and microbial extracts. Gel electrophoresis is a widely used analytical method that separates molecules based upon charge, size and shape. It is particularly useful for determining the size of biomolecules. Samples of proteins are loaded into wells made in the gel during casting. The gel, which consists of microscopic pores that act as a molecular sieve, is placed in a chamber containing a buffer solution and electrodes. Current is applied from a direct current (D.C.) power source. Since biomolecules are charged, they will migrate through the gel.

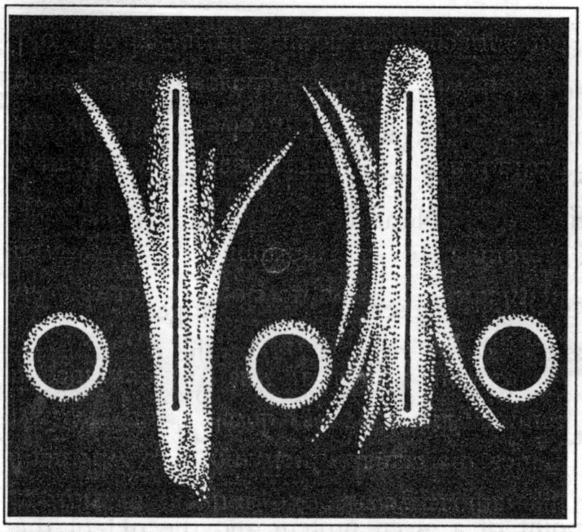

Fig. 25.4: Immunoelectrophoresis of protein

In immunoelectrophoresis, the proteins are first separated by horizontal agarose gel electrophoresis on the basis of their different charge-to-mass ratios (see Figure 25.4). The gel with the separated proteins (antigens) is then removed from the electric field and antibodies to the proteins are introduced into narrow troughs parallel to the separated antigens. Diffusion of both antigen antibody takes place and, at a particular locus, the equivalence point is reached resulting in precipitation. The purpose of this experiment is to demonstrate the use of immunoelectrophoresis to separate and characterize a mixture of serum proteins as well as to examine the specificity of the antigen-antibody interaction.

Applications

Immunoelectrophoresis has many different applications; examples include:

1. Determining the homogeneity of antigen-antibody systems
2. Diagnosing specific autoimmune disorders
3. Following the purification of an antigenic mixture
4. Elucidating the reactions among serologically related antigens

Radioimmunoassay (RIA)

One of the most sensitive techniques for detecting antigen or antibody is **radioimmunoassay (RIA)**. The technique was first developed in 1960 by two endocrinologists, S. A. Berson and Rosalyn Yalow, to determine levels of insulin–anti-insulin complexes in diabetics. Although their technique encountered some skepticism, it soon proved its value for measuring hormones, serum proteins, drugs, and vitamins at concentrations of 0.001 *micrograms* per milliliter or less. In 1977, some years after Berson's death, the significance of the technique was acknowledged by the award of a Nobel Prize to Yalow.

Prior to the development of the EIA/ELISA, the only option for conducting an immunoassay was radioimmunoassay, a technique using radioactively-labeled antigens or antibodies. In radioimmunoassay, the radioactivity provides the signal which indicates whether a specific antigen or antibody is present in the sample. The RIA was developed by Rosalyn Yalow and Solomon Aaron Berson in the 1950s. Radioimmunoassay was first described in a paper by Rosalyn Sussman Yalow and Solomon Berson published in 1960. Rosalyn Sussman Yalow received the 1977 Nobel Prize in Medicine for the development of the RIA for insulin. The precise measurement of minute amounts of such a hormone was a key event in endocrinology.

Radioimmunoassay (RIA) is a scientific method used to test antigens (for example, hormone levels in the blood) without the need to use a bioassay. It involves mixing known quantities of radioactive antigen (frequently labeled with gamma-radioactive isotopes of iodine attached to tyrosine) with antibody to that antigen, then adding unlabeled or "cold" antigen and measuring the amount of labeled antigen displaced. Initially, the radioactive antigen is bound to the antibodies. When "cold" (unlabeled, quest) antigen is added, the two compete for antibody binding sites—at higher concentrations of "cold" antigen, more of it binds to the antibody, displacing the radioactive variant. The bound antigens are separated from the unbound ones.

One method of separation used initially was the use of a second antibody directed against the first for precipitation and centrifugation. The use of charcoal suspension for precipitation was extended but replaced later by Drs. Werner and Acebedo at Columbia University for RIA of T3 and T4. An ultramicro RIA for human TSH was published in BBRC (1975) by Drs. Acebedo, Hayek. The free hormone stays in the supernatant, the radioactivity of which is measured and a binding curve is plotted. The technique is both extremely sensitive and specific, but it requires special precautions (because radioactive substances are used), sophisticated apparatus, and is expensive. The technique has been largely replaced by the ELISA method where instead of a radioactive signal, the antigen-antibody reaction is measured by colorometric signals.

Enzyme linked immunosorbent assay (ELISA)

Enzyme Immuno-Assay (EIA) is a synonym for the ELISA. Enzyme-Linked ImmunoSorbent Assay, or **ELISA**, is a biochemical technique used mainly in immunology to detect the presence of an antibody or an antigen in a sample. The ELISA has been used as a diagnostic tool in medicine and plant pathology, as well as a quality control check in various industries. In simple terms, in ELISA an unknown amount of antigen is affixed to a surface, and then a specific antibody is washed over the surface so that it can bind to the antigen. This antibody is linked to an enzyme, and in the final step a substance is added that the enzyme can convert to some detectable signal. Thus in the case of fluorescence ELISA, when light is shown upon the sample, any antigen/antibody complexes will fluoresce so that the amount of antigen in the sample can be measured.

Principle of ELISA

ELISA is advance technique over Radio Immune Assay, because radioactivity poses a health threat, a safer alternative was sought. A suitable alternative to radioimmunoassay would substitute a non-radioactive signal in place of the radioactive signal. When certain enzymes (such as peroxidase) react with appropriate substrates (such as ABTS or 3.3′, 5.5′-Tetramethylbenzidine), they can result in changes in color, which can be used as a signal. However, the signal has to be associated with the presence of antibody or antigen, which is why the enzyme has to be linked to an appropriate antibody. This linking process was independently developed by Stratis Avrameas and G.B. Pierce. Since it is necessary to remove any unbound antibody or antigen by washing, the antibody or antigen has to be fixed to the surface of the container, i.e. the *immunosorbent* has to be prepared. A technique to accomplish this was published by Wide and Porath in 1966.

The Enzyme-linked Immunosorbent Assay (ELISA) is a precise and sensitive antibody-based detection method for antigens such as bacterial and viral proteins. The reaction steps of the ELISA will be determined. None of the components have been prepared from human sources.

During an infection, an individual mounts an antibody response which eventually results in production of plasma IgG molecules that bind to various parts of the infectious agent. If these antibodies are present in the sample, they will bind to the adsorbed antigens in the well and remain there after washing, and will be detected by the ELISA technique.

All antibodies belong to a group of serum proteins known as globulins. Each antibody is made up of a heavy and light polypeptide chain. In general, antibodies are produced in response to the presence of a "non-self" antigenic response. Antibodies obtained from animals, such as rabbits, in response to an antigen are known as polyclonal antibodies.

Polyclonal antibodies are heterogeneous in structure and vary in their ability to bind to antigens. Antibodies that have a high affinity for non-specific antigens may give unwanted cross-reactions that can result in high backgrounds. Such antibodies can also give false negative results. By contrast, antibodies with weak binding constants may not be as sensitive. Enzyme linked immunosorbent assay (ELISA) tests were originally developed for antibody measurement but have also been adapted to successfully detect samples that contain antigens.

This ELISA experiment has been designed to detect an antibody directed against an antigen. ELISAs are done in microtiter plates usually made of polystyrene or polyvinyl chloride. The plates are somewhat transparent and contain many small wells, into which liquid samples are deposited.

Performing an ELISA involves at least one antibody with specificity for a particular antigen. The sample with an unknown amount of antigen is immobilized on a solid support (usually a polystyrene microtiter plate) either non-specifically (via adsorption to the surface) or specifically (via capture by another antibody specific to the same antigen, in a "sandwich" ELISA). After the antigen is immobilized the detection antibody is added, forming a complex with the antigen. The detection antibody can be covalently linked to an enzyme, or can itself be detected by a secondary antibody which is linked to an enzyme through bioconjugation. Between each step the plate is typically washed with a mild detergent solution to remove any proteins or antibodies that are not specifically bound. After the final wash step the plate is developed by adding an enzymatic substrate to produce a visible signal, which indicates the quantity of antigen in the sample. Older ELISAs utilize chromogenic substrates, though newer assays employ fluorogenic substrates with much higher sensitivity.

Method

The following are the basic steps of the ELISA technique.

1. The antigen is added to the wells where some remain adsorbed by hydrophobic association to the walls after washing away the excess. The antigens can be a lysate, a specific protein, or a mixture of the two. There is no specificity involved with the adsorption process, although some substances may exhibit low binding to the walls. In certain cases the antigens can be covalently cross-linked to the plastic using UV light.
2. After washing away unadsorbed material, the unoccupied sites on the walls of the plastic wells are blocked with proteins, typically gelatin or bovine serum albumin.
3. A solution that may or may not contain the primary antibody is added to the wells. If present in the solution, the primary antibody will bind to the adsorbed antigen in the well and remain after washing.
4. A solution containing the secondary antibody is then added to the wells. If the primary antibody has remained bound to the well, then the secondary antibody will bind to it and remain attached after washing. Secondary antibodies are purified and covalently cross linked to enzymes such as horseradish peroxidase. This coupling does not significantly affect the binding specificity and affinity of the antibody or the enzymatic activity of the peroxidase.
5. The wells are washed with buffer to remove unbound secondary antibody.
6. After washing the wells, substrate will be added to wells. The enzyme attached to the secondary antibody is a peroxidase. Peroxidase possesses a high catalytic activity and can exceed turnover rates of 106 per second. Consequently, amplification of a positive sample can occur over several orders of magnitude. Many hydrogen donor co-substrates can be used by peroxidase. These cosubstrates include o-diansidine,

aminoantipyrine, aminosalicylic acid and numerous phenolic compounds that develop color upon oxidation. Substrate 1 (S1) contains hydrogen peroxide and amino salicylate. The substrate solution added is nearly colorless. Peroxidase converts the peroxide to H_2O + O_2 using the salicylate as the hydrogen donor. The oxidized salicylate is brown and can be easily observed in wells that have received each of the components required for completion of reaction. Substrate contains hydrogen peroxide and azino-diethylbenzthiazoline sulfonate (ABTS). The substrate solution added is nearly colorless. Peroxidase converts the peroxide to H_2O + O_2 using the ABTS as the hydrogen donor. The oxidized ABTS is green and can be easily observed in positive wells. It should be noted that polyclonal antibody preparations to a given antigen can have variable binding affinities due to differences in the immunological responses between animals. Different immunizations with the same antigen in animals can also produce antibodies with variable binding affinities. The use of monoclonal antibodies directed against a single epitope eliminates this variability. Western blot analysis is usually used to confirm the ELISA results and to quantitate the size and amount of antigen. Western blots and ELISA-based tests are used as diagnostic tools.

This experiment demonstrates two important concepts. The first is the effect of the absence of the antigen or the primary antibody which results in the disruption of the ELISA reaction. The second is the demonstration that the substrate utilized by the enzyme attached to the secondary antibody can result in different positive well colors.

TYPES ELISA

Indirect ELISA

The indirect method is employed to detect a specific antibody in serum of suspected patient e.g. detection of HIV specific antibody in person suspected to suffering from AIDS. The steps of the general, "indirect," ELISA for determining serum antibody concentrations are : The first step in this procedure is to immobilize the known antigen concentration to a surface, often the well of a microtiter plate. The antigen is fixed to the surface to render it immobile. Simple adsorption of the protein to the plastic surface is usually sufficient. These samples of known antigen concentrations will constitute a standard curve used to calculate antigen concentrations of unknown samples. Note that the antigen itself may be an antibody. After incubation, the excess antigen is removed and wells are washed and dried. The plate wells or other surface are then coated with test serum samples of unknown antigen concentration, diluted into the same buffer used for the antigen standards. Since antigen immobilization in this step is due to non-specific adsorption, it is important for the total protein concentration to be similar to that of the antigen standards. A concentrated solution of non-interacting protein, such as Bovine Serum Albumin (BSA) or casein, is added to all plate wells. This step is known as blocking, because the serum proteins block non-specific adsorption of other proteins to the plate. The plate is washed, and a detection antibody specific to the antigen of interest is applied to all plate wells. This antibody will only bind to immobilized antigen on the well surface, not to other serum proteins or the blocking proteins. The plate is washed to remove any unbound detection antibody. After this wash, only the antibody-antigen complexes remain attached to the well. Secondary antibodies, which will bind to any remaining detection antibodies, are added to the wells. These secondary antibodies are conjugated to the substrate-specific enzyme. This step may be skipped if the detection antibody is conjugated to an enzyme. Wash the plate, so that excess unbound enzyme-antibody conjugates are removed. Apply a substrate which is converted by the enzyme to elicit a chromogenic or fluorogenic or electrochemical signal. View/quantify the result using a spectrophotometer, spectrofluorometer, or other optical/ electrochemical device.

Indirect ELISA

Antigen → Antibody → Antiantibody linked with enzyme → Substrate → Chromogen → Color product → **Positive reaction**

No Color → **Negative reaction**

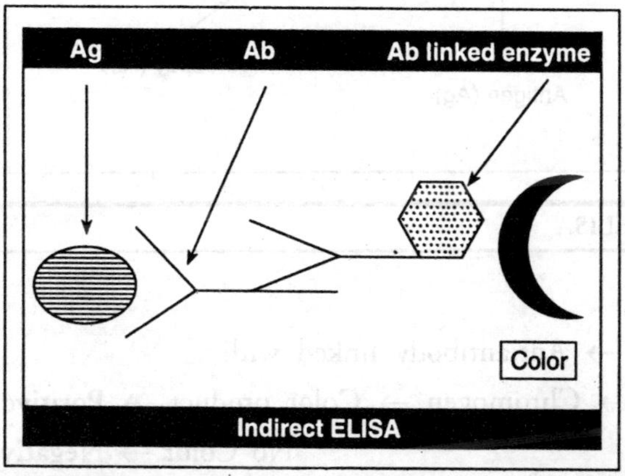

Fig. 25.5: Indirect ELISA

Sandwich ELISA

Antigen can be measure or detect by sandwich ELISA. A major disadvantage of the indirect ELISA is that the method of antigen immobilization is non-specific; any proteins in the sample will stick to the microtiter plate well, so small concentrations of analyte in serum must compete with other serum proteins when binding to the well surface.

The sandwich ELISA provides a solution to this problem. The sandwich ELISA may be run in a qualitative or quantitative format. Qualitative results provide a simple positive or negative result for a sample. The cutoff between positive and negative is determined by the analyst and may be statistical. Two or three times the standard deviation is often used to distinguish positive and negative samples. In quantitative ELISA, the optical density or fluorescent units of the sample is interpolated into a standard curve, which is typically a serial dilution of the target.

1. In this technique antibody is first incubated to microtiter plate. Plate is coated with a capture antibody

2. Antigen is added, and allow to react with immobilized antibody
3. After well is washed, second enzyme linked antibody is added, and allow to react bound antigen
4. Substrate is added, and colored reaction product is measure.

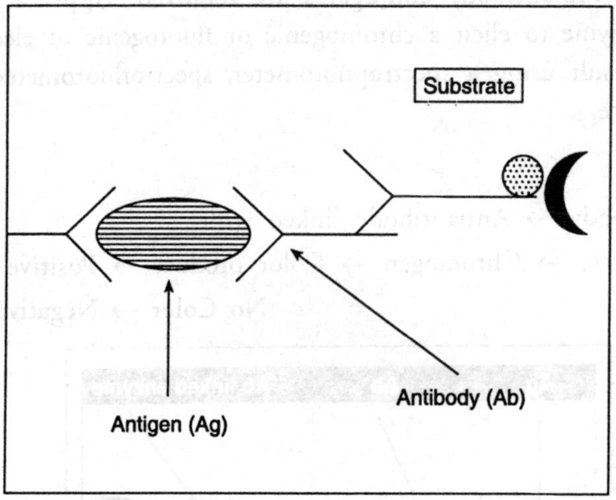

Fig. 25.6: Direct or sandwich ELISA

Sandwich ELISA

Antibody → Antigen → Antiantibody linked with
enzyme → Substrate → Chromogen → Color product → **Positive reaction**
No Color → **Negative reaction**

A less-common variant of this technique, called "sandwich" ELISA, is used to detect sample antigen. The steps are as follows: Prepare a surface to which a known quantity of capture antibody is bound. Block any non specific binding sites on the surface. Apply the antigen-containing sample to the plate. Wash the plate, so that unbound antigen is removed. Apply primary antibodies that bind specifically to the antigen. Apply enzyme-linked secondary antibodies which are specific to the primary antibodies. Wash the plate, so that the unbound antibody-enzyme conjugates are removed. Apply a chemical which is converted by the enzyme into a color or fluorescent or electrochemical signal. Measure the absorbance or fluorescence or electrochemical signal (e.g. current) of the plate wells to determine the presence and quantity of antigen. The image to the right includes the use of a secondary antibody conjugated to an enzyme, though technically this is not necessary if the capture antibody is conjugated to an enzyme.

However, use of a secondary-antibody conjugate avoids the expensive process of creating enzyme-linked antibodies for every antigen one might want to detect. By using an enzyme-linked antibody that binds the Fc region of other antibodies, this same enzyme-linked antibody can be used in a variety of situations. The major advantage of a sandwich ELISA is the ability to use crude or impure samples and still selectively bind any antigen that may be present. Without the first layer of "capture" antibody, any proteins in the sample (including serum proteins) may competitively adsorb to the plate surface, lowering the quantity of antigen immobilized.

Competitive ELISA

Another variation for measuring antigen is competitive ELISA. A third use of ELISA is through competitive binding. The steps for this ELISA are somewhat different than the first two examples:

1. Unlabeled antibody is incubated in the presence of its antigen.
2. These bound antibody/antigen complexes are then added to an antigen coated well.
3. The plate is washed, so that unbound antibody is removed. (The more antigen in the sample, the less antibody will be able to bind to the antigen in the well, hence "competition.")
4. The secondary antibody, specific to the primary antibody is added. This second antibody is coupled to the enzyme.
5. A substrate is added, and remaining enzymes elicit a chromogenic or fluorescent signal.

For competitive ELISA, the higher the original antigen concentration, the weaker the eventual signal.

Applications ELISA

1. The ELISA can be performed to evaluate either the presence of antigen or the presence of antibody in a sample, it is a useful tool both for determining serum antibody concentrations (such as with the HIV test or West Nile Virus) and also for detecting the presence of antigen.
2. It has also found applications in the food industry in detecting potential food allergens such as milk, peanuts, walnuts, almonds, and eggs.
3. The ELISA was the first screening test commonly employed for HIV. It has a high sensitivity. In an ELISA test, a person's serum is diluted 400-fold and applied to a plate to which HIV antigens have been attached. If antibodies to HIV are present in the serum, they may bind to these HIV antigens. The plate is then washed to remove all other components of the serum. A specially prepared "secondary antibody"—an antibody that binds to human antibodies—is then applied to the plate, followed by another wash. This secondary antibody is chemically linked in advance to an enzyme. Thus the plate will contain enzyme in proportion to the amount of secondary antibody bound to the plate. A substrate for the enzyme is applied, and catalysis by the enzyme leads to a change in color or fluorescence. ELISA results are

reported as a number; the most controversial aspect of this test is determining the "cut-off" point between a positive and negative result.

4. It is also used in testing cats for the FIV disease.

ELISPOT (Enzyme-Linked Immunospot)

The Enzyme-Linked Immunospot assay is a powerful tool for detecting and enumerating individual cells that secrete a particular biomarker of interest *ex vivo*. Originally developed for analysis of antibody secreting cells, the ELISPOT assay has been adapted to detect cells secreting multiple biomarkers such as cytokines, chemokines, and granzymes. The **Enzyme-linked immunosorbent spot (ELISPOT)** is a common method for monitoring immune responses in humans and animals. It was developed by Cecil Czerkinsky in 1983. The ELISPOT assay is based on, and was developed from a modified version of the ELISA immunoassay. ELISPOT assays were originally developed to enumerate B-cells secreting antigen-specific antibodies, and have subsequently been adapted for various tasks, especially the identification and enumeration of cytokine-producing cells at the single cell level. Simply put, at appropriate conditions the ELISPOT assay allows visualization of the secretory product of individual activated or responding cells. Each spot that develops in the assay represents a single reactive cell.

Thus, the ELISPOT assay provides both qualitative (type of immune protein) and quantitative (number of responding cells) information.

While ELISPOT assay techniques have existed for more than two decades now advancements are still being made in the assay. Modern ELISPOT analysis is typically performed using ELISPOT readers, which employ computer vision techniques to enumerate the actively producing cells. This allows much of the analysis process to be automated, and permits a greater level of accuracy than what can be achieved using manual inspection.

Principle

ELISpot assays employ the sandwich enzyme-linked immunosorbent assay (ELISA) technique. Either a monoclonal or polyclonal antibody specific for the chosen analyte is pre-coated onto a PVDF (polyvinylidene difluoride) backed microplate. Appropriately stimulated cells are pipetted into the wells and the microplate is placed into a humidified 37°C CO_2 incubator for a specified period of time. During this incubation period, the immobilized antibody, in the immediate vicinity of the secreting cells, binds secreted analyte. After washing away any cells and unbound substances, a biotinylated polyclonal antibody specific for the chosen analyte is added to the wells. Following a wash to remove any unbound biotinylated antibody, alkaline-phosphatase conjugated to streptavidin is added. Unbound enzyme is subsequently removed by washing and a substrate solution (BCIP/NBT) is added. A blue-black colored precipitate forms and appears as spots at the sites of cytokine localization, with each individual spot representing an individual analyte-secreting cell. The spots can be counted with an automated ELISPOT reader system or manually, using a stereomicroscope.

Method

In the ELISPOT procedure a color reaction is visualized as a spot for each individual cell producing the biomarker of interest (proteins, antibodies, cytokines, etc.). The specificity and the sensitivity of ELISPOT is attained by employing high affinity capture and detection antibodies and detection with enzyme-amplification, which allows a measurement of even very low frequency biomarker-producing cells. Plates are scanned and evaluated for the number of antigen specific cells using powerful imaging system—the latest model of ELISPOT reader.

The ELISPOT assays employ a technique very similar to the sandwich enzyme-linked immunosorbent assay (ELISA) technique. Either a monoclonal or polyclonal capture antibody is coated aseptically onto a PVDF (polyvinylidene fluoride) backed microplate. These antibodies are chosen for their specificity for the analyte in question. The plate is blocked, usually with a serum protein that is non-reactive with any of the antibodies in the assay. After this, cells of interest are plated out at varying densities, along with antigen or mitogen, and then placed in a humidified 37°C CO_2 incubator for a specified period of time.

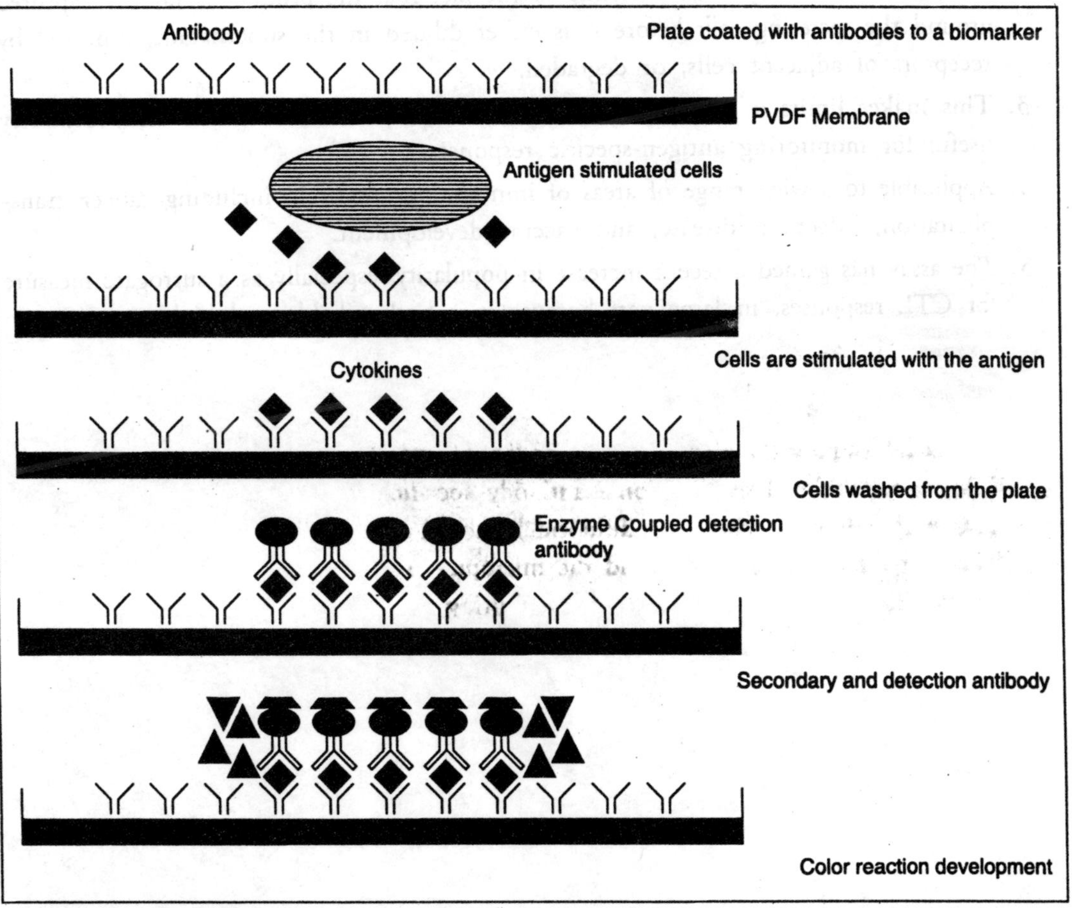

Fig. 25.7: ELISPOT Assay

Cytokine (or other cell product of interest) secreted by activated cells is captured locally by the coated antibody on the high surface area PVDF membrane. After washing the wells to remove cells, debris, and media components. A biotinylated polyclonal antibody specific for the chosen analyte is added to the wells. This antibody is reactive with a distinct epitope of the target cytokine and thus is employed to detect the captured cytokine. Following a wash to remove any unbound biotinylated antibody, the detected cytokine is then visualized using an avidin-HRP, and a precipitating substrate (e.g., AEC, BCIP/NBT). The colored end product (a spot, usually a blackish blue) typically represents an individual cytokine-producing cell. The spots can be counted manually (e.g., with a dissecting microscope) or using an automated reader to capture the microwell images and to analyze spot number and size.

Application

1. The ELISPOT assay is used for analysis of rare cell populations (e.g., antigen-specific responses) by virtue of exquisite sensitivity.
2. ELISPOT assays much more sensitive than conventional ELISA measurements. This exceptional sensitivity is in part due to the fact that the product is rapidly captured around the secreting cell: before it is either diluted in the supernatant, captured by receptors of adjacent cells, or degraded.
3. This makes limits of detection are below 1/100,000 rendering the assay uniquely useful for monitoring antigen-specific responses.
4. Applicable to a wide range of areas of immunology research, including cancer, transplantation, infectious disease, and vaccine development.
5. The assay has gained a recent increase in popularity, especially as a surrogate measure for CTL responses, in large part because it is both reliable and highly sensitive.

THE NOBEL PRIZE LAUREATES IN IMMUNOLOGY

- **1901** E.A. Von Behring (Germany) for the work on serum therapy especially its application against diphtheria.
- **1905** R. Koch (Germany) for the investigations concerning tuberculosis.
- **1908** E. Metchnikoff (Russia) and P. Ehrlich (Germany) for their work on immunity (respectively, phagocytosis/cellular theory and humoral theory).
- **1913** C.R. Richet (France) for the work on anaphylaxis.
- **1919** J. Bordet (Belgium) for the discoveries relating to immunity (complement).
- **1930** K. Landsteiner (Austria/USA) for the discovery of human blood groups.
- **1951** M. Theiler (South Africa) for the discoveries and developments concerning yellow fever.
- **1957** D. Bovet (Italy/Switzerland) for the discoveries related to histamine and compounds, which inhibit action of histamine and other substances on the vascular system and the skeleton muscles.
- **1960** Sir F. McFarlane Burnet (Australia) and Sir P.B. Medawar (Great Britain) for the discovery of acquired immunological tolerance.
- **1972** G.M. Edelman (USA) and R.R. Porter (Great Britain) for their discovery concerning the chemical structure of antibodies.
- **1977** R. Yalow (USA) for the development of radioimmunoassays of peptide hormones.
- **1980** B. Benacerraf (USA), J. Dausset (France) and G.D. Snell (USA) for their discoveries concerning genetically determined structures on the cell surface (major histocompatibility complex) that regulate immunological reactions.
- **1982** S.K. Bergstrom (Sweden), B.I. Samuelsson (Sweden) and J.R. Vane (UK) for their discoveries concerning prostaglandins and related biologically active substances.
- **1984** N.K. Jerne (Denmark/Switzerland) for theories concerning the specificity in development (lymphocyte clonality) and control of the immune system; +G.J.F. Köhler (Germany/Switzerland) and C. Milstein (Argentina/Great Britain) for the discovery of the principle for production of monoclonal antibodies.
- **1987** S. Tonegawa (Japan/USA) for the discovery of the genetic principle for generation of antibody diversity.
- **1990** J.E. Murray and E.D. Thomas (USA) for their discovery concerning organ and cell transplantation in the treatment of human diseases.
- **1996** P.C. Doherty (Australia/USA) and R.M. Zinkernagel (Switzerland) for their discoveries concerning the specificity of the cell mediated immune defense ("dual recognition").
- **1997** S.B. Prusiner (USA) for the discovery of prions as a new biological principle of infection.

1999 G. Blobel (USA) for discoveries concerning signal transduction.

AA: Arachidonic acid
ADCC: Antibody-dependent cell-mediated cytotoxicity
AICD: Activation-induced cell death
Allo: Allogeneic
APC: Antigen-presenting cell
BMC: Bone marrow cells
BMT: Bone marrow transplant
CAM: Cell adhesion molecules
CCR7: CC chemokine receptor 7
CD40L: CD40 ligand
CDR3: Complementarity-determining region 3
CAMP: Cyclic adenosine monophosphate
CLL: Chronic lymphocytic leukemia
CML: Cell-mediated lympholysis
CMV: Cytomegalovirus
COX: Cyclooxygenase
CsA: Cyclosporine
CTL: Cytotoxic T-lymphocyte
CTLp: Cytotoxic T-lymphocyte precursor
CXCR3: CXC chemokine receptor 3
DAF: Decay-accelerating factor
DC: Dendritic cell
DLI: Donor lymphocyte infusions
DST: Donor-specific transfusion
DT: Diphtheria endotoxin
DTH: Delayed-type hypersensitivity
ECF: Eosinophil chemotactic factors
ELISA: Enzyme-linked immunosorbent assay
ELISPOT: Enzyme-linked immunosorbent spot
ES cells: Embryonic stem cells
FasL: Fas ligand
FKBP: FK506-binding protein
GalT: Gal transferase
GM-CSF: Granulocyte-myelocyte colony stimulating factor
GVH: Graft versus host
GVHD: Graft-versus-host disease
GVL: Graft versus leukemia
H-2: Mouse MHC
HA: Histocompatibility antigen
HCT: Hematopoietic cell transplantation
HDAF: Human decay-accelerating factor
HEV: High endothelial venule
HLA: Human leukocyte antigen
IFN: Interferon
IP-10: Interferon gamma-induciple protein of 10 kD
ITIM: Immune receptor tyrosine based inhibitory motif
KGF: Keratinocyte growth factor
KIR: Killer inhibitory receptor
LAK: Lymphokine activated killer
LC: Langerhans cells
LTC: Leukotrienes
LT: Leukotrienes
Mabs: Monoclonal antibodies
MAC: Membrane attack complex
Mac1: Myeloid cell-surface marker CD11b
MALT: Mucosal associated lymphatic tissues
MCP: Membrane cofactor protein
MHC: Major histocompatibility complex
Mig: Monokine induced by IFN-& gamma
MIP1: Macrophage inflammatory protein 1
MLR: Mixed lymphocyte response
NCF: Neutrophil chemotactic factor
NK: Natural killer
NS: Natural suppressor
NSAIDs: Nonsteroidal anti-inflammatory drugs
PBMC: Peripheral blood mononuclear cell
PERV: Pig endogenous retrovirus
PGD: Prostaglandin
PMN: Polymorphonuclear leucocytes
PHSC: Pluripotent hematopoietic stem cell

PRA: Panel reactive antibody

Rag: Recombinase-activating gene

ROS: Reactive oxygen species

SCID: Severe combined immunodeficiency

SLA: Swine leukocyte antigen

SLE: Systemic lupus erythematosus

SRS-A: Slow reacting substance of anaphylaxis

STAT: Signal transducers and activators of transduction

TBI: Total body irradiation

TCR: T-cell receptor

TI: Thymic irradiation

TLI: Total lymphoid irradiation

TNFR1: Tumour necrosis factor receptor 1

Tr1: T-regulatory cells 1

PAF: Platelet activating factor

PAMPs: Pathogen-associated molecular patterns

TLR: Toll-like receptors

Tx: Thromboxanes

PCB: Polychlorinated-biphenyls

GVHD: Graft-versus-host disease

LAD: Leukocyte adhesion deficiency

PE: Pseudomonas endotoxin

Adjuvant: Any substance which nonspecifically enhances the immune response to antigen.

Affinity: The strength of binding between a receptor and ligand.

Agglutination: The aggregation of particles such as RBCs or bacteria by antibodies.

Allele: The variants of a polymorphic gene at a given genetic locus.

Allelic exclusion: The phenomenon whereby, following successful rearrangement of one allele of an antigen receptor gene, rearrangement of the other parental allele is suppressed, thereby ensuring each lymphocyte expresses only a single specificity of antigen receptor.

Allotypic: An allelic variant of an antigen which, because it is not present in all individuals, may be immunogenic in members of the same species which have a different version of the allele.

Anaphylotoxin: complement breakdown product (e.g., C3a, C4a or C5a) capable of directly triggering mast cell degranulation.

Anaphylaxis: An often fatal hypersensitivity reaction, triggered by IgE or anaphylatoxin-mediated mast cell degranulation, leading to anaphylactic shock due to vasodilatation and smooth muscle contraction

Antibody-dependent cellular cytotoxicity: A cytotoxic reaction in which an antibody-coated target cell is directly killed by an Fc receptor-bearing leukocyte, e.g., NK cell, macrophage or neutrophil.

Apoptosis: A form of programmed cell death, characterized by endonuclease digestion of DNA.

Atopic allergy: IgE-mediated hypersensitivity, i.e., asthma, eczema, hayfever and food allergy.

Avidity: The binding strength between two molecules taking into account the valency of the interaction. Thus, the avidity will always be equal to or greater than the intrinsic affinity.

Chemokines: A family of structurally related cytokines which selectively induce chemotaxis and activation of leukocytes.

Chemotaxis: The directional migration of cells towards an attractant.

Complementarity determining regions (CDR): The hypervariable amino acid sequences within antibody and T-cell receptor variable regions which interact with complementary amino acids on the antigen or peptide–MHC complex.

C-reactive protein: An acute phase protein which is able to bind to the surface of microorganisms where it functions as a stimulator of the classical pathway of complement activation, and as an opsonin for phagocytosis.

Cytokines: Low molecular weight proteins that stimulate or inhibit the differentiation, proliferation or function of immune cells.

Delayed-type hypersensitivity (DTH): A hypersensitivity reaction occurring within 48–72 h and mediated by cytokine release from sensitized T-cells.

Diversity (*D*) gene segments: Found in the immunoglobulin heavy chain gene and T-cell receptor b and d gene loci between the *V* and *J* gene segments. Encode part of the third hypervariable region in these antigen receptor chains.

Energy: Potentially reversible specific immunological tolerance in which the lymphocyte becomes functionally nonresponsive.

Endotoxin: Pathogenic cell wall-associated lipopolysaccharides of Gram-negative bacteria.

Epitope: That part of an antigen recognized by an antigen receptor.

Fab: Monovalent antigen-binding fragment obtained following papain digestion of immunoglobulin. Consists of an intact light chain and the N-terminal VH and CH1 domains of the heavy chain.

F(ab')2: Bivalent antigen-binding fragment obtained following pepsin digestion of immunoglobulin. Consists of both light chains and the N-terminal part of both heavy chains linked by disulfide bonds.

Fas: A member of the TNF receptor gene family. Engagement of Fas (CD95) on the surface of the cell by the Fas ligand (CD178) present on cytotoxic cells, can trigger apoptosis in the Fas-bearing target cell.

Fc: Crystallizable, non-antigen binding fragment of an immunoglobulin molecule obtained following papain digestion. Consists of the C-terminal portion of both heavy chains which is responsible for binding to Fc receptors and C1q.

Fluorescent antibody: An antibody conjugated to a fluorescent dye such as FITC.

Framework regions: The relatively conserved amino acid sequences which flank the hypervariable regions in immunoglobulin and T-cell receptor variable regions and maintain a common overall structure for all V-region domains.

Gammaglobulin: The serum proteins, mostly immunoglobulins, which have the greatest mobility towards the cathode during electrophoresis.

Germ line: The arrangement of the genetic material as transmitted through the gametes.

Germinal center: Discrete areas within lymph node and spleen where B-cell maturation and memory development occur.

Graft vs host (g.v.h.) reaction: Reaction occurring when T lymphocytes present in a graft recognize and attack host cells.

Granuloma: A tissue nodule containing proliferating lymphocytes, fibroblasts, and giant cells and epithelioid cells (both derived from activated macrophages), which forms due to inflammation in response to chronic infection or persistence of antigen in the tissues.

Granzymes: Serine esterases present in the granules of cytotoxic T lymphocytes and NK cells. They induce apoptosis in the target cell which they enter through perforin channels inserted into the target cell membrane by the cytotoxic lymphocyte.

Haplotype: The set of allelic variants present at a given genetic region.

Hapten: A low molecular weight molecule that is recognized by preformed antibody but is not itself immunogenic unless conjugated to a "carrier" molecule which provides epitopes recognized by helper T-cells.

Hematopoiesis: The production of erythrocytes and leukocytes.

High endothelial venule (HEV): Capillary venule composed of specialized endothelial cells allowing migration of lymphocytes into lymphoid organs.

HLA (*h*uman *l*eukocyte *a*ntigen): The human major histocompatibility complex (MHC).

Humanized antibody: A genetically engineered monoclonal antibody of non-human origin in which all but the antigen-binding CDR sequences have been replaced with sequences derived from human antibodies. This procedure is carried out to minimize the immunogenicity of therapeutic monoclonal antibodies.

Humoral: Pertaining to extracellular fluid such as plasma and lymph. The term humoral immunity is used to denote antibody mediated immune responses.

Hybridoma: Hybrid cell line obtained by fusing a lymphoid tumour cell with a lymphocyte which then has both the immortality of the tumour cell and the effector function (e.g., monoclonal antibody secretion) of the lymphocyte.

Hypervariable regions: Those amino acid sequences within the immunoglobulin and T-cell receptor variable regions which show the greatest variability and contribute most to the antigen or peptide–MHC binding site.

Idiotype: The complete set of idiotopes in the variable region of an antibody or T-cell receptor which react with an anti-idiotypic serum.

Idiotype network: A regulatory network based on interactions of idiotypes and anti-idiotypes present on antibodies and T-cell receptors.

Immune complex: Complex of antibody bound to antigen which may also contain complement components.

Immunofluorescence: Technique for detection of cell or tissue associated antigens by the use of a fluorescently-tagged ligand (e.g., an anti-immunoglobulin conjugated to fluorescein isothiocyanate).

Immunogen: Any substance which elicits an immune response. Whilst all immunogens are antigens, not all antigens are immunogens.

Immunoglobulin superfamily: Large family of proteins characterized by possession of "immunoglobulin-type" domains of approximately 110 amino acids folded into two b-pleated sheets. Members include immunoglobulins, T-cell receptors and MHC molecules.

Immunotoxin: A cytotoxic agent produced by chemically coupling a toxic agent to a specific monoclonal antibody.

Internal image: An epitope on an anti-idiotype which binds in a way that structurally and functionally mimics the antigen.

Isotype: An antibody constant region structure present in all normal individuals, i.e., antibody class or subclass.

ITAM: Immunoreceptor Tyrosine-based *A*ctivation *M*otifs are consensus sequences for src-family tyrosine kinases. These motifs are found in the cytoplasmic domains of several signalling molecules including the signal transduction units of lymphocyte antigen receptors and of Fc receptors.

J chain: Molecule which forms part of the structure of pentameric IgM and dimeric IgA. Joining (*J*) gene segments: Found in the immunoglobulin and T-cell receptor gene loci and, upon gene rearrangement, encode part of the third hypervariable region of the antigen receptors.

Lectins: A family of proteins, mostly of plant origin, which bind specific sugars on glycoproteins and glycolipids. Some lectins are mitogenic (e.g., PHA, ConA).

Linkage disequilibrium: The occurrence of two alleles being inherited together at a greater frequency than that expected from the product of their individual frequencies.

Lipopolysaccharide (LPS): Endotoxin derived from Gram-negative bacterial cell walls which has inflammatory and mitogenic actions.

Mannose binding lectin: A member of the collectin family of calcium-dependent lectins, and an acute phase protein. It activates the complement system by binding to mannose residues on the surface of microorganisms.

Marginal zone: The outer area of the splenic periarteriolar lymphoid sheath (PALS) which is rich in B-cells, particularly those responding to thymus-independent antigens.

Margination: Leukocyte adhesion to the endothelium of blood vessels in the early phase of an acute inflammatory reaction.

Membrane attack complex (MAC): Complex of complement components C5b–C9 which inserts as a pore into the membrane of target cells leading to cell lysis.

Memory cells: Clonally expanded T- and B-cells produced during a primary immune response and which are "primed" to mediate a secondary immune response to the original antigen.

MHC (major histocompatibility complex): A genetic region encoding molecules involved in antigen presentation to T-cells. Class I MHC molecules are present on virtually all nucleated cells and are encoded by HLA-A, B, and C in man, whilst class II MHC molecules are expressed on antigen-presenting cells (primarily macrophages, B-cells and interdigitating dendritic cells) and are encoded by HLA-DR, DQ, and DP in man. Allelic differences can be associated with the most intense graft rejection within a species.

MHC restriction: The necessity that T-cells recognize processed antigen only when presented by MHC molecules of the original haplotype associated with T-cell priming.

Mitogen: A substance which non-specifically induces lymphocyte proliferation.

Monoclonal antibody: Homogeneous antibody derived from a single B-cell clone and therefore all bearing identical antigen-binding sites and isotype.

Mucosal-associated lymphoid tissue (MALT): A system of lymphoid tissue found in the gastrointestinal and respiratory tracts.

Murine: Pertaining to mice.

Myeloma protein: Monoclonal antibody secreted by myeloma cells.

Negative selection: Deletion by apoptosis in the thymus of T-cells which recognize self peptides presented by self MHC molecules, thus preventing the development of autoimmune T-cells. Negative selection of developing B-cells is also thought to occur if they encounter high levels of self antigen in the bone marrow.

Nude mouse: Mouse which is T-cell deficient due to a homozygous gene defect (*nu*/*nu*) resulting in the absence of a thymus (and also lack of body hair).

Oncofetal antigen: Antigen whose expression is normally restricted to the fetus but which may be expressed during malignancy in adults.

Opsonin: Substance, e.g., antibody or C3b, which enhances phagocytosis by promoting adhesion of the antigen to the phagocyte.

Opsonization: Coating of antigen with opsonin to enhance phagocytosis.

Pathogen-associated molecular pattern (PAMP): Molecules such as lipopolysaccharide, peptidoglycan, lipoteichoic acids and mannans, which are widely expressed by microbial pathogens as repetitive motifs but are not present on host tissues. They are therefore utilized by the pattern recognition receptors (PRRs) of the immune system to distinguish pathogens from self antigens.

Pattern recognition receptor (PRR): Receptors on professional antigen-presenting cells and phagocytes which enable them to recognize pathogen-associated molecular patterns (PAMPs). Amongst the large number of different PRRs are the mannose receptor and the macrophage scavenger.

Respiratory burst: The increased oxidative metabolism which occurs in phagocytic cells following activation.

Rheumatoid factor: IgM, IgG and IgA autoantibodies to IgG, particularly the Fc region. These antibodies are usually found in the blood of patients with rheumatoid arthritis.

Secretory component: Proteolytic cleavage product of the poly-Ig receptor which remains associated with dimeric IgA in seromucus secretions.

Somatic hypermutation: The enhanced rate of point mutation in the immunoglobulin variable region genes which occurs following antigenic stimulation and acts as a mechanism for increasing antibody diversity and affinity.

Superantigen: An antigen which reacts with all the T-cells belonging to a particular T-cell receptor V region family, and which therefore stimulates (or deletes) a much larger number of cells than does conventional antigen.

TAP: The transporters associated with Antigen Processing (TAP-1 and TAP-2) are molecules which carry antigenic peptides from the cytoplasm into the lumen of the endoplasmic reticulum for incorporation into MHC class I molecules.

T-dependent antigen: An antigen which requires helper T-cells in order to elicit an antibody response.

Thymocyte: Developing T-cell in the thymus.

T-independent antigen: An antigen which is able to elicit an antibody response in the absence of T-cells.

Titer: Measure of the relative "strength" (a combination of amount and avidity) of an antibody or antiserum, usually given as the highest dilution which is still operationally detectable in, for example, an ELISA.

Toll-like receptors: A family of pattern-recognition receptors that recognize different molecules on bacterial surfaces. Activation of these receptors usually results in cytokine production and upregulation of the adaptive immune response.

Variable (V) gene segments: Genes that rearrange together with D (diversity) and J (joining) gene segments in order to encode the variable region amino acid sequences of immunoglobulins and T-cell receptors.

INDEX